LONDON MATHEMATICAL SOCIETY LECTURE NOTE SERIES

Managing Editor: Professor M. Reid, Mathematics Institute,
University of Warwick, Coventry CV4 7AL, United Kingdom

The titles below are available from booksellers, or from Cambridge University Press at
http://www.cambridge.org/mathematics

333 Synthetic differential geometry (2nd Edition), A. KOCK
334 The Navier–Stokes equations, N. RILEY & P. DRAZIN
335 Lectures on the combinatorics of free probability, A. NICA & R. SPEICHER
336 Integral closure of ideals, rings, and modules, I. SWANSON & C. HUNEKE
337 Methods in Banach space theory, J.M.F. CASTILLO & W.B. JOHNSON (eds)
338 Surveys in geometry and number theory, N. YOUNG (ed)
339 Groups St Andrews 2005 I, C.M. CAMPBELL, M.R. QUICK, E.F. ROBERTSON & G.C. SMITH (eds)
340 Groups St Andrews 2005 II, C.M. CAMPBELL, M.R. QUICK, E.F. ROBERTSON & G.C. SMITH (eds)
341 Ranks of elliptic curves and random matrix theory, J.B. CONREY, D.W. FARMER, F. MEZZADRI & N.C. SNAITH (eds)
342 Elliptic cohomology, H.R. MILLER & D.C. RAVENEL (eds)
343 Algebraic cycles and motives I, J. NAGEL & C. PETERS (eds)
344 Algebraic cycles and motives II, J. NAGEL & C. PETERS (eds)
345 Algebraic and analytic geometry, A. NEEMAN
346 Surveys in combinatorics 2007, A. HILTON & J. TALBOT (eds)
347 Surveys in contemporary mathematics, N. YOUNG & Y. CHOI (eds)
348 Transcendental dynamics and complex analysis, P.J. RIPPON & G.M. STALLARD (eds)
349 Model theory with applications to algebra and analysis I, Z. CHATZIDAKIS, D. MACPHERSON, A. PILLAY & A. WILKIE (eds)
350 Model theory with applications to algebra and analysis II, Z. CHATZIDAKIS, D. MACPHERSON, A. PILLAY & A. WILKIE (eds)
351 Finite von Neumann algebras and masas, A.M. SINCLAIR & R.R. SMITH
352 Number theory and polynomials, J. MCKEE & C. SMYTH (eds)
353 Trends in stochastic analysis, J. BLATH, P. MÖRTERS & M. SCHEUTZOW (eds)
354 Groups and analysis, K. TENT (ed)
355 Non-equilibrium statistical mechanics and turbulence, J. CARDY, G. FALKOVICH & K. GAWEDZKI
356 Elliptic curves and big Galois representations, D. DELBOURGO
357 Algebraic theory of differential equations, M.A.H. MACCALLUM & A.V. MIKHAILOV (eds)
358 Geometric and cohomological methods in group theory, M.R. BRIDSON, P.H. KROPHOLLER & I.J. LEARY (eds)
359 Moduli spaces and vector bundles, L. BRAMBILA-PAZ, S.B. BRADLOW, O. GARCÍA-PRADA & S. RAMANAN (eds)
360 Zariski geometries, B. ZILBER
361 Words: Notes on verbal width in groups, D. SEGAL
362 Differential tensor algebras and their module categories, R. BAUTISTA, L. SALMERÓN & R. ZUAZUA
363 Foundations of computational mathematics, Hong Kong 2008, F. CUCKER, A. PINKUS & M.J. TODD (eds)
364 Partial differential equations and fluid mechanics, J.C. ROBINSON & J.L. RODRIGO (eds)
365 Surveys in combinatorics 2009, S. HUCZYNSKA, J.D. MITCHELL & C.M. RONEY-DOUGAL (eds)
366 Highly oscillatory problems, B. ENGQUIST, A. FOKAS, E. HAIRER & A. ISERLES (eds)
367 Random matrices: High dimensional phenomena, G. BLOWER
368 Geometry of Riemann surfaces, F.P. GARDINER, G. GONZÁLEZ-DIEZ & C. KOUROUNIOTIS (eds)
369 Epidemics and rumours in complex networks, M. DRAIEF & L. MASSOULIÉ
370 Theory of p-adic distributions, S. ALBEVERIO, A.YU. KHRENNIKOV & V.M. SHELKOVICH
371 Conformal fractals, F. PRZYTYCKI & M. URBANSKI
372 Moonshine: The first quarter century and beyond, J. LEPOWSKY, J. MCKAY & M.P. TUITE (eds)
373 Smoothness, regularity and complete intersection, J. MAJADAS & A. G. RODICIO
374 Geometric analysis of hyperbolic differential equations: An introduction, S. ALINHAC
375 Triangulated categories, T. HOLM, P. JØRGENSEN & R. ROUQUIER (eds)
376 Permutation patterns, S. LINTON, N. RUŠKUC & V. VATTER (eds)
377 An introduction to Galois cohomology and its applications, G. BERHUY
378 Probability and mathematical genetics, N. H. BINGHAM & C. M. GOLDIE (eds)
379 Finite and algorithmic model theory, J. ESPARZA, C. MICHAUX & C. STEINHORN (eds)
380 Real and complex singularities, M. MANOEL, M.C. ROMERO FUSTER & C.T.C WALL (eds)
381 Symmetries and integrability of difference equations, D. LEVI, P. OLVER, Z. THOMOVA & P. WINTERNITZ (eds)
382 Forcing with random variables and proof complexity, J. KRAJÍČEK
383 Motivic integration and its interactions with model theory and non-Archimedean geometry I, R. CLUCKERS, J. NICAISE & J. SEBAG (eds)
384 Motivic integration and its interactions with model theory and non-Archimedean geometry II, R. CLUCKERS, J. NICAISE & J. SEBAG (eds)
385 Entropy of hidden Markov processes and connections to dynamical systems, B. MARCUS, K. PETERSEN & T. WEISSMAN (eds)
386 Independence-friendly logic, A.L. MANN, G. SANDU & M. SEVENSTER
387 Groups St Andrews 2009 in Bath I, C.M. CAMPBELL et al (eds)
388 Groups St Andrews 2009 in Bath II, C.M. CAMPBELL et al (eds)

389 Random fields on the sphere, D. MARINUCCI & G. PECCATI
390 Localization in periodic potentials, D.E. PELINOVSKY
391 Fusion systems in algebra and topology, M. ASCHBACHER, R. KESSAR & B. OLIVER
392 Surveys in combinatorics 2011, R. CHAPMAN (ed)
393 Non-abelian fundamental groups and Iwasawa theory, J. COATES *et al* (eds)
394 Variational problems in differential geometry, R. BIELAWSKI, K. HOUSTON & M. SPEIGHT (eds)
395 How groups grow, A. MANN
396 Arithmetic differential operators over the p-adic integers, C.C. RALPH & S.R. SIMANCA
397 Hyperbolic geometry and applications in quantum chaos and cosmology, J. BOLTE & F. STEINER (eds)
398 Mathematical models in contact mechanics, M. SOFONEA & A. MATEI
399 Circuit double cover of graphs, C.-Q. ZHANG
400 Dense sphere packings: a blueprint for formal proofs, T. HALES
401 A double Hall algebra approach to affine quantum Schur–Weyl theory, B. DENG, J. DU & Q. FU
402 Mathematical aspects of fluid mechanics, J.C. ROBINSON, J.L. RODRIGO & W. SADOWSKI (eds)
403 Foundations of computational mathematics, Budapest 2011, F. CUCKER, T. KRICK, A. PINKUS & A. SZANTO (eds)
404 Operator methods for boundary value problems, S. HASSI, H.S.V. DE SNOO & F.H. SZAFRANIEC (eds)
405 Torsors, étale homotopy and applications to rational points, A.N. SKOROBOGATOV (ed)
406 Appalachian set theory, J. CUMMINGS & E. SCHIMMERLING (eds)
407 The maximal subgroups of the low-dimensional finite classical groups, J.N. BRAY, D.F. HOLT & C.M. RONEY-DOUGAL
408 Complexity science: the Warwick master's course, R. BALL, V. KOLOKOLTSOV & R.S. MACKAY (eds)
409 Surveys in combinatorics 2013, S.R. BLACKBURN, S. GERKE & M. WILDON (eds)
410 Representation theory and harmonic analysis of wreath products of finite groups, T. CECCHERINI-SILBERSTEIN, F. SCARABOTTI & F. TOLLI
411 Moduli spaces, L. BRAMBILA-PAZ, O. GARCÍA-PRADA, P. NEWSTEAD & R.P. THOMAS (eds)
412 Automorphisms and equivalence relations in topological dynamics, D.B. ELLIS & R. ELLIS
413 Optimal transportation, Y. OLLIVIER, H. PAJOT & C. VILLANI (eds)
414 Automorphic forms and Galois representations I, F. DIAMOND, P.L. KASSAEI & M. KIM (eds)
415 Automorphic forms and Galois representations II, F. DIAMOND, P.L. KASSAEI & M. KIM (eds)
416 Reversibility in dynamics and group theory, A.G. O'FARRELL & I. SHORT
417 Recent advances in algebraic geometry, C.D. HACON, M. MUSTAŢĂ & M. POPA (eds)
418 The Bloch–Kato conjecture for the Riemann zeta function, J. COATES, A. RAGHURAM, A. SAIKIA & R. SUJATHA (eds)
419 The Cauchy problem for non-Lipschitz semi-linear parabolic partial differential equations, J.C. MEYER & D.J. NEEDHAM
420 Arithmetic and geometry, L. DIEULEFAIT *et al* (eds)
421 O-minimality and Diophantine geometry, G.O. JONES & A.J. WILKIE (eds)
422 Groups St Andrews 2013, C.M. CAMPBELL *et al* (eds)
423 Inequalities for graph eigenvalues, Z. STANIĆ
424 Surveys in combinatorics 2015, A. CZUMAJ *et al* (eds)
425 Geometry, topology and dynamics in negative curvature, C.S. ARAVINDA, F.T. FARRELL & J.-F. LAFONT (eds)
426 Lectures on the theory of water waves, T. BRIDGES, M. GROVES & D. NICHOLLS (eds)
427 Recent advances in Hodge theory, M. KERR & G. PEARLSTEIN (eds)
428 Geometry in a Fréchet context, C.T.J. DODSON, G. GALANIS & E. VASSILIOU
429 Sheaves and functions modulo p, L. TAELMAN
430 Recent progress in the theory of the Euler and Navier-Stokes equations, J.C. ROBINSON, J.L. RODRIGO, W. SADOWSKI & A. VIDAL-LÓPEZ (eds)
431 Harmonic and subharmonic function theory on the real hyperbolic ball, M. STOLL
432 Topics in graph automorphisms and reconstruction (2nd Edition), J. LAURI & R. SCAPELLATO
433 Regular and irregular holonomic D-modules, M. KASHIWARA & P. SCHAPIRA
434 Analytic semigroups and semilinear initial boundary value problems (2nd Edition), K. TAIRA
435 Graded rings and graded Grothendieck groups, R. HAZRAT
436 Groups, graphs and random walks, T. CECCHERINI-SILBERSTEIN, M. SALVATORI & E. SAVA-HUSS (eds)
437 Dynamics and analytic number theory, D. BADZIAHIN, A. GORODNIK & N. PEYERIMHOFF (eds)
438 Random walks and heat kernels on graphs, M.T. BARLOW
439 Evolution equations, K. AMMARI & S. GERBI (eds)
440 Surveys in combinatorics 2017, A. CLAESSON *et al* (eds)
441 Polynomials and the mod 2 Steenrod algebra I, G. WALKER & R.M.W. WOOD
442 Polynomials and the mod 2 Steenrod algebra II, G. WALKER & R.M.W. WOOD
443 Asymptotic analysis in general relativity, T. DAUDÉ, D. HÖFNER & J.-P. NICOLAS (eds)
444 Geometric and cohomological group theory, P.H. KROPHOLLER, I.J. LEARY, C. MARTÍNEZ-PÉREZ & B.E.A. NUCINKIS (eds)
445 Introduction to hidden semi-Markov models, J. VAN DER HOEK & R.J. ELLIOTT
446 Advances in two-dimensional homotopy and combinatorial group theory, W. METZLER & S. ROSEBROCK (eds)
447 New directions in locally compact groups, P.-E. CAPRACE & N. MONOD (eds)

London Mathematical Society Lecture Note Series: 447

New Directions
in Locally Compact Groups

Edited by

PIERRE-EMMANUEL CAPRACE
Université Catholique de Louvain, Belgium

NICOLAS MONOD
École Polytechnique Fédérale de Lausanne

CAMBRIDGE
UNIVERSITY PRESS

CAMBRIDGE
UNIVERSITY PRESS

University Printing House, Cambridge CB2 8BS, United Kingdom

One Liberty Plaza, 20th Floor, New York, NY 10006, USA

477 Williamstown Road, Port Melbourne, VIC 3207, Australia

314-321, 3rd Floor, Plot 3, Splendor Forum, Jasola District Centre, New Delhi - 110025, India

79 Anson Road, #06-04/06, Singapore 079906

Cambridge University Press is part of the University of Cambridge.

It furthers the University's mission by disseminating knowledge in the pursuit of education, learning and research at the highest international levels of excellence.

www.cambridge.org
Information on this title: www.cambridge.org/9781108413121
DOI: 10.1017/9781108332675

© Cambridge University Press 2018

First published 2018

A catalogue record for this publication is available from the British Library

ISBN 978-1-108-41312-1 Paperback

Contents

Foreword by George Willis   *page* vii

1   On the role of totally disconnected groups in the structure of
    locally compact groups   *Burger*   1

2   Locally compact groups as metric spaces   *Tessera*   9

3   A short primer on profinite groups   *Wilson*   17

4   Lectures on Lie groups over local fields   *Glöckner*   37

5   Abstract quotients of profinite groups,
    after Nikolov and Segal   *Klopsch*   73

6   Automorphism groups of trees: generalities and prescribed local
    actions   *Garrido–Glasner–Tornier*   92

7   Simon Smith's construction of an uncountable family of simple,
    totally disconnected, locally compact groups   *Reid–Willis*   117

8   The Neretin groups   *Garncarek–Lazarovich*   131

9   The scale function and tidy subgroups   *Brehm–Gheysens–Le Boudec–
    Rollin*   145

10 Contraction groups and the scale   *Wesolek*   161

11 The Bader–Shalom normal subgroup theorem   *Gal*   171

12 Burger–Mozes' simple lattices   *Bartholdi*   179

13 A lecture on invariant random subgroups   *Gelander*   186

14 L^2-Betti number of discrete and non-discrete groups   *Sauer*   205

15 Minimal normal closed subgroups in compactly generated tdlc
    groups   *Dumont–Gulko*   227

16 Elementary totally disconnected locally compact groups, after
    Wesolek   *Cesa–Le Maître*   236

17 **The structure lattice of a totally disconnected locally compact group** *Wilson* 258

18 **The centraliser lattice** *Hume–Stulemeijer* 267

19 **On the quasi-isometric classification of locally compact groups** *Cornulier* 275

20 **Future directions in locally compact groups: a tentative problem list** *Caprace–Monod* 343

Index 357

Foreword by George Willis

Totally disconnected locally compact (tdlc) groups are of interest for two reasons: on one hand, important classes of tdlc groups arise in combinatorial geometry, number theory and algebra and, on the other, an essential part of the task of describing the structure of general locally compact groups is understanding the totally disconnected case. Interest in these groups is currently very high because of the rapid progress being made with the general theory.

Advances in our understanding of the structure of tdlc groups are being made through three loosely related approaches:

- the *scale*, a positive integer-valued function defined on automorphisms of tdlc groups that relates to eigenvalues in algebraic representations of these groups and to translation distance in geometric representations;
- the *structure lattice* of locally normal subgroups of a tdlc group, which gives rise to a local theory underpinning a typology of simple tdlc groups; and
- a *decomposition theory* for tdlc groups that exploits methods for gauging their size and breaks a given group into smaller, and often simple, pieces.

These approaches are developing a conceptual framework that promises to support a comprehensive description of tdlc groups. At the same time, more examples filling out this framework are being found. There is still some way to go before this description could be regarded as complete however.

The current interest perhaps prompted the Oberwolfach Forschungsinstitut to ask P.-E. Caprace and N. Monod to organise the Arbeitsgemeinschaft *Totally Disconnected Groups* in October, 2014, with the aim of bringing together these approaches and the researchers and students involved. Lectures surveyed the background for the study of tdlc groups and introduced the main ideas and most recent developments in the three approaches described above. These notes, which cover the lectures as well as including a couple of other invited surveys, thus provide a valuable review of the current state of knowledge. It is to be hoped that they will serve as a reference for further work that goes toward completing the description of totally disconnected, locally compact groups.

1

On the role of totally disconnected groups in the structure of locally compact groups

Marc Burger

Abstract

We describe to what extent a general locally compact group decomposes into a totally disconnected part and a connected part that can be approximated by Lie groups. We also present a construction that highlights the relevance of groups acting on trees in the structure of general locally compact groups.

1.1	Van Dantzig's theorem	2
1.2	An application of the Gleason–Yamabe theorem	4
1.3	Totally disconnected groups and actions on trees	7
	References for this chapter	8

In the first part of this lecture we establish a fundamental theorem which says that if G is a totally disconnected locally compact group, every neighbourhood of $e \in G$ contains a compact open subgroup. Since such groups are profinite, this implies that the multiplication in a neighbourhood of e can be approximated with arbitrary accuracy by the multiplication in finite groups.

In the second part of the lecture we address the question to which extent a locally compact group decomposes into a totally disconnected part and a connected part; using the concept of amenable radical we will establish a decomposition result using the Gleason–Yamabe structure theorem.

In the third part we describe a construction which relates totally disconnected groups which are compactly generated to groups acting on trees via an appropriate Cayley–Abels graph ("Nebengruppenbild", see [3], §4).

Parts 2 and 3 of this lecture are taken from [1], 3.3–3.5, with some modifications. The heuristic principle is that certain questions, concerning for instance amenable actions of general compactly generated locally compact groups, can be analyzed by treating separately the case of semisimple Lie groups and closed cocompact subgroups of the automorphism group of a regular tree. For van Dantzig's theorem, one may consult [2], Chapter II, §7, and a good reference for Hilbert's 5th problem is [4].

1.1 Van Dantzig's theorem

Let us first fix some terminology: a topological space X is locally compact if every point admits a compact neighbourhood; if in addition X is T_2 (that is Hausdorff), then every point admits a fundamental system of compact neighbourhoods.

Let now G be a topological group; then the connected component G° of $e \in G$ is a normal subgroup and closed in addition; thus G/G° with quotient topology is a T_2 topological group and:

Lemma 1.1 *The topological group G/G° is totally disconnected.*

Recall

Definition 1.2 A non-empty topological space is totally disconnected if all its connected components are reduced to points.

Observe then that a topological group L is totally disconnected if and only if $L^\circ = (e)$.

Proof Let $\pi : G \to L := G/G^\circ$ be the canonical quotient homomorphism and assume that $\pi^{-1}(L^\circ) = F_1 \cup F_2$ where $F_i \subset G$ are closed disjoint subsets. Since $G^\circ \subset \pi^{-1}(L^\circ)$, we have for every $g \in \pi^{-1}(L^\circ)$ either $gG^\circ \subset F_1$ or $gG^\circ \subset F_2$. Thus, F_1 and F_2 are union of G°-cosets and the decomposition descends to the quotient. Since L° is connected, we conclude that $F_1 = \varnothing$ or $F_2 = \varnothing$. Hence $\pi^{-1}(L^\circ) \supset G^\circ$ is connected and hence $\pi^{-1}(L^\circ) = G^\circ$ which finally implies that $L^\circ = (e)$ and $L = G/G^\circ$ is totally disconnected. $\square$

Corollary 1.3 *If G is locally compact T_2, then G/G° is locally compact, T_2 and totally disconnected.*

Our aim is to establish the following result of D. van Dantzig [5].

Theorem 1.4 *If G is locally compact, T_2 and totally disconnected, then every compact neighbourhood of e contains a closed and open subgroup.*

In the sequel we will call "clopen" the subsets which are closed and open.

The following lemma is the essential ingredient:

Lemma 1.5 *Let X be a compact T_2 space. Then for every $x \in X$, the subset*

$$K_x := \bigcap \{U : U \ni x, U \text{ is clopen}\}$$

is connected.

Proof Assume that $K_x = K_1 \cup K_2$ is a disjoint union of closed subsets with $K_1 \ni x$; pick open subsets $U_i \supset K_i$ with $U_1 \cap U_2 = \varnothing$ and consider $F := \overline{U}_2 \backslash U_2$. Since $K_2 \subset U_2$, we have $F \cap K_2 = \varnothing$ and since $F \subset \overline{U}_2 \subset U_1^c$ we have $F \cap K_1 = \varnothing$. Thus, $F \cap K_x = \varnothing$ and by compactness of X, there are finitely many clopen subsets $U_i \ni x$, $1 \le i \le n$, with

$$\bigcap_{i=1}^{n} U_i \cap F = \varnothing.$$

Setting $V := \bigcap_{i=1}^{n} U_i$, we deduce from $V \cap (\overline{U}_2 \backslash U_2) = \varnothing$ that $V \cap U_2^c = V \cap \overline{U}_2^c$ which shows that $V \cap U_2^c$ is a clopen set containing x and avoiding K_2. Thus, $K_2 = \varnothing$ and K_x is connected. $\square$

With this at hand we obtain the following important information concerning the topology of totally disconnected spaces:

Lemma 1.6 *Let X be locally compact, T_2 and totally disconnected. The family of compact open subsets is a basis for the topology of X.*

Proof Let $x \in X$ and $C \ni x$ a compact neighbourhood of x. Since X is totally disconnected T_2, Lemma 1.5 implies

$$\{x\} = \bigcap \{U : U \subset C, \ U \text{ clopen in } C, \ U \ni x\}.$$

If $\overset{\circ}{C}$ denotes the interior of C, we have $(C \backslash \overset{\circ}{C}) \cap \{x\} = \varnothing$, hence there are $U_1, \ldots, U_n$ clopen in C with $x \in \bigcap_{i=1}^{n} U_i$ and $(\bigcap_{i=1}^{n} U_i) \cap (C \backslash \overset{\circ}{C}) = \varnothing$. Thus, $\bigcap_{i=1}^{n} U_i$ is open in X; since at any rate it is compact, this finishes the proof of the lemma. $\square$

Lemma 1.7 *Let G be a topological group and $C \subset U$ with C compact and U open. Then there is $V \ni e$ open, with $V = V^{-1}$ and $C \cdot V \subset U$.*

Proof By continuity of the multiplication and the inverse, there is for every $x \in C$ and open $V_x \ni e$ with $V_x = V_x^{-1}$ and $xV_x^2 \subset U$. In particular, $C \subset \bigcup_{x \in C} xV_x$

and hence there exist $x_1, \ldots, x_n$ in C with $C \subset \bigcap_{i=1}^{n} x_i V_{x_i}$. Now set $V := \bigcap_{i=1}^{n} V_{x_i}$; then $V = V^{-1} \ni e$ is open and

$$C \cdot V \subset \bigcup_{i=1}^{n} x_i V_{x_i} V \subset \bigcup_{i=1}^{n} x_i V_{x_i}^2 \subset U.$$

$\square$

Proof of Theorem 1.4. Let $U \ni e$ be a neighbourhood of e. By Lemma 1.6 there exists $C \subset G$ compact open subset with $e \in C \subset U$. By Lemma 1.7, since C is both compact and open, there exists $V \ni e$, $V = V^{-1}$ open with $CV \subset C$ and hence $CV = C$. Thus, the subgroup

$$L := \{g \in G : C \cdot g = C\}$$

contains $V \ni e$ and hence is open in G; since $C \ni e$, we have $L \subset C$, and since L is closed it is therefore compact.

$\square$

Thus, it follows from Theorem 1.4 that for such locally compact totally disconnected groups, the local multiplicative structure is completely encoded in compact groups. Now one can go one step further and apply Theorem 1.4 to compact groups:

Corollary 1.8 *A compact, T_2 totally disconnected group is a projective limit of finite groups.*

Proof By Theorem 1.4, the set $\mathscr{V}$ of all compact open subgroups of G form a fundamental system of neighbourhoods of e. Since G is compact, for every $H \in \mathscr{O}$, G/H is finite and thus $\bigcap_{x \in G} xHx^{-1}$ is still open and normal in G. Thus, $\mathscr{V} = \{H \in \mathscr{O} : H \triangleleft G\}$ also form a fundamental system of neighbourhoods of e. For every H, the group G/H is finite and we deduce from the above that the continuous homomorphism

$$G \longrightarrow \prod_{H \in \mathscr{V}} G/H$$

$$g \longmapsto (g \bmod H)$$

is injective, and hence provides a topological isomorphism from G to a closed subgroup of the above product of finite groups.

$\square$

1.2 An application of the Gleason–Yamabe theorem

An interesting application of Theorem 1.4 is the following: let G be locally compact totally disconnected; then any continuous homomorphism $\pi: G \to$

$GL(n,\mathbf{R})$ has open kernel. Indeed, let $U \ni Id$ be an open neighbourhood of Id in $GL(n,\mathbf{R})$ which does not contain any non-trivial subgroup; then $\pi^{-1}(U)$ is an open neighbourhood of e and hence contains a compact open subgroup L. But then $\pi(L) \subset U$ must be the trivial group. Thus, one of the possible obstructions to be a Lie group is to be non-discrete, totally disconnected.

For connected groups, examples which are not Lie groups include $(S^1)^{\mathbf{N}}$ or $\prod_{n \geq 1} U(n)$. The latter example is somehow universal, in that every separable compact T_2 group is isomorphic to a closed subgroup thereof; this is an easy consequence of the Peter–Weyl theorem and had already been observed by von Neumann.

One of the major structure theorems for locally compact groups is the following

Theorem 1.9 (Gleason–Yamabe)   *Let G be locally compact T_2. Then there is an open subgroup G' which is a projective limit of Lie groups.*

When G is connected we have automatically that $G = G'$ and we obtain

Corollary 1.10   *Let G be locally compact, T_2, connected. Then there is a normal compact subgroup $K \lhd G$ such that the quotient G/K is a Lie group.*

Our aim is to obtain a variant of Corollary 1.10 where one divides by a characteristic subgroup. This will involve the concept of amenable radical. Recall first

Definition 1.11   A topological group G is amenable if it fixes a point in every non-empty convex-compact G-space.

Here a convex-compact G-space is a convex compact subset S of a T_2 locally convex topological vector space on which G acts continuously, linearly, and preserving S.

Now we sketch the proof of the existence of the amenable radical:

Theorem 1.12   *Let G be locally compact T_2. Then there exists a unique maximal closed amenable normal subgroup $A(G)$ of G. It is topologically characteristic and*

$$A\big(G/A(G)\big) = (e).$$

Proof   Let $\mathcal{N}$ be the set of all closed normal amenable subgroups of G and let $A(G) = \overline{\langle N \colon N \in \mathcal{N} \rangle}$ be the closed subgroup generated by them. Clearly, $A(G)$ is topologically characteristic and we proceed to show that it is amenable. Let S be a convex-compact $A(G)$-space, and let $\{N_1, \ldots, N_r\} \subset \mathcal{N}$ be any finite collection. Observe that $N_1 \ldots N_{r-1}$ is a normal subgroup of G, in particular normalised by N_r; thus, N_r acts in the set $S^{N_1 \cdots N_{r-1}}$ of $N_1 \ldots N_{r-1}$-fixed points in S and

$$\left(S^{N_1,\ldots,N_{r-1}}\right)^{N_r} = \bigcap_{i=1}^{r} S^{N_i}.$$

Take first $r = 2$: since N_1 is amenable, $S^{N_1} \neq \varnothing$; since now S^{N_1} is a non-empty convex-compact N_2-space, we have $(S^{N_1})^{N_2} \neq \varnothing$, and hence $S^{N_1} \cap S^{N_2} \neq \varnothing$. Using the amenability of every N_i, one shows by induction that $(S^{N_1,\ldots,N_{r-1}})^{N_r} \neq \varnothing$, and hence $\bigcap_{i=1}^{r} S^{N_i} \neq \varnothing$. Thus, since S is compact, we conclude

$$\bigcap_{N \in \mathcal{N}} S^N \neq \varnothing,$$

and hence by continuity, $S^{A(G)} \neq \varnothing$. This shows that $A(G)$ is amenable. Finally, the fact that $A(G/A(G)) = e$ is an easy exercise. $\qquad\square$

We can now draw the following easy consequence from the existence of the amenable radical and Corollary 1.10.

Corollary 1.13 *Assume that G is locally compact, T_2 and connected. Then the quotient*

$$G/A(G) \simeq \prod_{i=1}^{n} S_i$$

is isomorphic to a direct product of connected, simple, centre-free, non-compact Lie groups.

Proof By Corollary 1.10 there exists $K \lhd G$ compact such that G/K is a Lie group. Since K is compact, in particular amenable, and normal in G, we must have that $A(G) \supset K$. Thus, $L := G/A(G)$ is a Lie group as well and in addition $A(L) = (e)$. Thus, L has trivial solvable radical as well as trivial centre; thus it is a direct product of simple adjoint Lie groups none of which can be compact. $\qquad\square$

We turn now to our main application of the results so far obtained. Let G be locally compact T_2 and G° as above its connected component of e; then the amenable radical $A(G^\circ)$ is a characteristic subgroup of G° and hence it is normal in G; let

$$L := G/A(G^\circ)$$

denote the quotient group.

Theorem 1.14 *The group $L°$ is a direct product of adjoint connected simple non-compact Lie groups. Its centraliser $Z_L(L°)$ in L is totally disconnected and the direct product $L° \cdot Z_L(L°)$ is open and of finite index in L.*

In particular, $G/A(G°)$ is virtually the direct product of its connected component of e with a totally disconnected normal subgroup. Observe that this totally disconnected group, that is $Z_L(L°)$, is locally isomorphic to $G/G°$, so that one can at least "lift" $G/G°$ locally to $G/A(G°)$.

Proof of Theorem 1.14. Observe (exercise!) that $L° = G°/A(G°)$; thus $A(L°) = (e)$ and the first assertion of the theorem follows from Corollary 1.13. For every $g \in L$, let $i(g): L \to L$ denote the automorphism given by conjugation $\ell \longmapsto g\ell g^{-1}$. By restriction to $L°$, we obtain a continuous homomorphism $L \to \mathrm{Aut}(L°)$, $g \longmapsto i(g)|_{L°}$. It follows then from the fact that $L°$ is connected semisimple with finite centre, that the group $\mathrm{Inn}(L°)$ of inner automorphisms of $L°$ is open of finite index in $\mathrm{Aut}(L°)$. As a result, the subgroup

$$L^* := \left\{ g \in L : i(g)\big|_{L°} \in \mathrm{Inn}(L°) \right\}$$

is open and finite index in L as well. Thus, for every $g \in L^*$ there exists an $h \in L°$ such that $g\ell g^{-1} = h\ell h^{-1}$ for every $\ell \in L°$, that is $h^{-1}g \in Z_L(L°)$, which shows that $L^* = L° \cdot Z_L(L°)$. Observe that this product is direct since $Z_L(L°) \cap L° \subset Z(L°) = (e)$. Finally, the subgroup $Z_L(L°)$ is totally disconnected since it is isomorphic to the open subgroup $L^*/L°$ of $L/L°$. $\qquad\square$

1.3 Totally disconnected groups and actions on trees

If G is a compactly generated locally compact group, then so is $G/A(G°)$ and hence the totally disconnected group $H := Z_L(L°)$ is compactly generated as well. Such groups are naturally related to groups acting cofinitely on trees in the following way.

Let H be totally disconnected generated by a compact set C and let $U < H$ be a compact open subgroup. Let $\mathscr{G} = (V, E)$ be the undirected simple graph with vertex set $V := G/U$ and edge set

$$E = \{(gU, gcU) : g \in H,\ c \in UCU\}.$$

Then $\mathscr{G}$ is regular of finite valency $d := |UCU/U|$ and H acts as a group of automorphisms on $\mathscr{G}$; the kernel of this action is $K := \bigcap_{g \in H} gUg^{-1}$ and hence compact. We call $\mathscr{G}$ a **Cayley–Abels graph**.

Let $H_1 := H/K$ and $\mathscr{T}_d \to \mathscr{G}$ the universal covering of $\mathscr{G}$. Then we obtain an exact sequence

$$(e) \to \pi_1(\mathscr{G}) \to \widehat{H} \to H_1 \to (e),$$

where $\widetilde{H} < \text{Aut}\,\mathscr{T}_d$ is the group of all automorphisms of the d-regular tree which cover elements from H_1.

Observe that $\widetilde{H}$ is a closed subgroup of $\text{Aut}\,\mathscr{T}_d$ acting transitively on the set of vertices of $\mathscr{T}_d$, in particular is compactly generated. When H_1 is non-compact and $\mathscr{G}$ is not a tree, $\pi_1(\mathscr{G})$ is a free group on countably many generators. In particular, if $\widetilde{H}_r := \widetilde{H}/[\pi_1, [\pi_1, \ldots]]$ denotes the quotient by the rth-term of the derived series of π_1, the amenable radical of the compactly generated group $\widetilde{H}_r$ contains the free solvable group of rank r on countably many generators.

References for this chapter

[1] M. Burger and N. Monod. Continuous bounded cohomology and applications to rigidity theory. *Geom. Funct. Anal.*, 12(2):219–280, 2002.

[2] E. Hewitt and K.A. Ross. *Abstract harmonic analysis*, volume I, 2nd edition, Vol. 115. Springer Verlag, Berlin, 1979.

[3] O. Schreier. Die Untergruppen der freien Gruppen. *Ann. Math. Sem. Univ. Hamburg*, 5(1):161–183, 1927.

[4] T. Tao. Hilbert's fifth problem and related topics. 2012, book available at `http://terrytao.files.wordpress.com/2012/03/hilbert-book.pdf`.

[5] D. van Dantzig, *Studien over topologische algebra* (proefschrift). Ph.D. thesis (Groningen, 1931).

2

Locally compact groups as metric spaces

Romain Tessera

Abstract

It is explained how the approach of coarse geometry, well known in the study of finitely generated groups, can be extended to the context of locally compact groups.

2.1	Introduction	9
2.2	Coarse structure on locally compact groups	10
2.3	Examples of adapted pseudo-metrics	11
2.4	Coarse connectedness and compact generation	11
2.5	Coarse simple connectedness and compact presentability	12
2.6	Geometric amenability	14
References for this chapter		16

2.1 Introduction

The aim of this lecture is to give a short introduction to the large-scale geometry of locally compact groups. The general idea is that one can attach to every σ-compact locally compact group an essentially unique coarse metric structure. Our main goal will be to show that certain algebraic properties such as compact generation and compact presentability can be characterized purely in terms of large-scale geometry. In a last section, we consider the notion of amenability from this metric point of view. There is a large literature on this topic, and we simply refer to the book [1] for references.

2.2 Coarse structure on locally compact groups

An **adapted metric** on a σ-compact locally compact group is a left-invariant pseudo-metric $d : G \times G \to [0, \infty)$, which is bounded on compact subsets of $G \times G$ and such that balls are relatively compact. Note that d is not assumed to be compatible with the topology. When the group G is generated by a compact subset S, i.e. is *compactly generated*, then one checks that the word metric d_S is adapted (although it induces the discrete topology on G).

Proposition 2.1 *Every σ-compact locally compact group admits adapted metrics.*

Proof Let $(S_n)_{n \geq 1}$ be an increasing sequence of compact symmetric subsets of G such that $\cup_n S_n = G$. For every n, let d_n be the left-invariant word metric associated to S_n, with the convention that $d_n(g, h) = \infty$ if $g^{-1}h$ does not belong to the subgroup G_n generated by S_n. Let us define a "weighted" word metric associated to the sequences S_n as follows. For all $g, h \in G$, $d(g, h)$ is defined as the infimum of $\sum_{i=0}^{k-1} j_i d_{j_i}(g_i, g_{i+1})$ over all sequences $g = g_0, \ldots, g_k = h$, and all sequences of integers $j_0, \ldots, j_{k-1}$. One easily sees that $d(g, h)$ is an adapted metric. The only non-obvious thing to check is properness: this follows from the fact that if $g^{-1}h$ does not belong to G_n, then $d(g, h) \geq n + 1$. Indeed, given $g = g_0, \ldots, g_k = h$ and $j_0, \ldots, j_{k-1}$, there must exist at least some $0 \leq i < k$ such that $g_i^{-1}g_{i+1}$ does not belong to G_n. But since for all $q \leq n$, $d_q(g_i, g_{i+1}) = \infty$, we deduce that $j_i d_{j_i}(g_i, g_{i+1}) \geq n + 1$. $\square$

Recall that a **coarse equivalence** between two (pseudo)-metric space is a map $F : X \to Y$ such that there exist functions $\rho_1, \rho_2 : [0, \infty) \to [0, \infty)$ such that $\lim_{t \to \infty} \rho_1(t) = \infty$, and a constant C, satisfying for all $x, x' \in X$

$$\rho_1(d(x, x')) \leq d(F(x), F(x')) \leq \rho_2(d(x, x'),$$

and for all $y \in Y$, there exists $x \in X$ such that $d(y, f(x)) \leq C$. One checks that being coarse equivalent defines an equivalence relation between (pseudo)-metric spaces, and that two adapted metrics on a locally compact group are coarse equivalent.

Recall that an isometric action of a group G on a metric space X is called **cobounded** if there exists a bounded subset $Y \subset X$ such that $X = GY$. Assuming G locally compact, the action is called **locally bounded** if for all compact subset $K \subset G$ and $x \in X$, Kx is bounded. It is called **metrically proper** if for all $R > 0$ and all $x \in X$, the set of elements of G mapping x at distance $\leq R$ is relatively compact in G. Finally an action is called **geometric** if it is isometric, cobounded, locally bounded and metrically proper. As we shall see in the next section, adapted metrics are often obtained in the following way: consider a geometric action of a locally compact group G on a metric space X and let $x \in X$, then the pseudo-metric on G defined by $\delta(g,h) := d(gx,hx)$ is adapted.

2.3 Examples of adapted pseudo-metrics

- **(Left-invariant Riemannian metrics)** An important class of examples is that of connected Lie groups, which can be endowed with left-invariant Riemannian metrics. In the case of semi-simple Lie groups, it is sometimes more fruitful to consider pseudo-metrics associated to their actions on their associated symmetric spaces, which in some sense reflects better their large-scale geometry: for instance these spaces are simply connected, and admit non-positively curved Riemannian metrics.
- **(Cayley graphs)** Finitely generated groups act geometrically on their Cayley graphs. However, as for semi-simple Lie groups, it is sometimes useful to consider pseudo-metrics associated to geometric actions on various classes of metric spaces (e.g. CAT(0) spaces).
- **(Cayley–Abels graphs)** More generally, every totally discontinuous locally compact, compactly generated group can be turned[1] into the isometry group of some vertex-transitive connected graph. This is the Cayley–Abels construction which can be outlined as follows: given a compact open subgroup K of G and a compact generating symmetric subset S such that $S = KSK$, the Cayley–Abels graph (G,K,S) is the quotient of the Cayley graph (G,S) by the K-action by right-translations. Its vertex set being $X = G/K$, and K being open, it follows that the G-action by left-translations on its Cayley–Abels graph is continuous (see §1.3 in Chapter 1 of this volume). A famous example of Cayley–Abels graph is the Bass–Serre tree of $\mathrm{GL}(2,\mathbf{Q}_p)$.

2.4 Coarse connectedness and compact generation

Definition 2.2 A metric space X is **coarsely connected** if there exists $C > 0$ such that every pair of points can be connected by a "discrete path" $x = x_1,\ldots,x_n = y$ such that $d(x_i,x_{i+1}) \leq C$. We shall call such a path a C-path (or a C-loop if in addition $x = y$).

The interest of this notion is illustrated by the following easy fact.

Proposition 2.3 *Let G be a σ-compact locally compact group. The following are equivalent.*

---

[1] up to taking the quotient by a compact normal subgroup.

(i) G is compactly generated;

(ii) G admits an adapted coarse connected (pseudo)-metric;

(iii) G acts geometrically on a geodesic metric space.

Proof **(i) implies (ii):** this follows from the obvious observation that the word metric associated to a compact symmetric generating subset is coarsely connected.

(ii) implies (i): Assume G not compactly generated, and let $d(g,h)$ be an adapted metric on G. For all integers n, let G_n be the subgroup generated by the ball $B(e,n)$. Clearly, the distance between two distinct cosets of G_n is at least n. In particular, given $C > 0$, if the elements g and h belong to distinct cosets of G_n for $n > C$, then these two elements cannot be joined by a C-path.

(i) implies (iii): Assume G compactly generated and let S be a compact symmetric generating subset. Then G acts geometrically on its Cayley graph (G,S), which is a geodesic metric space.

(iii) implies (ii) Fix a point $x \in X$, and consider the adapted pseudo-metric δ on G defined by $\delta(g,h) = d(gx,hx)$. The fact that X is geodesic easily implies that δ is coarsely connected.   □

2.5 Coarse simple connectedness and compact presentability

Let (X,d) be a pseudo-metric space and let $C > 0$. We shall say that two finite C-paths γ and γ' are C-elementarily homotopic if they have the same extremities, and one path can be obtained from the other by removing one point. Two C-paths γ and γ' are C-homotopic if there exists a sequence of C-paths $\gamma = \gamma_1, \ldots, \gamma_k = \gamma'$ such that for each $i = 1, \ldots, k-1$, γ_i and γ_{i+1} are C-elementarily homotopic.

Definition 2.4 A metric space X is **coarsely simply connected** if it is coarsely connected and if for each large enough $C > 0$, there exists $C' \geq C$ such that any C-loop (which in particular is a C'-loop) based at some $x \in X$ is C'-homotopic to the constant C-loop at x.

It can be checked that coarse simple connectedness is invariant under coarse equivalence. On the other hand, one has

Proposition 2.5 *A path connected simply connected metric space is coarsely simply connected.*

Proof Let $C > 0$, and let $\gamma = (x = x_0, \ldots, x_n = x)$ be a C-loop in X based at $x \in X$. Since X is path connected, one can extend γ to a continuous loop

$\tilde{\gamma} \colon \mathbf{R}/\mathbf{Z} \to X$, such that $\tilde{\gamma}(i/n) = x_i$ for all $0 \le i \le n$. Our first remark is that for all large enough $A \in \mathbf{N}$, the finite sequence $\gamma' = (\tilde{\gamma}(i/(An)))_{0 \le i \le An}$ is also a C-loop, and is $2C$-homotopic to γ.

As X is simply connected, there is a continuous map $h \colon \mathbf{R}/\mathbf{Z} \times [0,1] \to X$, such that $h(\cdot,0) = \tilde{\gamma}$, and $h(\cdot,1) = x$. Using that h is uniformly continuous, taking N and A large enough, we have that

- for all $0 \le k \le N$, $\gamma_k := (h(i/(An),k/N))_{0 \le i \le An}$ is a C-loop;
- for every $0 \le k \le N-1$, and every $0 \le i \le An$,

$$d(\gamma_k(i), \gamma_{k+1}(i)) = d(h(i/(An),k/N), h(i/(An),(k+1)/N)) \le C.$$

It easily follows from the second point that γ_k and γ_{k+1} are $2C$-homotopic. Observe that $\gamma_0 = \gamma'$ and that $\gamma_N = \{x\}$. Hence, combining this with our first remark, we deduce that γ is $2C$-homotopic to the trivial loop based at x. This finishes the proof that X is coarsely simply connected. ☐

Recall that a group is **compactly presentable** if it admits a presentation $\langle S;R \rangle$, where S is a compact generating subset of G, and R is a set of words in S of length bounded by some constant k. These two notions are related via the following result.

Theorem 2.6 *Let G be a σ-compact locally compact group. The following are equivalent.*

  (i) G is compactly presentable;
 (ii) G admits an adapted coarse simply connected (pseudo)-metric;
(iii) G acts geometrically on a geodesic simply connected metric space.

Proof Except for one of them, we leave the proofs as exercises, only providing some indications.
(i) implies (iii): Let $\langle S;R \rangle$ be a compact presentation for G. Up to enlarging the generating set S, one can assume that all relators have length 3. Let X be the simplicial 2-complex, whose 1-skeleton is the Cayley graph (G,S), and where a Euclidean equilateral triangle is glued along each left-translate of a loop corresponding to a relator. One checks that this space is simply connected.
(iii) implies (ii): Follows from Proposition 2.5.
(ii) implies (i): Let us give more details for this one. Assume that d is an adapted C-coarsely simply connected pseudo-metric on G. Using that G is C-coarsely connected, one checks that $S = \overline{B(e,C)}$ is a (compact symmetric) generating subset of G. Assume that G is (C,NC)-simply connected for some $N \ge 1$. We define R to be the set of all words of length $3N$ in S that correspond to the trivial element of G. Let us prove that $\langle S;R \rangle$ is a presentation of G. This

amounts to proving that any *relation*, i.e. every reduced word in $\langle S \rangle$ corresponding to the trivial element in G, can be written as a product of conjugates of *relators* (elements of R). In order to simplify notation (without altering the main idea of the proof), we shall assume in the sequel that $N = 1$.

Note that a reduced word $w = s_1 \ldots s_n$ in $\langle S \rangle$ defines a C-path

$$(e = w(0), w(1) = s_1, w(2) = s_1 s_2, \ldots)$$

issued from the neutral element in G. Let $w = s_1 \ldots s_n$ and $w' = s'_1 \ldots s'_m$ be relations whose corresponding C-loops are C-elementarily homotopic. Up to permuting them, we can assume that w' is obtained from w by removing a point. In other words, $n = m+1$ and the C-loop corresponding to w' is

$$(e = w(0) \ldots w(i-1), w(i+1), \ldots w(n) = e)$$

for some $1 \leq i \leq n$. This translates into the fact that $w' = s_1 \ldots s_{i-1} s'_i s_{i+2} s_n$, where $s'_i = s_i s_{i+1}$. Observe that $r = s'_i s_{i+1}^{-1} s_i^{-1} \in R$, and that in the free group $\langle S \rangle$, $w'w^{-1} = (s_1 \ldots s_{i-1}) r (s_1 \ldots s_{i-1})^{-1}$. Since every C-loop is C-homotopic to the trivial loop, we deduce that every relation is a product of conjugates of elements of R. Hence $\langle S; R \rangle$ is a presentation of G. $\qquad\square$

2.6 Geometric amenability

A locally compact group is called **amenable** (in Følner's sense) if, for all $\varepsilon > 0$ and all compact subset $Q \subset G$, there exists a compact subset F such that $\mu(gF \triangle F) \leq \varepsilon \mu(F)$, where μ denotes a *left* Haar measure on G. Amenability has several other characterizations: for instance, a locally compact group is amenable if and only if every continuous action on a metrizable compact set preserves a Borel probability measure on that set. Amenability is stable under taking closed subgroups, quotients, extensions, and direct limits. Compact groups and locally compact solvable groups are amenable. Non-compact semisimple algebraic groups over a local field $\mathbf{k}$ (e.g. $SL(2, \mathbf{k})$) are non-amenable. We refer to [2, Appendix G] for a concise and very informative account of amenability.

Let us consider the following natural question: is amenability a geometric property? By *geometric property*, we mean a property characterized in terms of metric measure space. Moreover, we expect such a property to be invariant under coarse equivalence. This is known to be true for discrete groups. On the other hand, every connected Lie group (or every algebraic group over a local field) admits a co-compact amenable subgroup (take for instance a maximal solvable subgroup) and therefore is always coarse equivalent to an amenable group. A concrete example is as follows: for every local field $\mathbf{k}$, the subgroup T of $GL(2, \mathbf{k})$ consisting of upper triangular matrices is closed and cocompact, hence coarse equivalent to $GL(2, \mathbf{k})$. For $\mathbf{k}$ non-archimedean, T acts properly vertex-transitively on the Bass–Serre tree of $GL(2, \mathbf{k})$. Therefore in this case, T is coarse equivalent to a finitely generated non-abelian free group. On the other hand, T is solvable, hence amenable. As we shall see below, this apparent contradiction results from the fact that T is non-unimodular: in other words, its left Haar measure is not invariant under right multiplication.

To simplify the discussion let us assume that G is compactly generated, and that S is a compact symmetric generating subset of G containing 1. Følner's criterion is then equivalent to the existence of a sequence of compact subsets F_n of positive measure such that

$$\mu(SF_n \setminus F_n)/\mu(F_n) \to 0,$$

as n tends to infinity.

Given a graph X and a set of vertices A, we define the boundary of A as the set of vertices in the complement of A, which are related by an edge to a vertex of A. In other words,

$$\partial A := \{x \in [A]_1 \setminus A\},$$

where $[A]_1 = \{x \in V(X),\ d(x, A) \le 1\}$.

Definition 2.7 Consider the Cayley graph $X = (G, S)$. Let us say that G is **geometrically amenable** if there exists a sequence of compact subsets F_n of positive measure such that

$$\mu(\partial F_n)/\mu(F_n) \to 0.$$

It is easy to see that this definition does not depend on S. More generally, we say that a locally compact group G is geometrically amenable if all its compactly generated subgroups are geometrically amenable. It can be shown that geometric amenability is invariant under coarse equivalence (among σ-compact locally compact groups).

In order to compare this notion with Følner's criterion, observe that $[F_n]_1 = F_n S$, so that $\partial F_n = F_n S \setminus F_n$. Therefore, geometric amenability corresponds to Følner's criterion where left-translation is replaced with right-translation by S. It follows that if G is unimodular, then the two notions are equivalent. Moreover, it is easy to see that unimodularity is a necessary condition for being geometrically amenable. Indeed, let $\delta : G \to \mathbf{R}_+^*$ be the modular function. Let S be a compact generating set of G. If G is not unimodular, then one can find $s_0 \in S$

such that $\delta(s_0) > 1$. Then, for all compact subset $F \subset G$, $\mu(Fs_0) = \delta(s_0)\mu(F)$. It follows that

$$\mu(\partial F) \geq (\delta(s_0) - 1)\mu(F).$$

Let us summarize this discussion as follows.

Proposition 2.8 *Let G be a σ-compact locally compact group equipped with a left Haar measure. Then G is amenable and unimodular if and only if it is geometrically amenable. In particular being amenable and unimodular is invariant under coarse equivalence.*

References for this chapter

[1] Yves de Cornulier and Pierre de la Harpe. *Metric geometry of locally compact groups*, EMS Tracts in Mathematics, 25. European Mathematical Society, Zürich, 2016.

[2] Bachir Bekka, Pierre de la Harpe and Alain Valette. *Kazhdan's property (T)*, New Mathematical Monographs, 11. Cambridge University Press, Cambridge, 2008.

3

A short primer on profinite groups

John S. Wilson

Abstract

A short introduction to profinite groups is given, with a particular focus on areas of current activity.

3.1	Introduction	17
3.2	Infinite products and their closed subgroups	19
3.3	Galois groups	19
3.4	Profinite plagiarism	20
3.5	Subgroups of finite index	20
3.6	Baire's category theorem and applications	22
3.7	Profinite completions	23
3.8	Free groups	25
3.9	Small pro-p groups	26
3.10	Probability	27
3.11	The profinite topology on abstract groups; separability properties	29
3.12	Cohomology	30
3.13	Presentations of pro-p groups	32
	References for this chapter	34

3.1 Introduction

The aim of this chapter is to introduce profinite groups in some of their various guises and to proceed rapidly to a few topics of recent interest. The choice of topics is subjective and a rather incomplete representation of the many exciting developments in recent years.

Profinite groups may be defined in any of the following equivalent ways:

- as inverse limits of finite groups;
- as the compact Hausdorff totally disconnected groups;
- as the closed subgroups of Cartesian products of finite groups;
- as the Galois groups of algebraic field extensions.

Among the profinite groups, pro-p groups play a distinguished role, like the role of finite p-groups among finite groups. They can be defined as follows:

- as the inverse limits of finite p-groups;
- as the compact Hausdorff totally disconnected groups such that $x^{p^n} \to 1$ as $n \to \infty$ for all elements x;
- as the closed subgroups of Cartesian products of finite p-groups;
- as the Galois groups of algebraic field extensions such that each finite sub-extension has p-power degree.

Profinite groups arise in analysis and the general theory of topological groups as the quotients of compact groups modulo the connected component of the identity. Every totally disconnected locally compact group has a basis consisting of profinite groups by van Dantzig's theorem (see Chapter 1 in this volume). Profinite groups also arise naturally in the following contexts:

- finite group theory: as objects reflecting properties of infinite families of groups and allowing concise formulations of asymptotic results;
- infinite group theory: as examples, and as completions;
- algebraic number theory: as Galois groups; etc.
- model theory and combinatorics: as groups of automorphisms of rooted trees and other structures (in connection for example with the small index property).

Techniques from *all* of the above areas have been used in the study of profinite groups.

Since profinite groups are compact and Hausdorff, their subgroups of principal interest, those that are profinite in the subspace topology, are just the closed subgroups. Some of the standard terminology reflects this fact. We say that a subset X *topologically* generates a profinite group G if the abstract subgroup generated by X is dense, and we often omit the word 'topologically'. A profinite group is called *finitely generated* if it is generated by some finite set, in other words, if it has an abstract finitely generated dense subgroup.

3.2 Infinite products and their closed subgroups

Let $C = \prod_{\lambda \in \Lambda} G_\lambda$ be the Cartesian product of a family of groups $\{G_\lambda \mid \lambda \in \Lambda\}$. If each G_λ is a topological group, then so is C, with subbase of open sets

$$\{\pi_\lambda^{-1}(U_\lambda) \mid \lambda \in \Lambda, U_\lambda \text{ open in } G_\lambda\},$$

where $\pi_\lambda : C \to G_\lambda$ is the projection map for each λ. Tychonoff's theorem implies that if each G_λ is compact then so is C. Hence if each G_λ is compact and Hausdorff, then so is each closed subgroup of C.

In particular, we may take each G_λ to be a finite group with the discrete topology. The profinite groups are precisely the groups (topologically) isomorphic to closed subgroups of such products; and the countably based profinite groups are precisely the (groups isomorphic to) closed subgroups of the Cartesian product $\prod_{n \in \mathbf{N}} \mathrm{Sym}(n)$. Indeed, each countably based profinite group can be embedded in $\prod_{n \geqslant 5} A_n$, which is a 2-generator profinite group; see [25].

3.3 Galois groups

Let K/k be an algebraic field extension; thus K is a field, k is a subfield, and for each $a \in K$ there is a polynomial $f \in k[X] \setminus \{0\}$ such that $f(a) = 0$. The extension K/k is called a *Galois extension* if for each irreducible polynomial f over k the number $|\{a \in K \mid f(a) = 0\}|$ is 0 or $\deg f$. The *Galois group* $\mathrm{Gal}(K/k)$ consists of all field automorphisms of K that act trivially on k.

A *between field* of K/k is a subfield of K that contains k. Write

$$\mathcal{N} = \{\mathrm{Gal}(K/L) \mid L \text{ a between field}, L/k \text{ a finite Galois extension}\}.$$

We make $\mathrm{Gal}(K/k)$ into a topological group with respect to the *Krull topology*, in which $\mathcal{N}$ is taken as a base of neighbourhoods of 1. Then $\mathrm{Gal}(K/k)$ becomes a profinite group. Every profinite group is isomorphic to a Galois group $\mathrm{Gal}(K/k)$. Every countably based profinite group is isomorphic to a group $\mathrm{Gal}(K/k)$ with K countable.

In the context of arbitrary profinite groups, the fundamental theorem of Galois theory is the following assertion:

Theorem 3.1 *If K/k is a Galois extension of fields then $\gamma: F \mapsto \mathrm{Gal}(K/F)$ is an inclusion-reversing bijection from the set of between fields F of K/k to the set of closed subgroups of $G = \mathrm{Gal}(K/k)$; moreover*

(1) *$\gamma(F)$ is open if and only if $[F : k]$ is finite; then $|G : \gamma(F)| = [F : k]$;*
(2) *$\gamma(F) \triangleleft G$ if and only if F/k is Galois; then $G/\mathrm{Gal}(K/F) \cong \mathrm{Gal}(F/k)$.*

3.4 Profinite plagiarism

Some ideas concerning finite groups extend immediately to profinite groups.

Let π be a set of primes, and π' the complementary set. A positive integer n is called a π-number if all of its prime divisors are elements of π.

Now let G be a finite group, and H a subgroup. Then H is called a π-group if $|H|$ is a π-number, and H is called a Hall π-subgroup of G if also $|G:H|$ is a π'-number. (Thus if $\pi = \{p\}$ for some prime p then a Hall π-subgroup is just a Sylow p-subgroup.)

These definitions can be transferred to profinite groups G: a closed subgroup H of G is called

(1) a pro-π-subgroup if HN/N is a π-group for each $N \triangleleft_O G$;
(2) a Hall pro-π-subgroup if HN/N is a Hall π-subgroup of G/N for each $N \triangleleft_O G$.

Proposition 3.2 *Let G be a profinite group. If G/N has Hall π-subgroups for all $N \triangleleft_O G$ then G has Hall π-subgroups; if in addition the Hall π-subgroups of G/N are conjugate for each N then the Hall π-subgroups of G are all conjugate.*

The proof is an easy application of Zorn's lemma, or compactness, or non-emptiness of inverse images of finite sets. These three approaches can often be used interchangeably in the proof of results of this type.

Many elementary results about finite groups have obvious counterparts in the profinite setting. For example, a profinite group is pronilpotent (i.e. is an inverse limit of finite nilpotent groups) if and only if it is the Cartesian product of its Sylow subgroups. Moreover the *Frattini subgroup* $\Phi(G)$ of each profinite group G, that is, the intersection of all maximal open subgroups of G, is pronilpotent. If G is a pro-p group then $\Phi(G)$ is the closed subgroup (topologically) generated by $\{x^p \mid x \in G\} \cup \{[x,y] \mid x,y \in G\}$; then a subset X generates G if and only if it maps to a generating set in $G/\Phi(G)$ and so, if G is finitely generated, we have $|G/\Phi(G)| = p^d$ where d is the smallest size of a generating set for G.

3.5 Subgroups of finite index

Proposition 3.3 *Let G be profinite.*

(a) *Each open subgroup has finite index and is closed.*

(b) *Each closed subgroup of finite index is open.*

(c) *Each open subset is a union of cosets Nt with $N \lhd G$ open, $t \in T$.*

Assertions (a), (b) above are clear; the proof of assertion (c) relies on the general fact that a (locally) compact Hausdorff space is totally disconnected if and only if its compact open subsets form a base for its topology (see Lemma 1.6 in Chapter 1 of this volume).

Write $N \lhd_O G$ if N is an open normal subgroup of G and $N \lhd_f G$ if N is a normal subgroup of finite index in G.

Easy examples show that subgroups of finite index in profinite groups need not be open. Consider a product $C = \prod_{i \in \mathbf{N}} F_i$ with each F_i isomorphic to a non-trivial finite group F. From the definition of the product topology it follows that C has $\aleph_0$ open normal subgroups. However it is a fact that C has $2^{2^{\aleph_0}}$ subgroups of index $|F|$. When $F \cong \mathbf{Z}/p\mathbf{Z}$ for a prime p this is clear, since $\dim_{\mathbf{F}_p} C = 2^{\aleph_0}$, and the dual of a space of infinite dimension d has dimension 2^d.

It is natural to ask what profinite groups G have the property that all subgroups of finite index are closed. Here are some results on this question.

Theorem 3.4 (Saxl and Wilson [39], also Martinez and Zel'manov [30]) *Let $C = \prod_{i \in \mathbf{N}} F_i$ with each F_i a finite simple group. Then all subgroups of finite index are open if and only if there are only finitely many groups F_i of each isomorphism type.*

Theorem 3.5 (Smith and Wilson [41]) *Let G be profinite. Then all subgroups of G of finite index are open if and only if G has only countably many subgroups of finite index, and this is the case if and only if G has only finitely many subgroups of index n for each integer n.*

In the 1970s Serre observed that all subgroups of finite index in finitely generated pro-p groups are open and he speculated about whether the same holds for profinite groups. This matter has now been resolved:

Theorem 3.6 (Nikolov and Segal [31]) *If G is a finitely generated profinite group then all subgroups of finite index are open in G.*

The proof of this important result is long and difficult. An attempt to describe the strategy of the proof can be found in [50]; see also Chapter 5 in this volume.

3.6 Baire's category theorem and applications

Baire's category theorem allows us to reduce many problems about profinite groups to problems about finite groups. It plays a role in the proofs of all theorems in the previous section, For profinite groups, it can be stated as follows:

if G is a profinite group and $(C_n)_{n \in \mathbf{N}}$ is a countable family of closed subsets whose union contains a non-empty open set, then some C_n contains a non-empty open set; i.e. C_n contains an open coset Nt.

We shall give two illustrations of the use of Baire's category theorem.

Application 1. Suppose that G is profinite, and that the abstract commutator subgroup $G' = \langle [x,y] \mid x,y \in G \rangle$ is closed. The map

$$G \times \cdots \times G \to G: \quad (x_1, x_2, \ldots, x_{2n}) \mapsto [x_1, x_2][x_3, x_4] \ldots [x_{2n-1}, x_{2n}]$$

is continuous, and so its image C_n is closed in G, hence closed in G'. However $G' = \bigcup C_n$, and so some C_n contains a coset Nt with $N \lhd_O G'$, $t \in G'$. Therefore $N = Nt(t^{-1}) \subseteq C_{2n}$ and $G' = C_r$ for $r = 2n + |G'/N|$.

A similar argument shows that $G' = C_r$ for some r if it is assumed that $|G : G'|$ is countable instead of that G' is closed. Similar conclusions also hold if we replace the commutator word $[x,y]$ by any other word w from a free (abstract) group.

Application 2. Let G be a profinite torsion group (i.e. every element of G has finite order). For each n the set $D_n := \{g \in G \mid g^n = 1\}$ is the preimage of 1 under the continuous map $g \mapsto g^n$, and so is closed. Thus since $G = \bigcup D_n$ we can find $n \in \mathbf{N}$, $N \lhd_O G$ and $t \in G$ such that $Nt \subseteq D_n$.

Application 2 is used in the proofs of the following statements.

Theorem 3.7 (Wilson [43]). *If G is a compact torsion group then G has a finite series*

$$1 = G_0 \leqslant G_1 \leqslant \cdots \leqslant G_n = G$$

of closed characteristic subgroups with each G_i/G_{i-1} either (i) *a p-group for some prime p or* (ii) *(isomorphic to)* $\prod_{\lambda \in \Lambda}(S)$ *for some finite simple group S and some set* Λ.

Theorem 3.8 (Zel'manov [55]). *Every compact torsion group is locally finite (i.e. each finite subset generates a finite abstract subgroup).*

A subset of a group is said to have **finite exponent** if its elements have bounded orders. The example of finite dihedral 2-groups shows that the existence of a coset of given exponent and given index does not bound the exponent of a group. The following question remains open.

Question. Do all profinite torsion groups have finite exponent?

The answer is positive for abelian groups G: in the context and notation of Application 2, if $Nt \subseteq D_n$ then $x^n = x^n t^n = (xt)^n = 1$ for all $x \in N$ and so the elements of G have order dividing $n |G: N|$. The general case reduces to the case of profinite p-groups, by the first theorem of this section. Some progress has been made in a special case.

Theorem 3.9 (Khukhro and Shumyatsky [19]). *If G is a compact torsion group having an elementary abelian subgroup with finite centraliser, then G has finite exponent.*

Such questions are hard because one cannot in general deduce the existence of identities on subgroups from the existence of identities on open cosets. In a few cases this can be done (even with a weaker condition than a single identity on cosets). This was the case with Zel'manov's theorem above. We give another example. The Engel words are recursively defined by $[x,_0 y] = x$, $[x,_r y] = [x,_{r-1} y]$ for $r \geqslant 1$, and G is called an Engel group if for all $g, h \in G$ there is an $n \in \mathbf{N}$ such that $[g,_n h] = 1$.

Theorem 3.10 (Wilson and Zel'manov [52]). *Every finitely generated profinite Engel group is nilpotent.*

The start of the proof is an application of Baire's category theorem to find open cosets Nx_0, Ny_0 on which one of the words $[x,_n y]$ is a law, and then to pass to an associated Lie algebra L, and to deduce that a genuine Lie algebra identity holds on a large subalgebra of L.

3.7 Profinite completions

Let G be an *abstract* group and let $\{K_\lambda \mid \lambda \in \Lambda\}$ be the family of normal subgroups of finite index in G. Write $C = \prod_{\lambda \in \Lambda} (G/K_\lambda)$ and consider the map $g \mapsto (gK_\lambda)$ from G to C; let $\widehat{G}$ be the closure of the image of G and j the map $G \to \widehat{G}$.

The group $\widehat{G}$ is called the *profinite completion* of G; the image of G in $\widehat{G}$ is dense, and $\ker j = \bigcap_{\lambda \in \Lambda} K_\lambda$. The profinite completion has the following universal property:

if H is a profinite group and $\theta : G \to H$ a (continuous) homomorphism then there exists a unique (continuous) homomorphism $\bar{\theta}$ such that the following diagram commutes:

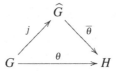

Examples. (1) The profinite completion of $\mathbf{Z}$ is $\prod_p \mathbf{Z}_p$, where $\mathbf{Z}_p$ is the group of p-adic integers.

(2) Kassabov and Nikolov [18] proved that if C is a Cartesian product of alternating groups and is finitely generated (as a profinite group) then C is the profinite completion of a subgroup that is finitely generated as an abstract group.

The universal property implies that all subgroups of finite index in a profinite group G are open if and only if the natural map from G (regarded as an abstract group) to its profinite completion is an isomorphism.

Pro-p completions are defined in an analogous way to profinite completions: instead of taking all normal subgroups of finite index in the construction one takes all normal subgroups of p-power index, and the universal property concerns maps to pro-p groups. Other completions can be obtained by taking different families $\mathscr{F}$ of normal subgroups K with the property that whenever $K_1, K_2 \in \mathscr{F}$ there is a subgroup $K_3 \in \mathscr{F}$ with $K_3 \leqslant K_1 \cap K_2$.

Completions via Cauchy sequences

Let G be an abstract group, and $(K_n)_{n \in \mathbf{N}}$ subgroups with $K_n \triangleleft_f G$, $K_{n+1} \leqslant K_n$, $\bigcap K_n = 1$. Then G is a metric space with metric defined by

$$d(x,y) = \inf\{n^{-1} \mid xy^{-1} \in K_n\}.$$

Moreover G becomes a metric group; that is, the map $(x,y) \mapsto xy^{-1}$ is continuous. We may now form the completion $\overline{G}$ of G as a metric space: the elements of $\overline{G}$ are equivalence classes of Cauchy sequences under the relation defined by

$$(a_n) \sim (b_n) \quad \text{if and only if} \quad d(a_n, b_n) \to 0.$$

It is easy to check that $\overline{G}$ is a profinite group.

If each subgroup of finite index contains some subgroup K_n then $\overline{G}$ is the profinite completion of G. Corresponding statements hold for the pro-p completion of G and other completions. For example, taking $G = \mathbf{Z}$ and $K_n = p^n \mathbf{Z}$ we have $\overline{G} = \mathbf{Z}_p$.

In performing calculations with profinite groups it is often helpful to think of elements of profinite groups as (represented by) limits of Cauchy sequences.

3.8 Free groups

The abstract free group F on a set X is defined by its universal property:

if H is a group and $\theta \colon X \to H$ is a function then there exists a unique homomorphism $\bar{\theta}$ such that the following diagram commutes:

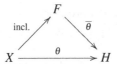

For X finite, the free profinite group on X is simply obtained by restricting to profinite groups H and requiring the maps $\bar{\theta}$ to be continuous. The free pro-p group on X is defined similarly. These groups can be constructed as the profinite and pro-p completions of the abstract free group on X.

For infinite sets X a small modification is required. A map $\theta \colon X \to H$ with H profinite is called 1-*convergent* if $\{x \mid \theta(x) \notin N\}$ is finite for each $N \triangleleft_O H$; in the definition of the free profinite group $\widehat{F}$ on X we insist that the inclusion map $X \to \widehat{F}$ is 1-*convergent* and restrict to 1-convergent maps θ to profinite groups H.

It is easy to see that the free pro-p group and free profinite group on a set with one element are respectively $\mathbf{Z}_p$ and $\prod_p \mathbf{Z}_p$. A number of Galois groups of interest are free, or have the property that all Sylow subgroups are free. For example, if k is a field of non-zero characteristic p and K is the union of all Galois extensions of finite p-power degree in an algebraic closure of k, then $\mathrm{Gal}(K/k)$ is a free pro-p group.

Closed subgroups of free pro-p groups are free pro-p groups. However the subgroup structure of free profinite groups is more complex. The maximal abelian subgroups are as diverse as they could be:

Theorem 3.11 (Haran and Lubotzky [15], Herfort and Ribes [16]) *Each non-abelian free profinite group has, among its maximal abelian subgroups, copies of the groups $\prod_{p \in L} \mathbf{Z}_p$ for all sets L of primes.*

The proof of this result in [15] used Galois-theoretic methods.

The free pro-p group on a finite set has a rather explicit description, given by Lazard [21]. Let $\mathbf{F}_p[\![y_1, \ldots, y_n]\!]$ be the formal power series ring in non-commuting indeterminates $y_1, \ldots, y_n$. Then clearly each element $x_i := 1 + y_i$ is invertible. Lazard proved that the closure of $\langle x_1, \ldots, x_n \rangle$ (in the obvious topology on the ring) is the free pro-p group E on $\{x_1, \ldots, x_n\}$, and moreover, that $\mathbf{F}_p[\![y_1, \ldots, y_n]\!]$ is the completed group algebra of E.

3.9 Small pro-p groups

Consider the group $\mathrm{SL}_n(\mathbf{Z}_p)$ of invertible $n \times n$ matrices of determinant 1 over the ring $\mathbf{Z}_p$ of p-adic integers, where $n \geqslant 2$, and write G for the kernel of the map from $\mathrm{SL}_n(\mathbf{Z}_p)$ to $\mathrm{SL}_n(\mathbf{F}_p)$ induced by the quotient map from $\mathbf{Z}_p$ to $\mathbf{F}_p = \mathbf{Z}_p/p\mathbf{Z}_p$. The group G is a pro-p group, and it has the structure of an analytic group over $\mathbf{Z}_p$. This group is in some senses the prototype for all p-adic analytic groups.

A pro-p group G is called *powerful* if its commutator subgroup is contained in the closed subgroup generated by all qth powers, where $q = p$ if p is odd and $q = 4$ if $p = 2$. The *lower Frattini series* of G is defined as follows:

$$\Phi_0(G) = G, \quad \text{and } \Phi_{r+1}(G) = \Phi(\Phi_r(G)) \text{ for } r \geqslant 0.$$

A pro-p group G is called *uniform* if it is finitely generated and powerful and if $|\Phi_r(G) : \Phi_{r+1}(G)| = |G : \Phi(G)|$ for all r. In [21], in a major and comprehensive study of p-adic analytic groups, Lazard proved, among other things, that a pro-p group has the structure of a p-adic analytic group if and only if it has an open uniform subgroup. More light was shed on p-adic analytic groups by Lubotzky and Mann in [24]: they characterized the p-adic analytic groups as the pro-p groups whose subgroups are all finitely generated and have bounded generating number. A completely different treatment of Lazard's theorem and a full account of the results of Lubotzky and Mann and many related results is given in [9]. The p-adic analytic groups are perhaps the best understood infinite profinite groups.

In [21] Lazard also proved that every pro-p group having an open uniform subgroup can be embedded in $\mathrm{SL}_n(\mathbf{Z}_p)$ for some integer n. Thus these groups have many subgroups, but, as we shall see, sometimes they have few quotient groups.

Simple profinite groups are clearly finite. Among the infinite profinite groups, those playing a role most similar to that of simple groups are the *just infinite* groups. These are the infinite profinite groups all of whose non-trivial closed normal subgroups are open; thus they have as few closed normal subgroups as possible. Examples are $\mathbf{Z}_p$, and the quotient of $\mathrm{SL}_n(\mathbf{Z}_p)$ modulo its centre for $n \geqslant 2$.

Further examples of just infinite pro-p groups are certain extensions G of finite direct powers A of $\mathbf{Z}_p$ by finite p-groups H, namely those such as the tensor product $A \otimes_{\mathbf{Z}_p} \mathbf{Q}_p$ is an irreducible $\mathbf{Q}_p H$-module on which H acts faithfully, with respect to the action of H on G induced by conjugation; here, $\mathbf{Q}_p$ is the field of fractions of $\mathbf{Z}_p$. These just infinite groups have played a profound role in the analysis of finite p-groups by coclass, based on the remarkable conjectures (now proved) of Leedham-Green and Newman [23]; see [22] for a discussion of the analysis. This approach to finite p-groups is one of the most striking examples of the importance of profinite groups in encoding and elucidating properties of finite groups.

Another just infinite pro-p group that has received much attention is the *Nottingham group* $\mathrm{Nott}(p)$: this can be described as the group of (continuous) automorphisms of the formal power series ring $\mathbf{F}_p[[t]]$ that induce the identity map in $\mathbf{F}_p[[t]]/t^2\mathbf{F}_p[[t]]$. It is a 2-generator pro-p group and in [5] Camina proved that every countably based pro-p group can be embedded in it. Like the groups $\mathrm{SL}_n(\mathbf{Z}_p)$, it is *hereditarily just infinite*; that is, each of its open subgroups is just infinite.

The just infinite profinite groups fall into three classes. First, there are those that have an open subgroup isomorphic to a finite power of $\mathbf{Z}_p$ for some prime p. The second comprises the just infinite groups not in the first class which have an open normal subgroup isomorphic to a finite direct power of a hereditarily just infinite group. The third class consists of the so-called just infinite *branch groups*. The dense abstract subgroups of such groups have provided many important counter-examples in abstract group theory. The just infinite analytic pro-p groups fall into the first two classes. In [48] examples were given of hereditarily just infinite profinite groups that have no open pro-p subgroup for any prime p.

3.10 Probability

Let G be a profinite group. The family of $\mathscr{B}$ of Borel sets in G is the smallest family of subsets that contains all closed subsets and that is closed with respect to taking complements and countable unions. The *normalised Haar measure* on G is the (unique) function $m: \mathscr{B} \to [0, 1]$ with the following properties: $m(\varnothing) = 0$, $m(G) = 1$, $m\left(\bigcup_{i=1}^{\infty} B_i\right) = \sum_{i=1}^{\infty} m(B_i)$ for each countable family $\{B_i\}_{i=1}^{\infty}$ of disjoint open members of $\mathscr{B}$, and $m(gB) = m(Bg) = m(B)$ for all $B \in \mathscr{B}$, $g \in G$. With respect to this measure, G can be regarded as a probability space. (See Fried and Jarden [12], pp. 201–206.) This fact has given rise to many interesting investigations. In [27], Mann introduced the notion of a positively finitely generated profinite group. For each profinite group G and integer

$k > 0$, write $P_G(k)$ for the probability that a random ordered k-tuple of elements of G generates G. The group G is said to be positively finitely generated (PFG) if $P_G(k) > 0$ for some k. In [29] it is proved that G is PFG if and only if the number of maximal subgroups of index n in G grows at most polynomially as a function of n. It is known from [4] that a finitely generated profinite group G is PFG unless every finite group is a continuous image of some open subgroup of G.

Now let G be finitely generated, and consider the series $\sum \mu_G(H)|G : H|^{-s}$, where H runs through all open subgroups of G, μ_G is the Möbius function, defined by requiring that $\mu_G(G) = 1$ and $\sum_{K \geqslant H} \mu_G(K) = 0$ for each open subgroup H of G, and s is a complex variable. Collecting summands corresponding to subgroups of equal index, one obtains a Dirichlet series $\sum_n a_n n^{-s}$.

Mann has shown in [27], [28] that this series converges in a half-plane for any prosoluble group G, and indeed for any profinite group G for which (i) $\mu_G(H)$ is bounded by a polynomial function of $|G : H|$ and (ii) the number of subgroups H of index n satisfying $\mu_G(H) \neq 0$ grows at most polynomially in n. If the series converges for a natural number k then the sum is $P_G(k)$, and accordingly the Dirichlet series is denoted by $P_G(s)$. The formal inverse of $P_G(s)$ is called the *probabilistic zeta function* of G; when $G = \widehat{\mathbf{Z}}$ it is the ordinary Riemann zeta function. In Detomi and Lucchini [8] it is shown that G is prosoluble if and only if the coefficients of the series $P_G(s) = \sum_n a_n n^{-s}$ satisfy $a_{ij} = a_i a_j$ whenever i and j are relatively prime, or, equivalently, if and only if $P_G(s)$ can be written as a product over all primes p of power series in p^{-s}.

We mention two further results on probability, and a question. As a consequence of more general results, Abért [1] has proved that n randomly chosen elements of a just infinite profinite branch group abstractly generate an abstract free subgroup of rank n with probability 1; he also obtained the same conclusion for groups having open subgroups mapping continuously to all finite alternating groups. An asymptotic result proved in [47], concerning generators for soluble subgroups of finite groups, can be succinctly formulated as follows: a profinite group has an open prosoluble subgroup if and only if with probability greater than 0 two randomly chosen elements (topologically) generate a prosoluble group.

In results concerning profinite groups, it is sometimes natural to ask whether a hypothesis on all subsets of n elements can be weakened to the hypothesis on almost all (with respect to the Haar measure) such subsets. This seems to lead to difficult problems. For example the following question is open; it reduces to a question about pro-p groups:

Question. Suppose that G is a profinite group and that with probability 1 a randomly chosen element of G has finite order. Does every element of G have finite order?

3.11 The profinite topology on abstract groups; separability properties

Let G be an *abstract* residually finite group. The topology on $\widehat{G}$ induces a topology on G: in this topology, the open sets are unions of cosets Nt with $|G: N|$ finite.

Despite the simplicity of this definition, the profinite topology leads quickly to difficult problems. Even in simple cases it has subtle properties that are not quite immediate. For example, in $\mathbf{Z}$, the set $\{n^2 \mid n \in \mathbf{Z}\}$ is closed but $\{6n^2 - 5n \mid n \in \mathbf{Z}\}$ is not closed; and the set of Fibonacci numbers (indexed by positive and negative integers) is closed. (See [42].)

The profinite topology is relevant to separability properties arising in connection with decision problems, such as the conjugacy problem and generalized word problem.

An abstract group G is called *conjugacy separable* if any two non-conjugate elements of G have non-conjugate images in some finite image of G, and G is *subgroup separable* (or *LERF*) if each finitely generated subgroup is the intersection of all subgroups of finite index containing it. It is easy to see that G is conjugacy separable if and only if each conjugacy class is closed (in the profinite topology on G), and that G is subgroup separable if and only if each finitely generated subgroup is closed.

Separability properties are subtle and hard to establish. For example, it is unknown whether if $|G: H| = 2$ then the conjugacy separability of one of G or H entails the conjugacy separability of the other.

To establish separability results for groups, one can either attempt to work within the groups themselves or pass to their profinite completions. Here are some results proved using the first approach.

Theorem 3.12

(a) (Fine–Rosenberger [11]) *Fuchsian groups are conjugacy separable.*

(b) (Wilson–Zalesskii [51]) *Many abstract branch groups are conjugacy separable.*

(c) (Grigorchuk–Wilson [14]) *The Grigorchuk group is subgroup separable.*

The second approach, involving passage from the group to its profinite completion, was pioneered by Ribes and Zalesskii.

For example, in [33] they established the conjugacy separability of certain groups $G_1 *_{\mathbf{Z}} G_2$ with G_1, G_2 conjugacy separable; suitable candidates for G_1, G_2 are virtually free groups and finitely generated nilpotent groups.

Theorem 3.13 (Ribes and Zalesskii [32]) *Let F be a free abstract group, $H_1, \ldots, H_n$ finitely generated subgroups of F. Then the product*

$$H_1 H_2 \ldots H_n = \{h_1 h_2 \ldots h_n \mid h_i \in H_i\}$$

is closed in F.

This very striking result solved a 20-year-old problem in semigroup theory; it has also connections with automata theory. The proof used a deep and powerful theory of profinite group actions on profinite trees, resembling the Bass–Serre theory for abstract groups, developed by Gildenhuis and Ribes in [13] and by Mel'nikov and Zalesskii in [53], [54]. A connected account of this theory appears in [34]. Now there are other proofs of the above theorem: for example a proof due to Coulbois [7] (building on work of Lascar and Herwig [17]) using ideas of model theory concerning extensions of partial isomorphisms of structures, and a constructive proof by Auinger and Steinberg [2] using inverse automata.

3.12 Cohomology

Let G be a profinite group and M a continuous G-module; that is, M is an abelian topological group and there is a continuous map $M \times G \to M$ making M into a G-module. Define the cohomology group $\mathrm{H}^n(G, M)$ in the usual way, but using only *continuous* cocycles from the set $G^{(n)}$ of n-tuples of elements of G to M.

Thus we let $C^n = C^n(G, M)$ be the additive group of continuous maps $G^{(n)} \to M$ for $n \geq 1$ and $C^0 = M$. We introduce certain group homomorphisms

$$\partial_n \colon C^{n-1} \to C^n$$

with $\partial_{n+1} \partial_n = 0$ for $n \geq 1$. So $B_n = \mathrm{im}\, \partial_n \leq Z_n = \ker \partial_{n+1}$, and we define $\mathrm{H}^n(G, M) = Z^n / B^n$ for each n. The group $\mathrm{H}^0(G, M)$ is the submodule M^G of fixed points of G.

There are two important special cases:

(1) M is compact and Hausdorff. This case becomes important in the study of

extensions

$$1 \to M \to H \to G \to 1$$

with H profinite and M abelian. For $n \leqslant 2$ there are the usual interpretations of $H^n(G,M)$ in terms of splittings over M and conjugacy classes of complements.

(2) M is discrete. In this case, by compactness, every continuous map $G^{(n)} \to M$ factors through $(G/N)^{(n)}$ for some $N \triangleleft_O G$. From this it follows fairly easily that

$$H^n(G,M) = \varinjlim_{N \triangleleft_O G} H^n(G/N, M^N);$$

here M^N is the submodule of points of M fixed by all elements of N.

In both of these cases there are long exact sequences associated with short exact sequences, and the cohomology is determined completely by the properties of these sequences and the vanishing of cohomology groups on a suitable family of modules.

The special case when G is a pro-p and $\mathbf{F}_p$ is regarded as a trivial module for G is of particular importance. In this case, all groups $H^n(G, \mathbf{F}_p)$ are vector spaces over $\mathbf{F}_p$. We have $\dim H^0(G, \mathbf{F}_p) = 1$, and $\dim H^1(G, \mathbf{F}_p)$ is the smallest number $d(G)$ of elements that can (topologically) generate G. One proof of a result in the next section uses the following elementary characterization of $H^2(G, \mathbf{F}_p)$ for a pro-p group G.

Lemma 3.14 *Suppose that $G = H/K$ with H a finitely generated pro-p group. Then*

$$d(H) - d_H(K) \geqslant d(G) - \dim(H^2(G, \mathbf{F}_p)),$$

with equality if H is a free pro-p group.

Here, for $K \triangleleft G$ we write $d_G(K)$ for the smallest cardinality of a set Y that generates K as a normal subgroup; thus $d_G(K)$ is the smallest number of G-conjugacy classes that can topologically generate K. The inequality has the natural interpretation if $d(G)$ is finite but some of the other terms are not.

Finite cohomological dimension.

The cohomological dimension $\operatorname{cd} G$ of a profinite group G is defined by writing $\operatorname{cd} G \leqslant n$ if $H^r(G,M) = 0$ for all $r > n$ and all discrete G-modules M. Thus $\operatorname{cd} G$ is a non-negative integer or ∞. Consideration of procyclic groups and use of the fact that $\operatorname{cd} H \leqslant \operatorname{cd} G$ if $H \leqslant G$ shows that $\operatorname{cd} G$ can only be finite if G is torsion-free, and by a theorem of Tate, $\operatorname{cd} G \leqslant 1$ if and only if all Sylow subgroups of G are free. (See [40].)

Many important examples of field extensions of local and global fields have Galois groups that either themselves have finite cohomological dimension or have Sylow subgroups of finite cohomological dimension (see [40]). For example, if l, p are primes and k, K are respectively a finite extension of $\mathbf{Q}_l$ and the algebraic closure of $\mathbf{Q}_l$, then the Sylow pro-p subgroups of $\mathrm{Gal}(K/k)$ have cohomological dimension at most 2. The groups P of this type that are not free constitute the important class of *Demushkin groups*, and they were completely classified by Labute [20].

The torsion-free p-adic analytic groups have finite cohomological dimension. Indeed, for uniform pro-p groups, a much stronger result holds. For any pro-p group there is a natural product operation, the *cup product*, that makes the direct sum

$$H(G, \mathbf{F}_p) = \bigoplus_{n \geqslant 0} H^n(G, \mathbf{F}_p)$$

into a ring.

Theorem 3.15 (Lazard, [21]) *Let G be a pro-p group. Then $H(G, \mathbf{F}_p)$ is the exterior algebra of a d-dimensional vector space V over $\mathbf{F}_p$ if and only if G is a uniform pro-p group such that $G/\Phi(G) \cong V$.*

3.13 Presentations of pro-p groups

Let G be a pro-p group. Again we write $d(G)$ for the size of a minimal generating set of G, and for $K \triangleleft G$ we write $d_G(K)$ for the smallest cardinality of a set that generates K as a normal subgroup.

A *presentation* of a pro-p group G is an epimorphism $\pi \colon F \to G$ from a free pro-p group F; it is a finite presentation if both $d(F)$ and $d_E(\ker \pi)$ are finite, and it has n generators and r relators if $d(F) = n$ and $d_F(\ker \pi) = r$. The generators are understood to be free generators of F and the relators are the members of a set Y with cardinality $d_F(\pi)$ that generates $\ker \pi$ as a normal subgroup.

The p-adic analytic groups are finitely presented. The Demushkin groups G mentioned in the previous section have finite presentations with one relator, and so $\dim H^2(G, \mathbf{F}_p) = 1$; they are precisely the one-relator groups for which the cup product bilinear map $H^1(G, \mathbf{F}_p) \times H_2(G, \mathbf{F}_p) \to H(G, \mathbf{F}_p)$ is non-singular. The work of Labute shows that for $p \neq 2$, they have a presentation $\pi \colon F \to G$ where F is free with basis $x_1, \ldots, x_{2d}$ for some $d \in \mathbf{N}$ and with relator

$$[x_1, x_2] \ldots [x_{2d-1}, x_{2d}] \quad \text{or} \quad x_1^q [x_1, x_2] \ldots [x_{2d-1}, x_{2d}]$$

for some power q of p.

Ershov proved [10] that for p odd the Nottingham pro-p group Nott(p) is finitely presented. The following question is open.

Question. Do there exist finitely presented just infinite branch pro-p groups?

We shall discuss a result concerning groups having presentations with fewer relations than generators. The following would be a reasonable exercise for undergraduate students of linear algebra:

Exercise 3.16 Let V be a vector space over k with a basis of n elements. Let R be a subspace generated by m elements where $m < n$. Let $S \subseteq V$ map to a generating set of V/R. Prove that there are $n - m$ elements of S that map to a basis of a subspace of V/R.

One might give students the hint that they should recall Steinitz' exchange lemma. One could also ask what exactly are the requirements on k in the Exercise. It is certainly sufficient for it to be a division ring.

It is perhaps remarkable that the statement of the Exercise remains valid when one passes from the category of vector spaces to the category of abstract groups or the category of pro-p groups:

Theorem 3.17 (Wilson [46]) *Let F be a free (abstract or pro-p) group of rank n. Let R be a normal subgroup of F that is generated as a normal subgroup by m elements where $m \leqslant n$. Let $S \subseteq F$ map to a generating set of F/R. Then there are $n - m$ elements of S that map to a free basis of a free subgroup of F/R.*

The result can be restated as follows: *if G has a presentation with n generators and m relators where $m \leqslant n$, and if S generates G, then some $n - m$ elements of S generate a free subgroup.*

The special case of the above theorem in which F is an abstract group, S is a free basis of F and $m = 1$ is essentially Magnus' famous Freiheitssatz [26]. The result in the case when S a free basis of F is deep and it is due to Romanovskii; for abstract groups it was proved in [35], and for pro-p groups in [36]. The case of the Theorem when F is an abstract group follows easily from the pro-p case and the fact that the pro-p completion of F is the free pro-p group on the same set.

The proof in [46] was an application of Romanovskii's deep result from [36] and easy cohomology. A more direct proof is given in [49]. The main ingredients are (a) the Exercise above, (b) a version of the Magnus embedding for pro-p groups and (c) in the pro-p case, the fact, proved by Romanovskii [36] using results of Cohn [6], that certain completed group algebras embed in division rings.

Further results of a similar character for abstract groups were proved in [38] and it is probable that the proofs can be modified to give analogous statements about pro-p groups. However, not all results for abstract presentations can be transferred easily to pro-p presentations.

Question. Let G be a finitely presented pro-p group with d generators and m relators, where $m \leqslant r - 2$. Is it true that some subgroup of finite index in G maps surjectively to a free pro-p group of rank 2?

The corresponding result for abstract groups is a well known theorem of Baumslag and Pride [3]. The proof relies crucially on the finiteness of words in the free generators, while in a pro-p group one only knows that the elements can be represented by infinite words, whose finite left segments form a Cauchy sequence giving approximations modulo terms of a central series. A proof in the case of pro-p groups would require entirely new ideas.

References for this chapter

[1] M. Abért. Group laws and free subgroups in topological groups. *Bull. London Math. Soc.* **37** (2005), 525–534.

[2] K. Auinger and B. Steinberg. A constructive version of the Ribes–Zalesskii product theorem. *Math. Z.* **250** (2005), 287–297.

[3] B. Baumslag and S. J. Pride. Groups with two more generators than relators. *J. London Math. Soc.* (2) **17** (1978), 425–426.

[4] A. V. Borovik, L. Pyber and A. Shalev. Maximal subgroups in finite and profinite groups. *Trans. Amer. Math. Soc.* **348** (1996), 3745–3761.

[5] R. Camina. The Nottingham group. In *New horizons in pro-p groups* (Birkhäuser, 2000), pp. 205–221.

[6] P. M. Cohn. On the embedding of rings in skew fields. *Proc. London Math. Soc.* (3) **11** (1961), 511–530.

[7] T. Coulbois. Free product, profinite topology and finitely generated subgroups. *Internat. J. Algebra Comput.* **11** (2001), 171–184.

[8] E. Detomi and A. Lucchini. Profinite groups with multiplicative probabilistic zeta function. *J. London Math. Soc.* (2) **70** (2004), 165–181.

[9] J. D. Dixon, M. P. F. du Sautoy, A. Mann and D. Segal. Analytic pro-p groups, 2nd edition (Cambridge University Press, 1999).

[10] M. V. Ershov. The Nottingham group is finitely presented. *J. London Math. Soc.* (2) **71** (2005), 362–378.

[11] B. Fine and G. Rosenberger. Conjugacy separability of Fuchsian groups and related questions. In *Combinatorial group theory, Contemp. Math.* **109** (1990), 11–18.

[12] M. D. Fried and M. Jarden. *Field arithmetic* (Springer-Verlag, 1986).

[13] D. Gildenhuis and L. Ribes. Profinite groups and Boolean graphs. *J. Pure Appl. Algebra* **12** (1978), 21–47.

[14] R. I. Grigorchuk and J. S. Wilson. A structural property concerning abstract commensurability of subgroups. *J. London Math. Soc.* (2) **68** (2003), 671–682.

[15] D. Haran and A. Lubotzky. Maximal abelian subgroups of free profinite groups. *Math. Proc. Cambridge Philos. Soc.* **97** (1985), 51–55.

[16] W. N. Herfort and L. Ribes. Torsion elements and centralizers in free products of profinite groups. *J. Reine Angew. Math.* **358** (1985), 155–161.

[17] B. Herwig and D. Lascar. Extending partial automorphisms and the profinite topology on free groups. *J. Reine Angew. Math.* **358** (1985), 155–161.

[18] M. Kassabov and N. Nikolov. Cartesian products as profinite completions. *Int. Math. Res. Not.* **2006**, Art. ID 72947, 1–17.

[19] E. I. Khukhro and P. Shumyatsky. Bounding the exponent of a finite group with automorphisms. *J. Algebra* **212** (1999), 363–374.

[20] J. Labute. Classification of Demushkin groups. *Canad. J. Math.* **19** (1967), 106–132.

[21] M. Lazard. Groupes analytiques p-adiques. *Inst. Hautes Études Sci. Publ. Math.* **26** (1965), 389–603.

[22] C. R. Leedham-Green and S. M. McKay. *The structure of groups of prime power order.* London Mathematical Society Monographs, New Series, 27 (Oxford University Press, 2002).

[23] C. R. Leedham-Green and M. F. Newman. Space groups and groups of prime-power order I. *Arch. Math. (Basel)* **35** (1980), 193–202.

[24] A. Lubotzky and A. Mann. Powerful p-groups. I. finite groups, II. p-adic analytic groups. *J. Algebra* **105** (1987), 484–505 and 506–515.

[25] A. Lubotzky and J. S. Wilson. An embedding theorem for profinite groups. *Arch. Math. (Basel)* **42** (1984), 397–399.

[26] W. Magnus. Über diskontinuierliche Gruppen mit einer definierenden Relation (Der Freiheitssatz). *J. Reine Angew. Math.* **163** (1930), 141–165.

[27] A. Mann. Positively finitely generated groups. *Forum Math.* **8** (1996), 429–459.

[28] A. Mann. A probabilistic zeta function for arithmetic groups. *Internat. J. Algebra. Comput.* **15** (2005), 1053–1059.

[29] A. Mann and A. Shalev. Simple groups, maximal subgroups, and probabilistic aspects of profinite groups. *Israel J. Math.* **96** (1996), 449–468.

[30] C. Martinez López and E. I. Zel'manov. Products of powers in finite simple groups. *Israel J. Math.* **96** (1996), 449–468.

[31] N. Nikolov and D. Segal. On finitely generated profinite groups. I. Strong completeness and uniform bounds, II. Products in quasisimple groups. *Ann. of Math.* (2) **165** (2007), 171–238 and 239–273.

[32] L. Ribes and P. A. Zalesskii. On the profinite topology on a free group. *Bull. London Math. Soc.* **25** (1993), 37–43.

[33] L. Ribes and P. A. Zalesskii. Conjugacy separability of amalgamated free products of groups. *J. Algebra* **179** (1996), 751–774.

[34] L. Ribes and P. A. Zalesskii. Pro-p trees and applications. In *New horizons in pro-p groups* (Birkhaüser, 2000), pp. 75–119.

[35] N. S. Romanovskii. Free subgroups of finitely presented groups. *Algebra and Logic* **16** (1977), 62-68.

[36] N. S. Romanovskii. A generalized theorem on freedom for pro-p groups. *Siberian Math. J.* **27** (1986), 267–280.

[37] N. S. Romanovskii. Shmel'kin embeddings for abstract and profinite groups. *Algebra and Logic* **38** (1999), 326–334.

[38] N. S. Romanovskii and J. S. Wilson. Free product decompositions in certain images of free products of groups. *J. Algebra* **310** (2007), 57–69.

[39] J. Saxl and J. S. Wilson. A note on powers in simple groups. *Math. Proc. Cambridge Philos. Soc.* **122** (1997), 91–94.

[40] J.-P. Serre. *Galois cohomology* (Springer-Verlag, 1965).

[41] M. G. Smith and J. S. Wilson. On subgroups of finite index in compact Hausdorff groups. *Arch. Math. (Basel)* **80** (2003), 123–129.

[42] J. S. Wilson. Polycyclic groups and topology. *Rend. Sem. Math. Fis. Milano* **51** (1981), 17–28.

[43] J. S. Wilson. On the structure of compact torsion groups. *Monatsh. Math.* **96** (1983), 57–66.

[44] J. S. Wilson. *Profinite groups* (Clarendon Press, Oxford, 1998).

[45] J. S. Wilson. On abstract and profinite just infinite groups. In *New horizons in pro-p groups* (Birkhäuser, 2000), pp. 181–203.

[46] J. S. Wilson. On growth of groups with few relators. *Bull. London Math. Soc.* **36** (2004), 1–2.

[47] J. S. Wilson. The probability of generating a soluble subgroup of a finite group. *J. London Math. Soc.* (2) **75** (2007), 431–446.

[48] J. S. Wilson. Large hereditarily just infinite groups. *J. Algebra* **324** (2010), 248–255.

[49] J. S. Wilson. Free subgroups in groups with few relators. *Enseign. Math.* **56** (2010), 173–185.

[50] J. S. Wilson. Finite index subgroups and verbal subgroups in profinite groups. *Astérisque* **339** (2011), 387–408.

[51] J. S. Wilson and P. A. Zalesskii. Conjugacy separability of certain torsion groups. *Arch. Math. (Basel)* **68** (1997) 441–449.

[52] J. S. Wilson and E. I. Zel'manov. Identities for Lie algebras of pro-p groups. *J. Pure Appl. Algebra* **81** (1992), 103–109.

[53] P. A. Zalesskii and O. V. Mel'nikov. Subgroups of profinite groups acting on trees. *Math. USSR-Sb.* **63** (1989), 405–424.

[54] P. A. Zalesskii and O. V. Mel'nikov. Fundamental groups of graphs of groups. *Leningrad Math. J.* **1** (1989), 921–940.

[55] E. I. Zel'manov. On periodic compact groups. *Israel J. Math.* **77** (1992), 83–95.

4

Lectures on Lie groups over local fields

Helge Glöckner

Abstract

The goal of these notes is to provide an introduction to p-adic Lie groups and Lie groups over fields of Laurent series, with an emphasis on the dynamics of automorphisms and the specialization of Willis' theory to this setting. In particular, we shall discuss the scale, tidy subgroups and contraction groups for automorphisms of Lie groups over local fields. Special attention is paid to the case of Lie groups over local fields of positive characteristic.

4.1	Introduction	38
4.2	Generalities	40
4.3	Basic facts concerning p-adic Lie groups	49
4.4	Lazard's characterization of p-adic groups	53
4.5	Iteration of linear automorphisms	54
4.6	Scale and tidy subgroups for automorphisms of p-adic Lie groups	57
4.7	p-adic contraction groups	58
4.8	Pathologies in positive characteristic	61
4.9	Tools from non-linear analysis: invariant manifolds	62
4.10	The scale, tidy subgroups and contraction groups in positive characteristic	64
4.11	The structure of contraction groups in the case of positive characteristic	67
4.12	Further results that can be adapted to positive characteristic	68
References for this chapter		70

4.1 Introduction

Lie groups over local fields (notably p-adic Lie groups) are among those totally disconnected, locally compact groups which both are well accessible and arise frequently. For example, p-adic Lie groups play an important role in the theory of pro-p-groups (i.e., projective limits of finite p-groups), where they are called "analytic pro-p-groups." In ground-breaking work in the 1960s, Michel Lazard obtained deep insights into the structure of analytic pro-p-groups and characterized them within the class of pro-p-groups [32] (see §3.9 in Chapter 3 of this volume; see also [10] and [11] for more details and later developments).

It is possible to study Lie groups over local fields from various points of view and on various levels, taking more and more structure into account. At the most basic level, they can be considered as mere topological groups. Next, we can consider them as analytic manifolds, enabling us to use ideas from Lie theory and properties of analytic functions. At the highest level, one can focus on those Lie groups which are linear algebraic groups over local fields (in particular, reductive or semi-simple groups), and study them using techniques from the theory of algebraic groups and algebraic geometry.

While much of the literature focusses either on pro-p-groups or on algebraic groups, our aims are complementary: We emphasize aspects at the borderline between Lie theory and the structure theory of totally disconnected, locally compact groups, as developed in [46], [47] and [3]. In particular, we shall discuss the scale, tidy subgroups and contraction groups for automorphisms of Lie groups over local fields.

For each of these topics, an algebraic group structure does not play a role, but merely the Lie group structure.

For the definition and basic properties of the scale function *and of* tidy subgroups, *see Chapter 9 in this volume; compare also 4.21 below.*

Scope and structure of the lectures. After an introduction to some essentials of Lie theory and calculus over local fields, we discuss the scale, contraction groups, tidy subgroups (and related topics) in three stages:

- In a first step, we consider linear automorphisms of finite-dimensional vector spaces over local fields.

A sound understanding of this special case is essential.

- Next, we turn to automorphisms of p-adic Lie groups.
- Finally, we discuss Lie groups over local fields of positive characteristic and their automorphisms.

For various reasons, the case of positive characteristic is more complicated than the p-adic case, and it is one of our goals to explain which additional ideas are needed to tackle this case.

As usual in an expository article, we shall present many facts without proof. However, we have taken care to give at least a sketch of proof for all central results, and to explain the main ideas. In particular, we have taken care to explain carefully the ideas needed to deal with Lie groups over local fields of positive characteristic, and found it appropriate to give slightly more details when they are concerned (the more so because not all of the results are available yet in the published literature).

Mention should be made of what these lectures do not strive to achieve.

First of all, we do not intend to give an introduction to linear algebraic groups over local fields, nor to Bruhat–Tits buildings or to the representation theory of p-adic groups (and harmonic analysis thereon). The reader will find nothing about these topics here.

Second, we shall hardly speak about the theory of analytic pro-p-groups, although this theory is certainly located at a very similar position in the spectrum ranging from topological groups to algebraic groups (one of Lazard's results will be recalled in Section 4.4, but no later developments). One reason is that we want to focus on aspects related to the structure theory of totally disconnected, locally compact groups—which is designed primarily for the study of non-compact groups.

It is interesting to note that also in the area of pro-p-groups, Lie groups over fields (and suitable pro-p-rings) of positive characteristic are attracting more and more attention. The reader is referred to the seminal work [33] and subsequent research (like [1], [2], [7], [28] and [29]; cf. also [11]).

Despite the importance of p-adic Lie groups in the area of pro-p-groups, it is not fully clear yet whether p-adic Lie groups (or Lie groups over local fields) play a fundamental role in the theory of general locally compact groups (comparable to that of real Lie groups, as in [25]). Two recent studies indicate that this may be the case, at least in the presence of group actions. They describe situations where Lie groups over local fields are among the building blocks for general structures:

Contraction groups Let G be a totally disconnected, locally compact group and $\alpha\colon G \to G$ be an automorphism which is **contractive**, i.e.,

$$\lim_{n\to\infty} \alpha^n(x) = 1,$$

for each $x \in G$ (where $1 \in G$ is the neutral element). Then the torsion elements form a closed subgroup $\mathrm{tor}(G)$ and

$$G = \mathrm{tor}(G) \times G_{p_1} \times \cdots \times G_{p_n}$$

for certain α-stable p-adic Lie groups G_p (see [22]).

$\mathbf{Z}^n$-actions If $\mathbf{Z}^n$ acts with a dense orbit on a locally compact group G via automorphisms, then G has a compact normal subgroup K invariant under the action such that

$$G/K \cong \mathbf{L}_1 \times \cdots \times \mathbf{L}_m,$$

where $\mathbf{L}_1,\ldots,\mathbf{L}_m$ are local fields of characteristic 0 and the action is diagonal, via scalar multiplication by field elements [9].

In both studies, Lazard's theory of analytic pro-p-groups accounts for the occurrence of p-adic Lie groups.

4.2 Generalities

This section compiles elementary definitions and facts concerning local fields, analytic functions and Lie groups.

Basic information on local fields can be found in many books, e.g., [44] and [39]. Our main sources for Lie groups over local fields are [40] and [6].

Local fields

By a **local field**, we mean a totally disconnected, locally compact, non-discrete topological field $\mathbf{K}$. Each local field admits an **ultrametric absolute value** $|.|$ defining its topology, i.e.,

(a) $|t| \geq 0$ for each $t \in \mathbf{K}$, with equality if and only if $t = 0$;
(b) $|st| = |s| \cdot |t|$ for all $s, t \in \mathbf{K}$;
(c) The **ultrametric inequality** holds, i.e., $|s+t| \leq \max\{|s|, |t|\}$ for all $s, t \in \mathbf{K}$.

An example of such an absolute value is what we call the **natural absolute value**, given by $|0| := 0$ and

$$|x| := \Delta_{\mathbf{K}}(m_x) \quad \text{for } x \in \mathbf{K} \setminus \{0\},$$

where $m_x \colon \mathbf{K} \to \mathbf{K}, y \mapsto xy$ is scalar multiplication by x and $\Delta_{\mathbf{K}}(m_x)$ its module, the definition of which is recalled in 4.3 (cf. [44, Chapter II, §2]).[1]

It is known that every local field $\mathbf{K}$ either is a field of formal Laurent series over some finite field (if $\text{char}(\mathbf{K}) > 0$), or a finite extension of the field of p-adic numbers for some prime p (if $\text{char}(\mathbf{K}) = 0$). Let us fix our notation concerning these basic examples.

Example 4.1 Given a prime number p, the field $\mathbf{Q}_p$ of p-adic numbers is the completion of $\mathbf{Q}$ with respect to the p-adic absolute value,

$$\left| p^k \frac{n}{m} \right|_p := p^{-k} \quad \text{for } k \in \mathbf{Z} \text{ and } n, m \in \mathbf{Z} \setminus p\mathbf{Z}.$$

We use the same notation, $|.|_p$, for the extension of the p-adic absolute value to $\mathbf{Q}_p$. Then the topology coming from $|.|_p$ makes $\mathbf{Q}_p$ a local field, and $|.|_p$ is the natural absolute value on $\mathbf{Q}_p$. Every non-zero element x in $\mathbf{Q}_p$ can be written uniquely in the form

$$x = \sum_{k=n}^{\infty} a_k p^k$$

with $n \in \mathbf{Z}$, $a_k \in \{0, 1, \dots, p-1\}$ and $a_n \neq 0$. Then $|x|_p = p^{-n}$. The elements of the form $\sum_{k=0}^{\infty} a_k p^k$ form the subring $\mathbf{Z}_p = \{x \in \mathbf{Q}_p \colon |x|_p \leq 1\}$ of $\mathbf{Q}_p$, which is open and also compact, because it is homeomorphic to $\{0, 1, \dots, p-1\}^{\mathbf{N}_0}$ via $\sum_{k=0}^{\infty} a_k p^k \mapsto (a_k)_{k \in \mathbf{N}_0}$.

Example 4.2 Given a finite field $\mathbf{F}$ (with q elements), we let $\mathbf{F}((X)) \subseteq \mathbf{F}^{\mathbf{Z}}$ be the field of formal Laurent series $\sum_{k=n}^{\infty} a_k X^k$ with $a_k \in \mathbf{F}$. Here addition is pointwise, and multiplication is given by the Cauchy product. We endow $\mathbf{F}((X))$ with the topology arising from the ultrametric absolute value

$$\left| \sum_{k=n}^{\infty} a_k X^k \right| := q^{-n} \quad \text{if } a_n \neq 0. \tag{4.1}$$

Then the set $\mathbf{F}[[X]]$ of formal power series $\sum_{k=0}^{\infty} a_k X^k$ is a compact and open subring of $\mathbf{F}((X))$, and thus $\mathbf{F}((X))$ is a local field. Its natural absolute value is given by (4.1).

---

[1] Note that if $\mathbf{K}$ is an extension of $\mathbf{Q}_p$ of degree d, then $|p| = p^{-d}$ depends on the extension.

Beyond local fields, we shall occasionally use **ultrametric fields** $(\mathbf{K}, |.|)$. Thus $\mathbf{K}$ is a field and $|.|$ an ultrametric absolute value on $\mathbf{K}$ which defines a non-discrete topology on $\mathbf{K}$. For example, we occasionally use an algebraic closure $\overline{\mathbf{K}}$ of a local field $\mathbf{K}$ and exploit that an ultrametric absolute value $|.|$ on $\mathbf{K}$ extends uniquely to an ultrametric absolute value on $\overline{\mathbf{K}}$ (see, e.g., [39, Theorem 16.1]). The same notation, $|.|$, will be used for the extended absolute value. An ultrametric field $(\mathbf{K}, |.|)$ is called **complete** if $\mathbf{K}$ is a complete metric space with respect to the metric given by $d(x,y) := |x-y|$.

Basic consequences of the ultrametric inequality

Let $(\mathbf{K}, |.|)$ be an ultrametric field and $(E, \|.\|)$ be a normed $\mathbf{K}$-vector space whose norm is ultrametric in the sense that $\|x+y\| \leq \max\{\|x\|, \|y\|\}$ for all $x, y \in E$.

Since $\|x\| = \|x+y-y\| \leq \max\{\|x+y\|, \|y\|\}$, it follows that $\|x+y\| \geq \|x\|$ if $\|y\| < \|x\|$ and hence

$$\|x+y\| = \|x\| \quad \text{for all } x,y \in E \text{ such that } \|y\| < \|x\| \tag{4.2}$$

("the strongest wins," [38, p. 77]). Hence, small perturbations have a much smaller impact in the ultrametric case than they would have in the real case (a philosophy which will be made concrete below).

The ultrametric inequality has many other useful consequences. For example, consider the balls

$$B_r^E(x) := \{y \in E : \|y-x\| < r\}$$

and

$$\overline{B}_r^E(x) := \{y \in E : \|y-x\| \leq r\}$$

for $x \in E$, $r \in]0, \infty[$. Then $B_r^E(0)$ and $\overline{B}_r^E(0)$ are *subgroups* of $(E, +)$ with non-empty interior (and hence are both open and closed). Specializing to $E = \mathbf{K}$, we see that

$$\mathbf{O} := \{t \in \mathbf{K} : |t| \leq 1\} \tag{4.3}$$

is an open subring of $\mathbf{K}$, its so-called **valuation ring**. If $\mathbf{K}$ is a local field, then $\mathbf{O}$ is a compact subring of $\mathbf{K}$ (which is maximal and independent of the choice of absolute value).

Calculation of indices

Indices of compact subgroups inside others are omnipresent in the structure theory of totally disconnected groups. We therefore perform some basic calculations of indices now. Haar measure λ on a locally compact group G will be used as a tool and also the notion of the **module** of an automorphism, which measures the distortion of Haar measure. We recall:

4.3 *Let G be a locally compact group and $\mathbf{B}(G)$ be the σ-algebra of Borel subsets of G. We let $\lambda : \mathbf{B}(G) \to [0,\infty]$ be a Haar measure on G, i.e., a non-zero Radon measure which is left-invariant in the sense that $\lambda(gA) = \lambda(A)$ for all $g \in G$ and $A \in \mathbf{B}(G)$. It is well known that a Haar measure always exists, and that it is unique up to multiplication with a positive real number (cf. [24]). If $\alpha : G \to G$ is a (bicontinuous) automorphism, then also*

$$\mathbf{B}(G) \to [0,\infty], \qquad A \mapsto \lambda(\alpha(A))$$

is a left-invariant non-zero Radon measure on G and hence a multiple of Haar measure: There exists $\Delta(\alpha) > 0$ such that $\lambda(\alpha(A)) = \Delta(\alpha)\lambda(A)$ for all $A \in \mathbf{B}(G)$. If $U \subseteq G$ is a relatively compact, open, non-empty subset, then

$$\Delta(\alpha) = \frac{\lambda(\alpha(U))}{\lambda(U)} \tag{4.4}$$

(cf. [24, (15.26)], where however the conventions differ). We also write $\Delta_G(\alpha)$ instead of $\Delta(\alpha)$, if we wish to emphasize the underlying group G.

Remark 4.4 Let U be a compact open subgroup of G. If $U \subseteq \alpha(U)$, with index $[\alpha(U) : U] =: n$, we can pick representatives $g_1, \ldots, g_n \in \alpha(U)$ for the left cosets of U in $\alpha(U)$. Exploiting the left-invariance of Haar measure, (4.4) turns into

$$\Delta(\alpha) = \frac{\lambda(\alpha(U))}{\lambda(U)} = \sum_{j=1}^{n} \frac{\lambda(g_j U)}{\lambda(U)} = [\alpha(U) : U]. \tag{4.5}$$

If $\alpha(U) \subseteq U$, applying (4.5) to α^{-1} instead of α and $\alpha(U)$ instead of U, we obtain

$$\Delta(\alpha^{-1}) = [U : \alpha(U)]. \tag{4.6}$$

In the following, the group $\mathrm{GL}_n(\mathbf{K})$ of invertible $n \times n$-matrices will frequently be identified with the group $\mathrm{GL}(\mathbf{K}^n)$ of linear automorphisms of $\mathbf{K}^n$.

Lemma 4.5 *Let $\mathbf{K}$ be a local field, $|.|$ its natural absolute value and $A \in \mathrm{GL}_n(\mathbf{K})$. Then*

$$\Delta_{\mathbf{K}^n}(A) = |\det A| = \prod_{i=1}^{n} |\lambda_i|, \tag{4.7}$$

where $\lambda_1, \ldots, \lambda_n$ are the eigenvalues of A in an algebraic closure $\overline{\mathbf{K}}$ of $\mathbf{K}$.

Proof Let $\lambda_{\mathbf{K}^n}$ and λ be Haar measures on $(\mathbf{K}^n, +)$ and $(\mathbf{K}, +)$, respectively, such that $\lambda_{\mathbf{K}^n} = \lambda \otimes \cdots \otimes \lambda$. Let $\mathbf{O} \subseteq \mathbf{K}$ be the valuation ring and $U := \mathbf{O}^n$. Since both $A \mapsto \Delta_{\mathbf{K}^n}(A)$ and $A \mapsto |\det A|$ are homomorphisms from $\mathrm{GL}_n(\mathbf{K})$ to the multiplicative group $]0, \infty[$, it suffices to check the first equality in (4.7) for diagonal matrices and elementary matrices of the form $\mathbf{1} + e_{ij}$ with matrix units e_{ij}, where $i \neq j$ (because these generate $\mathrm{GL}_n(\mathbf{K})$ as a group). For such an elementary matrix A, we have $A(U) = U$ and thus

$$\Delta_{\mathbf{K}^n}(A) = \frac{\lambda_{\mathbf{K}^n}(A(U))}{\lambda_{\mathbf{K}^n}(U)} = 1 = |\det A|.$$

If A is diagonal with diagonal entries $t_1, \ldots, t_n$, we also have

$$\Delta_{\mathbf{K}^n}(A) = \frac{\lambda_{\mathbf{K}^n}(A(U))}{\lambda_{\mathbf{K}^n}(U)} = \prod_{i=1}^{n} \frac{\lambda(t_i \mathbf{O})}{\lambda(\mathbf{O})} = \prod_{i=1}^{n} \Delta_{\mathbf{K}}(m_{t_i}) = \prod_{i=1}^{n} |t_i| = |\det A|,$$

as required. The final equality in (4.7) is clear if all eigenvalues lie in $\mathbf{K}$. For the general case, pick a finite extension $\mathbf{L}$ of $\mathbf{K}$ containing the eigenvalues, and let $d := [\mathbf{L} : \mathbf{K}]$ be the degree of the field extension. Then $\Delta_{\mathbf{L}^n}(A) = (\Delta_{\mathbf{K}^n}(A))^d$. Since the extended absolute value is given by

$$|x| = \sqrt[d]{\Delta_{\mathbf{L}}(m_x)}$$

(see [27, Chapter 9, Theorem 9.8] or [39, Exercise 15.E]), the final equality follows from the special case already treated (applied now to $\mathbf{L}$). $\qquad\square$

4.6 *For later use, consider a ball $B_r(0) \subseteq \mathbf{Q}_p^n$ with respect to the maximum norm, where $r \in]0, \infty[$. Let $m \in \mathbf{N}$ and $A \in \mathrm{GL}_n(\mathbf{Q}_p)$ be the diagonal matrix with diagonal entries $p^m, \ldots, p^m$. Then (4.6) and (4.7) imply that*

$$[B_r(0) : B_{p^{-m}r}(0)] = [B_r(0) : A.B_r(0)] = |\det(A^{-1})| = p^{mn}. \qquad (4.8)$$

Analytic functions, manifolds and Lie groups

Given a complete ultrametric field $(\mathbf{K}, |.|)$ and $n \in \mathbf{N}$, we equip $\mathbf{K}^n$ with the maximum norm, $\|(x_1, \ldots, x_n)\| := \max\{|x_1|, \ldots, |x_n|\}$ (the choice of norm does not really matter because all norms are equivalent; see [39, Theorem 13.3]). If $\alpha \in \mathbf{N}_0^n$ is a multi-index, we write $|\alpha| := \alpha_1 + \cdots + \alpha_n$. We mention that confusion with the absolute value $|.|$ is unlikely; the intended meaning of $|.|$ will always be clear from the context. If $\alpha \in \mathbf{N}_0^n$ and $y = (y_1, \ldots, y_n) \in \mathbf{K}^n$, we abbreviate $y^\alpha := y_1^{\alpha_1} \cdots y_n^{\alpha_n}$, as usual. See [40] for the following concepts.

Definition 4.7 Given an open subset $U \subseteq \mathbf{K}^n$, a map $f\colon U \to \mathbf{K}^m$ is called **analytic**[2] if it is given locally by a convergent power series around each point $x \in U$, i.e.,

$$f(x+y) = \sum_{\alpha \in \mathbf{N}_0^n} a_\alpha y^\alpha \quad \text{for all } y \in \overline{B}_r^{\mathbf{K}^n}(0),$$

with $a_\alpha \in \mathbf{K}^m$ and some $r > 0$ such that $\overline{B}_r^{\mathbf{K}^n}(x) \subseteq U$ and

$$\sum_{\alpha \in \mathbf{N}_0^n} \|a_\alpha\| r^{|\alpha|} < \infty.$$

It can be shown that compositions of analytic functions are analytic [40, Theorem, p. 70]. We can therefore define an n-dimensional analytic manifold M over a complete ultrametric field $\mathbf{K}$ in the usual way, namely as a Hausdorff topological space M, equipped with a set $\mathscr{A}$ of homeomorphisms ϕ from open subsets of M onto open subsets of $\mathbf{K}^n$ (the so-called **charts**) such that the transition map $\psi \circ \phi^{-1}$ is analytic, for all $\phi, \psi \in \mathscr{A}$.

Analytic mappings between analytic manifolds are defined as usual (by checking analyticity in local charts).

A **Lie group** over a complete ultrametric field $\mathbf{K}$ is a group G, equipped with a (finite-dimensional) analytic manifold structure which turns the group multiplication

$$m\colon G \times G \to G, \quad m(x,y) := xy$$

and the group inversion

$$j\colon G \to G, \quad j(x) := x^{-1}$$

into analytic mappings.

Besides the additive groups of finite-dimensional $\mathbf{K}$-vector spaces, the most obvious examples of $\mathbf{K}$-analytic Lie groups are general linear groups.

Example 4.8 $\mathrm{GL}_n(\mathbf{K}) = \det^{-1}(\mathbf{K}^\times)$ is an open subset of the space $M_n(\mathbf{K}) \cong \mathbf{K}^{n^2}$ of $n \times n$-matrices and hence is an n^2-dimensional $\mathbf{K}$-analytic manifold. The group operations are rational maps and hence analytic.

Beyond this, one can show (cf. [34, Chapter I, Proposition 2.5.2]):

Example 4.9 Every (group of $\mathbf{K}$-rational points of a) linear algebraic group defined over a local field $\mathbf{K}$ is a $\mathbf{K}$-analytic Lie group, viz. every subgroup

---

[2] In other parts of the literature related to rigid analytic geometry, such functions are called **locally analytic** to distinguish them from functions which are globally given by a power series.

$G \leq \mathrm{GL}_n(\mathbf{K})$ which is the set of joint zeros of a set of polynomial functions $M_n(\mathbf{K}) \to \mathbf{K}$. For example, $\mathrm{SL}_n(\mathbf{K}) = \{A \in \mathrm{GL}_n(\mathbf{K}) \colon \det(A) = 1\}$ is a $\mathbf{K}$-analytic Lie group.

Remark 4.10 Not every Lie group is linear. For instance, in Remark 4.50 we shall encounter an analytic Lie group G over $\mathbf{K} = \mathbf{F}_p((X))$ which does not admit a faithful, continuous linear representation $G \to \mathrm{GL}_n(\mathbf{K})$ for any n.

The Lie algebra functor. If G is a $\mathbf{K}$-analytic Lie group, then its tangent space $L(G) := T_1(G)$ at the identity element can be made a Lie algebra via the identification of $x \in L(G)$ with the corresponding left invariant vector field on G (noting that the left invariant vector fields form a Lie subalgebra of the Lie algebra $\mathscr{V}^\omega(G)$ of all analytic vector fields on G).

If $\alpha\colon G \to H$ is an analytic group homomorphism between $\mathbf{K}$-analytic Lie groups, then the tangent map $L(\alpha) := T_1(\alpha)\colon L(G) \to L(H)$ is a linear map and actually a Lie algebra homomorphism (cf. [6, Chapter 3, §3.7 and §3.8], Lemma 5.1 on p. 129 in [40, Part II, Chapter V.1]). An **analytic automorphism** of a Lie group G is an invertible group homomorphism $\alpha\colon G \to G$ such that both α and α^{-1} are analytic.

Ultrametric inverse function theorem

Since small perturbations do not change the size of a given non-zero vector in the ultrametric case (as "the strongest wins"), the ultrametric inverse function theorem has a much nicer form than its classical real counterpart. Around a point with invertible differential, an analytic map behaves on balls simply like an affine-linear map (namely its linearization).

In the following three propositions, we let $(\mathbf{K}, |.|)$ be a complete ultrametric field and equip $\mathbf{K}^n$ with any ultrametric norm $\|.\|$ (e.g., the maximum norm). Given $x \in \mathbf{K}^n$ and $r > 0$, we abbreviate $B_r(x) := B_r^{\mathbf{K}^n}(x)$. The total differential of f at x is denoted by $f'(x)$. Now the ultrametric inverse function theorem (for analytic functions) reads as follows:[3]

Proposition 4.11 *Let $f\colon U \to \mathbf{K}^n$ be an analytic map on an open set $U \subseteq \mathbf{K}^n$ and $x \in U$ such that $f'(x) \in \mathrm{GL}_n(\mathbf{K})$. Then there exists $r > 0$ such that $B_r(x) \subseteq U$,*

$$f(B_s(y)) = f(y) + f'(x).B_s(0) \quad \text{for all } y \in B_r(x) \text{ and } s \in]0, r],$$

---

[3] A proof is obtained, e.g., by combining [14, Proposition 7.1 (b)′] (a C^k-analogue of Proposition 4.11) with the version of the inverse function theorem for analytic maps from [40, p. 73].

and $f|_{B_s(y)}$ is an analytic diffeomorphism onto its open image. If $f'(x)$ is an isometry, then so is $f|_{B_r(x)}$ for small r. $\qquad\square$

It is useful that r can be chosen uniformly in the presence of parameters. As a special case of [14, Theorem 8.1 (b)'] (which only requires that f be C^k in a suitable sense), an 'ultrametric inverse function theorem with parameters' is available:[4]

Proposition 4.12 *Let $P \subseteq \mathbf{K}^m$ and $U \subseteq \mathbf{K}^n$ be open, $f: P \times U \to \mathbf{K}^n$ be a $\mathbf{K}$-analytic map, $p \in P$ and $x \in U$ such that $f'_p(x) \in \mathrm{GL}_n(\mathbf{K})$, where $f_p :=$ $f(p,\bullet): U \to \mathbf{K}^n$. Then there exists an open neighbourhood $Q \subseteq P$ of p and $r > 0$ such that $B_r(x) \subseteq U$ and*

$$f_q(B_s(y)) = f_q(y) + f'_p(x).B_s(0) \tag{4.9}$$

for all $q \in Q$, $y \in B_r(x)$ and $s \in\ ]0,r]$. If $f'_p(x)$ is an isometry, then also $f_q|_{B_r(x)}$ is an isometry for all $q \in Q$, if Q and r are chosen sufficiently small. $\qquad\square$

We mention that the group of linear isometries is large (an open subgroup).

Proposition 4.13 *The group $\mathrm{Iso}(\mathbf{K}^n)$ of linear isometries is open in $\mathrm{GL}_n(\mathbf{K})$. If $\mathbf{K}^n$ is equipped with the maximum norm, then*

$$\mathrm{Iso}(\mathbf{K}^n) = \mathrm{GL}_n(\mathbf{O}) = \{A \in \mathrm{GL}_n(\mathbf{K}): A, A^{-1} \in M_n(\mathbf{O})\},$$

where $\mathbf{O}$ (as in (4.3) from Section 4.2) is the valuation ring of $\mathbf{K}$.

Proof The subgroup $\mathrm{Iso}(\mathbf{K}^n)$ will be open in $\mathrm{GL}_n(\mathbf{K})$ if it is an identity neighbourhood. The latter is guaranteed if we can prove that $1 + A \in \mathrm{Iso}(\mathbf{K}^n)$ for all $A \in M_n(\mathbf{K})$ of operator norm

$$\|A\|_{\mathrm{op}} := \sup\left\{\frac{\|Ax\|}{\|x\|}: 0 \neq x \in \mathbf{K}^n\right\} < 1.$$

However, for each $0 \neq x \in \mathbf{K}^n$ we have $\|Ax\| \leq \|A\|_{\mathrm{op}}\|x\| < \|x\|$ and hence $\|(1+A)x\| = \|x + Ax\| = \|x\|$ by (4.2) from Section 4.2, as the strongest wins.

Now assume that $\|.\|$ is the maximum norm on $\mathbf{K}^n$, and $A \in \mathrm{GL}_n(\mathbf{K})$. If A is an isometry, then $\|Ae_i\| = 1$ for the standard basis vector e_i with j-th component δ_{ij}, and hence $A \in M_n(\mathbf{O})$. Likewise, $A^{-1} \in M_n(\mathbf{O})$ and hence $A \in \mathrm{GL}_n(\mathbf{O})$. If $A \in \mathrm{GL}_n(\mathbf{O})$, then $\|Ax\| \leq \|x\|$ and $\|x\| = \|A^{-1}Ax\| \leq \|Ax\|$, whence $\|Ax\| = \|x\|$ and A is an isometry. $\qquad\square$

---

[4] Combining the cited theorem and Proposition 4.11, one can also show that, for Q and r small enough, $f_q|_{B_r(x)}: B_r(x) \to f_q(x) + f'_p(x).B_r(0) = f_p(x) + f'_p(0).B_r(0)$ is an analytic diffeomorphism for each $q \in Q$. However, this additional information shall not be used in the current work. We also mention that the results from [14] were strengthened further in [18].

We mention that Proposition 4.13 and its proof remain valid if $(\mathbf{K}, |.|)$ is an arbitrary (not necessarily complete) ultrametric field and the norm $\|.\|$ on $\mathbf{K}^n$ is equivalent to the maximum norm. This is useful for the proof of Lemma 4.30.

Construction of small open subgroups

Let G be a Lie group over a local field $\mathbf{K}$, and $|.|$ be an absolute value on $\mathbf{K}$ defining its topology. Fix an ultrametric norm $\|.\|$ on $L(G)$ and abbreviate $B_r(x) := B_r^{L(G)}(x)$ for $x \in L(G)$ and $r > 0$. Using an analytic diffeomorphism $\phi : G \supseteq U \to V \subseteq L(G)$ such that $1 \in U$ and such that $\phi(1) = 0$, the group multiplication gives rise to a multiplication $\mu : W \times W \to V$, $\mu(x,y) = x * y$ for an open 0-neighbourhood $W \subseteq V$, via

$$x * y := \phi(\phi^{-1}(x)\phi^{-1}(y)).$$

It is easy to see that the first order Taylor expansions of multiplication and inversion in local coordinates read

$$x * y = x + y + \cdots \tag{4.10}$$

and

$$x^{-1} = -x + \cdots \tag{4.11}$$

(compare [40, p. 113]). Applying the ultrametric inverse function theorem with parameters to the maps $(x,y) \mapsto x * y$ and $(x,y) \mapsto y * x$, we find $r > 0$ with $B_r(0) \subseteq W$ such that

$$x * B_s(0) = x + B_s(0) = B_s(0) * x \tag{4.12}$$

for all $x \in B_r(0)$ and $s \in \,]0,r]$ (exploiting that the map $\mu(0,.) = \mu(.,0) = \mathrm{id}$ has derivative id at 0, which is an isometry). In particular, (4.12) implies that $B_s(0) * B_s(0) = B_s(0)$ for each $s \in \,]0,r]$, and thus also $y^{-1} \in B_s(0)$ for each $y \in B_s(0)$.

Summing up:

Lemma 4.14 $(B_s(0), *)$ is a group for each $s \in \,]0,r]$ and hence $\phi^{-1}(B_s(0))$ is a compact open subgroup of G, for each $s \in \,]0,r]$. Moreover, $B_s(0)$ is a normal subgroup of $(B_r(0), *)$. □

Thus small balls in $L(G)$ correspond to compact open subgroups in G. These special subgroups are very useful for many purposes. In particular, we shall see later that for suitable choices of the norm $\|.\|$, the groups $\phi^{-1}(B_s(0))$ will be tidy (see 4.21 below) for a given automorphism α, as long as α is well behaved (exceptional cases where this goes wrong will be pinpointed as well).

Remark 4.15 Note that (4.12) entails that the indices of $B_s(0)$ in $(B_r(0), +)$ and $(B_r(0), *)$ coincide (as the cosets coincide). This observation will be useful later.

4.3 Basic facts concerning p-adic Lie groups

For each local field $\mathbf{K}$ of characteristic 0, the exponential series $\sum_{k=0}^{\infty} \frac{1}{k!} A^k$ converges on some 0-neighbourhood $U \subseteq M_n(\mathbf{K})$ and defines an analytic mapping $\exp \colon U \to \mathrm{GL}_n(\mathbf{K})$. More generally, for such $\mathbf{K}$, it can be shown that every $\mathbf{K}$-analytic Lie group G admits an exponential function in the following sense:

Definition 4.16 An analytic map $\exp_G \colon U \to G$ on an open 0-neighbourhood $U \subseteq L(G)$ with $\mathbf{O}U \subseteq U$ is called an **exponential function** if $\exp_G(0) = 1$, $T_0(\exp_G) = \mathrm{id}$ and

$$\exp_G((s+t)x) = \exp_G(sx)\exp_G(tx) \quad \text{for all } x \in U \text{ and } s, t \in \mathbf{O},$$

where $\mathbf{O} := \{t \in \mathbf{K} \colon |t| \leq 1\}$ is the valuation ring.

Since $T_0(\exp_G) = \mathrm{id}$, after shrinking U one can assume that $\exp_G(U)$ is open and $\exp_G$ is a diffeomorphism onto its image. By Lemma 4.14, after shrinking U further if necessary, we may assume that $\exp_G(U)$ is a subgroup of G. Hence also U can be considered as a Lie group. The Taylor expansion of multiplication with respect to the logarithmic chart $\exp_G^{-1}$ is given by the Baker–Campbell–Hausdorff (BCH-) series

$$x * y = x + y + \frac{1}{2}[x, y] + \cdots \tag{4.1}$$

(all terms of which are nested Lie brackets with rational coefficients), and hence $x * y$ is given by this series for small U (cf. Lemma 3 and Theorem 2 in [6, Chapter 3, §4, no. 2]). In this case, we call $\exp_G(U)$ a **BCH-subgroup of G**.

For later use, we note that

$$x^n = x * \cdots * x = nx$$

for all $x \in U$ and $n \in \mathbf{N}_0$ (since $[x, x] = 0$ in (4.1)). As a consequence, the closed subgroup of $(U, *)$ generated by $x \in U$ is of the form

$$\overline{\langle x \rangle} = \mathbf{Z}_p x. \tag{4.2}$$

We also note that

$$\exp_U := \mathrm{id}_U \tag{4.3}$$

is an exponential map for U.

Next, let us consider homomorphisms between Lie groups. We first recall that if $\alpha\colon G \to H$ is an analytic homomorphism between real Lie groups, then the diagram

$$
\begin{array}{ccc}
G & \xrightarrow{\alpha} & H \\
\exp_G \uparrow & & \uparrow \exp_H \\
L(G) & \xrightarrow{L(\alpha)} & L(H)
\end{array}
$$

commutes (a fact referred to as the "naturality of exp"). If $\alpha\colon G \to H$ is an analytic homomorphism between Lie groups over a local field $\mathbf{K}$ of characteristic 0, we can still choose $\exp_G\colon U_G \to G$ and $\exp_H\colon U_H \to H$ with $L(\alpha).U_G \subseteq U_H$ and

$$
\exp_H \circ L(\alpha)|_{U_G} = \alpha \circ \exp_G \tag{4.4}
$$

(see Proposition 8 in [6, Chapter 3, §4, no. 4]).

The following fact (see [6, Chapter 3, Theorem 1 in §8, no. 1]) is essential.

Proposition 4.17 *Every continuous homomorphism between p-adic Lie groups is analytic.*

As a consequence, there is at most one p-adic Lie group structure on a given topological group. Following the general custom, we call a topological group a p-adic Lie group if it admits a p-adic Lie group structure.

We record standard facts; for the proofs, the reader is referred to Proposition 7 in §1, no. 4; Proposition 11 in §1, no. 6; and Theorem 2 in §8, no. 2 in [6, Chapter 3].

Proposition 4.18 *Closed subgroups, finite direct products and Hausdorff quotient groups of p-adic Lie groups are p-adic Lie groups.* □

It is also known that a topological group which is an extension of a p-adic Lie group by a p-adic Lie group is again a p-adic Lie group. In fact, the case of compact p-adic Lie groups is known in the theory of analytic pro-p-groups (one can combine [48, Proposition 8.1.1 (b)] with Proposition 1.11 (ii) and Corollary 8.33 from [10]). The general case then is a well known consequence (see, e.g., [22, Lemma 9.1 (a)]).

The following fact is essential for us (cf. Step 1 in the proof of Theorem 3.5 in [43]).

Proposition 4.19 *Every p-adic Lie group G has an open subgroup U which satisfies the ascending chain condition on closed subgroups.*

Proof We show that every BCH-subgroup U has the desired property. It suffices to discuss $U \subseteq L(G)$ with the BCH-multiplication. It is known that the Lie algebra $L(H)$ of a closed subgroup $H \leq U$ can be identified with the set of all $x \in L(U) = L(G)$ such that

$$\exp_U(Wx) \subseteq H$$

for some 0-neighbourhood $W \subseteq \mathbf{Q}_p$ (see Corollary 1 (ii) in [6, Chapter 3, §4, no. 4]). Since $\exp_U = \mathrm{id}_U$ here (see (4.3)) and $\mathbf{Z}_p x = \overline{\langle x \rangle} \subseteq H$ for each $x \in H$ (by (4.2)), we deduce that

$$L(H) = \mathrm{span}_{\mathbf{Q}_p}(H)$$

is the linear span of H in the current situation. Now consider an ascending series $H_1 \leq H_2 \leq \cdots$ of closed subgroups. We may assume that each H_n has the same dimension and thus $\mathfrak{h} := L(H_1) = L(H_2) = \cdots$ for each n. Then $H := \mathfrak{h} \cap U$ is a compact group in which H_1 is open, whence $[H : H_1]$ is finite and the series has to stabilize. $\qquad\square$

We record an important consequence:

Proposition 4.20 *If G is a p-adic Lie group and $H_1 \subseteq H_2 \subseteq \cdots$ an ascending sequence of closed subgroups of G, then also $H := \bigcup_{n \in \mathbf{N}} H_n$ is closed in G.*

Proof Let $V \subseteq G$ be an open subgroup satisfying an ascending chain condition on closed subgroups. Then

$$V \cap H_1 \subseteq V \cap H_2 \subseteq \cdots$$

is an ascending sequence of closed subgroups of V and thus stabilizes, say at $V \cap H_m$. Then

$$V \cap H = V \cap H_m$$

is closed in G. Being locally closed, the subgroup H is closed (compare [24, Theorem (5.9)]). $\qquad\square$

Two important applications are now described. The second one (Corollary 4.24) is part of [43, Theorem 3.5 (ii)]. The first application might also be deduced from the second using [3, Theorem 3.32].

4.21 *We recall from [46] and [47]: If G is a totally disconnected, locally compact group and α an automorphism of G, then a compact open subgroup $V \subseteq G$ is called **tidy for** α if it has the following properties:*

TA *$V = V_+ V_-$, where $V_+ := \bigcap_{n \in \mathbf{N}_0} \alpha^n(V)$ and $V_- := \bigcap_{n \in \mathbf{N}_0} \alpha^{-n}(V)$; and*

TB *The ascending union $V_{++} := \bigcup_{n\in\mathbf{N}_0} \alpha^n(V_+)$ is closed in G.*

If V satisfies **TA**, *it is also called* **tidy above**. *If V satisfies* **TB**, *it is also called* **tidy below**.

Corollary 4.22 *Let G be a p-adic Lie group, $\alpha\colon G \to G$ be an automorphism and $V \subseteq G$ be a compact open subgroup. Then V is tidy below.*

Proof The subgroup $\bigcup_{n\in\mathbf{N}_0}\alpha^n(V_+)$ is an ascending union of closed subgroups of G and hence closed, by Proposition 4.20. □

The second application of Proposition 4.20 concerns contraction groups (see Chapter 10 in this volume for a detailed discussion of that notion).

Definition 4.23 Given a topological group G and automorphism $\alpha\colon G \to G$, we define the **contraction group**[5] **of** α via

$$U_\alpha := \{x \in G\colon \alpha^n(x) \to 1 \text{ as } n \to \infty\}. \tag{4.5}$$

Corollary 4.24 *Let G be a p-adic Lie group. Then the contraction group U_α is closed in G, for each automorphism $\alpha\colon G \to G$.*

Proof Let $V_1 \supseteq V_2 \supseteq \cdots$ be a sequence of compact open subgroups of G which form a basis of identity neighbourhoods (cf. Lemma 4.14). Then an element $x \in G$ belongs to U_α if and only if

$$(\forall n \in \mathbf{N})\,(\exists m \in \mathbf{N})\,(\forall k \geq m)\ \alpha^k(x) \in V_n.$$

Since $\alpha^k(x) \in V_n$ if and only if $x \in \alpha^{-k}(V_n)$, we deduce that

$$U_\alpha = \bigcap_{n\in\mathbf{N}}\bigcup_{m\in\mathbf{N}}\bigcap_{k\geq m}\alpha^{-k}(V_n).$$

Note that $W_n := \bigcup_{m\in\mathbf{N}}\bigcap_{k\geq m}\alpha^{-k}(V_n)$ is an ascending union of closed subgroups of G and hence closed, by Proposition 4.20. Consequently, the contraction group $U_\alpha = \bigcap_{n\in\mathbf{N}} W_n$ is closed. □

Remark 4.25 We mention that contraction groups of the form U_α arise in many contexts: In representation theory in connection with the Mautner phenomenon (see [34, Chapter II, Lemma 3.2] and (for the p-adic case) [43]); in probability theory on groups (see [23], [41], [42] and (for the p-adic case) [8]); and in the structure theory of totally disconnected, locally compact groups [3].

[5] Some authors may prefer to call α the contraction **subgroup** of α. Another recent notation for U_α is con(α).

4.4 Lazard's characterization of *p*-adic groups

Lazard [32] obtained various characterizations of *p*-adic Lie groups within the class of locally compact groups, and many more have been found since in the theory of analytic pro-*p*-groups (see [10, pp. 97–98]). These characterizations (and the underlying theory of analytic pro-*p*-groups) are of great value for the structure theory of totally disconnected groups. For example, they explain the occurrence of *p*-adic Lie groups in the areas of $\mathbf{Z}^n$-actions and contractive automorphisms mentioned in the introduction. We recall one of Lazard's characterizations in a form recorded in [40, p. 157]:

Theorem 4.26 *A topological group G is a p-adic Lie group if and only if it has an open subgroup U with the following properties:*

(a) *U is a pro-p-group;*

(b) *U is topologically finitely generated, i.e., $U = \overline{\langle F \rangle}$ for a finite subset $F \subseteq U$;*

(c) *$[U,U] := \langle xyx^{-1}y^{-1} : x,y \in U \rangle \subseteq \{x^{p^2} : x \in U\}$.*

Although the proof of the sufficiency of conditions (a)–(c) is non-trivial, their necessity is clear. In fact, (a) is immediate from (4.8) in Section 4.2 and Remark 4.15, while (c) can easily be proved using the second order Taylor expansion of the commutator map and the inverse function theorem (applied to the map $x \mapsto x^{p^2}$). To obtain (b), one picks an exponential map $\exp_G : V \to G$ as well as a basis $x_1, \ldots, x_d \in V$ of $L(G)$, and notes that the analytic map

$$\phi : (\mathbf{Z}_p)^d \to G, \quad (t_1, \ldots, t_d) \mapsto \exp_G(t_1 x_1) \cdot \ldots \cdot \exp_G(t_d x_d)$$

has an invertible differential at the origin (the map $(t_1, \ldots, t_d) \mapsto \sum_{j=1}^d t_j x_j$), and hence restricts to a diffeomorphism from $p^n \mathbf{Z}_p^d$ (for some $n \in \mathbf{N}_0$) onto an open identity neighbourhood U, which can be chosen as subgroup of G (by Lemma 4.14). Replacing the elements x_j with $p^n x_j$, we can always achieve that ϕ is a diffeomorphism from $\mathbf{Z}_p^d$ onto a compact open subgroup U of G (occasionally, one speaks of "coordinates of the second kind" in this situation). Then

$$\overline{\langle \exp_G(x_1), \ldots, \exp_G(x_d) \rangle} = U,$$

establishing (b).

4.5 Iteration of linear automorphisms

A good understanding of linear automorphisms of vector spaces over local fields is essential for an understanding of automorphisms of general Lie groups.

Decomposition of E and adapted norms

For our first consideration, we let $(\mathbf{K}, |.|)$ be a complete ultrametric field, E be a finite-dimensional $\mathbf{K}$-vector space and $\alpha\colon E \to E$ be a linear automorphism. We let $\overline{\mathbf{K}}$ be an algebraic closure of $\mathbf{K}$, and use the same symbol, $|.|$, for the unique extension of the given absolute value to $\overline{\mathbf{K}}$ (see [39, Theorem 16.1]). We let $R(\alpha)$ be the set of all absolute values $|\lambda|$, where $\lambda \in \overline{\mathbf{K}}$ is an eigenvalue of the automorphism $\alpha_{\overline{\mathbf{K}}} := \alpha \otimes \mathrm{id}_{\overline{\mathbf{K}}}$ of $E_{\overline{\mathbf{K}}} := E \otimes_{\mathbf{K}} \overline{\mathbf{K}}$ obtained by extension of scalars. For each $\lambda \in \overline{\mathbf{K}}$, we let

$$(E_{\overline{\mathbf{K}}})_{(\lambda)} := \{x \in E_{\overline{\mathbf{K}}}\colon (\alpha_{\overline{\mathbf{K}}} - \lambda)^d x = 0\}$$

be the generalized eigenspace of $\alpha_{\overline{\mathbf{K}}}$ in $E_{\overline{\mathbf{K}}}$ corresponding to λ (where d is the dimension of the $\mathbf{K}$-vector space E). Given $\rho \in R(\alpha)$, we define

$$(E_{\overline{\mathbf{K}}})_\rho := \bigoplus_{|\lambda|=\rho} (E_{\overline{\mathbf{K}}})_{(\lambda)} \subseteq E_{\overline{\mathbf{K}}},$$

where the sum is taken over all $\lambda \in \overline{\mathbf{K}}$ such that $|\lambda| = \rho$. As usual, we identify E with $E \otimes 1 \subseteq E_{\overline{\mathbf{K}}}$.

The following fact (cf. (1.0) on p. 81 in [34, Chapter II]) is essential:[6]

Lemma 4.27 *For each $\rho \in R(\alpha)$, the vector subspace $(E_{\overline{\mathbf{K}}})_\rho$ of $E_{\overline{\mathbf{K}}}$ is defined over $\mathbf{K}$, i.e., $(E_{\overline{\mathbf{K}}})_\rho = (E_\rho)_{\overline{\mathbf{K}}}$, where $E_\rho := (E_{\overline{\mathbf{K}}})_\rho \cap E$. Thus*

$$E = \bigoplus_{\rho \in R(\alpha)} E_\rho, \tag{4.1}$$

and each E_ρ is an α-invariant vector subspace of E. $\square$

It is useful for us that certain well-behaved norms exist on E (cf. [12, Lemma 3.3] and its proof for the p-adic case).

Definition 4.28 A norm $\|.\|$ on E is **adapted** to α if the following holds:

A1 $\|.\|$ is ultrametric;
A2 $\left\|\sum_{\rho \in R(\alpha)} x_\rho\right\| = \max\{\|x_\rho\|\colon \rho \in R(\alpha)\}$
for each $(x_\rho)_{\rho \in R(\alpha)} \in \prod_{\rho \in R(\alpha)} E_\rho$; and
A3 $\|\alpha(x)\| = \rho\|x\|$ for each $\rho \in R(\alpha)$ and $x \in E_\rho$.

[6] In [34, p. 81], $\mathbf{K}$ is a local field, but the proof works also for complete ultrametric fields.

Proposition 4.29 *Let E be a finite-dimensional vector space over a complete ultrametric field* $(\mathbf{K}, |.|)$ *and* $\alpha \colon E \to E$ *be a linear automorphism. Then E admits a norm* $\|.\|$ *adapted to* α.

This follows from the next lemma.

Lemma 4.30 *For each* $\rho \in R(\alpha)$, *there exists an ultrametric norm* $\|.\|_\rho$ *on* E_ρ *such that* $\|\alpha(x)\|_\rho = \rho\|x\|_\rho$ *for each* $x \in E_\rho$.

Proof Assume $\rho \geq 1$ first. We choose a $\overline{\mathbf{K}}$-basis $w_1, \ldots, w_m$ of $(E_{\overline{\mathbf{K}}})_\rho =: V$ such that the matrix A_ρ of $\alpha_{\overline{\mathbf{K}}}|_V^V \colon V \to V$ with respect to this basis has Jordan canonical form. We let $\|.\|$ be the maximum norm on V with respect to this basis.

If $\rho = 1$, then $A_\rho \in \mathrm{GL}_m(\overline{\mathbf{O}})$, where $\overline{\mathbf{O}} = \{t \in \overline{\mathbf{K}} \colon |t| \leq 1\}$ is the valuation ring of $\overline{\mathbf{K}}$, and hence $\alpha_{\overline{\mathbf{K}}}|_V^V$ is an isometry with respect to $\|.\|$ (by Proposition 4.13). Thus $\|\alpha_{\overline{\mathbf{K}}}(x)\| = \|x\| = \rho\|x\|$ for all $x \in V$.

If $\rho > 1$, we note that for each $k \in \{1, \ldots, m\}$, there exists an eigenvalue μ_k such that $w_k \in (E_{\overline{\mathbf{K}}})_{(\mu_k)}$. Then $\alpha_{\overline{\mathbf{K}}}(w_k) = \mu_k w_k + (\alpha_{\overline{\mathbf{K}}}(w_k) - \mu_k w_k)$ for each k, with $\|\mu_k w_k\| = \rho > 1 \geq \|\alpha_{\overline{\mathbf{K}}}(w_k) - \mu_k w_k\|$. As a consequence, $\|\alpha_{\overline{\mathbf{K}}}(x)\| = \rho\|x\|$ for each $x \in V$, using (4.2) from Section 4.2.

In either of the preceding cases, we define $\|.\|_\rho$ as the restriction of $\|.\|$ to E_ρ. To complete the proof, assume $\rho < 1$ now. Then $E_\rho = E_{\rho^{-1}}(\alpha^{-1})$, where $\rho^{-1} > 1$. Thus, by what has already been shown, there exists an ultrametric norm $\|.\|_\rho$ on E_ρ such that $\|\alpha^{-1}(x)\|_\rho = \rho^{-1}\|x\|_\rho$ for each $x \in E_\rho$. Then $\|\alpha(x)\|_\rho = \rho\|x\|_\rho$ for each $x \in E_\rho$, as required. $\square$

Proof of Proposition 4.29. For each $\rho \in R(\alpha)$, we choose a norm $\|.\|_\rho$ on E_ρ as described in Lemma 4.30. Then

$$\left\| \sum_{\rho \in R(\alpha)} x_\rho \right\| := \max \left\{ \|x_\rho\|_\rho \colon \rho \in R(\alpha) \right\} \quad \text{for } (x_\rho)_{\rho \in R(\alpha)} \in \prod_{\rho \in R(\alpha)} E_\rho$$

defines a norm $\|.\| \colon E \to [0, \infty[$ which, by construction, is adapted to α. $\square$

Contraction group and Levi factor for a linear automorphism

For α a linear automorphism of a finite-dimensional vector space E over a local field $\mathbf{K}$, we now determine the contraction group (as introduced in Definition 4.23) and the associated Levi factor, defined as follows:

Definition 4.31 Let G be a topological group and $\alpha \colon G \to G$ be an automorphism. Following [3], we define the **Levi factor** M_α as the set of all $x \in G$ such that the two-sided orbit $\alpha^{\mathbf{Z}}(x)$ is relatively compact in G.

It is clear that both U_α and M_α are subgroups. If G is locally compact and totally disconnected, then M_α is closed, as can be shown using tidy subgroups as a tool (see [3, p. 224]).

Proposition 4.32 *Let α be a linear automorphism of a finite-dimensional* **K**-*vector space E. Then*

$$U_\alpha = \bigoplus_{\substack{\rho \in R(\alpha) \\ \rho < 1}} E_\rho, \quad M_\alpha = E_1 \quad and \quad U_{\alpha^{-1}} = \bigoplus_{\substack{\rho \in R(\alpha) \\ \rho > 1}} E_\rho.$$

Furthermore, $E = U_\alpha \oplus M_\alpha \oplus U_{\alpha^{-1}}$ as a **K**-*vector space.*

Proof Using an adapted norm on E, the characterizations of U_α, M_α and $U_{\alpha^{-1}}$ are clear. That E is the indicated direct sum follows from (4.1). $\qquad\square$

Tidy subgroups and the scale for a linear automorphism

In the preceding situation, define

$$E_+ := \bigoplus_{\rho \geq 1} E_\rho \quad and \quad E_- := \bigoplus_{\rho \leq 1} E_\rho. \tag{4.2}$$

Assume that **K** is endowed with its natural absolute value $|.|$.

Proposition 4.33 *For each $r > 0$ and norm $\|.\|$ on E adapted to α, the ball $B_r := \{x \in E \colon \|x\| < r\}$ is tidy for α, with*

$$(B_r)_\pm = B_r \cap E_\pm. \tag{4.3}$$

The scale of α is given by

$$s_E(\alpha) = \Delta_{E_+}(\alpha|_{E_+}^{E_+}) = |\det(\alpha|_{E_+}^{E_+})| = \prod_{\substack{j \in \{1,\ldots,d\} \\ |\lambda_j| \geq 1}} |\lambda_j|,$$

where $\lambda_1, \ldots, \lambda_d$ are the eigenvalues of $\alpha_{\overline{\mathbf{K}}}$ (occurring with multiplicities).

Proof Let $x \in B_r$. It is clear from the definition of an adapted norm that $\alpha^{-n}(x) \in B_r$ for all $n \in \mathbf{N}_0$ (i.e., $x \in (B_r)_+$) if and only if $x \in \bigoplus_{\rho \geq 1} E_\rho = E_+$. Thus $(B_r)_+$ is given by (4.3), and $(B_r)_-$ can be discussed analogously. It follows from property **A2** in the definition of an adapted norm that

$$B_r = \bigoplus_{\rho \in R(\alpha)} (B_r \cap E_\rho), \tag{4.4}$$

and thus $B_r = (B_r)_+ + (B_r)_-$. Hence B_r is tidy above. Since

$$(B_r)_{++} = \bigcup_{n \in \mathbf{N}_0} \alpha^n((B_r)_+) = (B_r \cap E_1) \oplus \bigoplus_{\rho > 1} E_\rho$$

is closed, B_r is also tidy below and thus B_r is tidy.

Since $K := (B_r)_+$ is open in E_+, compact and $K \subseteq \alpha(K)$, we obtain from (4.5) in Section 4.2 that

$$\Delta_{E_+}\left(\alpha|_{E_+}^{E_+}\right) = [\alpha(K) : K] = s_E(\alpha).$$

Here

$$\Delta_{E_+}\left(\alpha|_{E_+}^{E_+}\right) = \left|\det\left(\alpha|_{E_+}^{E_+}\right)\right| = \prod_{\substack{j \in \{1, \dots, d\} \\ |\lambda_j| \geq 1}} |\lambda_j|,$$

by (4.7) in Lemma 4.5. ☐

4.6 Scale and tidy subgroups for automorphisms of p-adic Lie groups

In this section, we determine tidy subgroups and calculate the scale for automorphisms of a p-adic Lie group.

4.34 *Applying (4.4) from Section 4.3 to an automorphism α of a p-adic Lie group G, we find an exponential function $\exp_G \colon V \to G$ which is a diffeomorphism onto its image, and an open 0-neighbourhood $W \subseteq V$ such that $L(\alpha).W \subseteq V$ and*

$$\exp_G \circ L(\alpha)|_W = \alpha \circ \exp_G|_W. \tag{4.1}$$

Thus α is locally linear in a suitable (logarithmic) chart, which simplifies an understanding of the dynamics of α. Many aspects can be reduced to the dynamics of the linear automorphism $\beta := L(\alpha)$ of $L(G)$, as discussed in Section 4.5. After shrinking W, we may assume that also $L(\alpha^{-1}).W \subseteq V$ and

$$\exp_G \circ L(\alpha^{-1})|_W = \alpha^{-1} \circ \exp_G|_W. \tag{4.2}$$

To construct subgroups tidy for α, we pick a norm $\|.\|$ on $E := L(G)$ adapted to $\beta := L(\alpha)$, as in Proposition 4.29. There is $r > 0$ such that $B_r := B_r^E(0) \subseteq W$ and

$$V_t := \exp_G(B_t)$$

is a compact open subgroup of G, for each $t \in \,]0, r]$ (see Lemma 4.14). Abbreviate $Q_t := B_t \cap \bigoplus_{\rho < 1} E_\rho$. Since $B_t = (B_t)_+ \oplus Q_t$ by (4.4) from Section 4.5 (applied to the linear automorphism β), the ultrametric inverse function theorem

(Proposition 4.11) and (4.10) from Section 4.2 imply that $V_t = \exp_G((B_t)_+)\exp_G(Q_t)$ and thus

$$V_t = \exp_G((B_t)_+)\exp_G((B_t)_-) \tag{4.3}$$

for all $t \in \,]0, r]$, after shrinking r if necessary. Then we have (as first recorded in [15, Theorem 3.4 (c)]):

Theorem 4.35 *The subgroup V_t of G is tidy for α, for each $t \in \,]0, r]$. Moreover,*

$$s_G(\alpha) = s_{L(G)}(L(\alpha)),$$

where $s_{L(G)}(L(\alpha))$ can be calculated as in Proposition 4.33.

Proof Since $\beta((B_t)_-) \subseteq (B_t)_-$ and $\beta^{-1}((B_t)_+) \subseteq (B_t)_+$, it follows with (4.1), (4.2) and induction that

$$\alpha^{\mp n}(\exp_G((B_t)_\pm)) = \exp_G(\beta^{\mp n}((B_t)_\pm)) \subseteq \exp_G((B_t)_\pm)$$

for each $n \in \mathbf{N}_0$ and hence

$$\exp_G((B_t)_\pm) \subseteq (V_t)_\pm.$$

Therefore $V_t = (V_t)_+(V_t)_-$ (using (4.3)) and thus V_t is tidy above. By Corollary 4.22, V_t is also tidy below and thus V_t is tidy for α.

Closer inspection shows that $L((V_t)_\pm) = L(G)_\pm$ (see [12, Theorem 3.5] or proof of Theorem 4.45 below), where $L(G)_\pm$ is defined as in (4.2) from Section 4.5, using β. Since the module of a Lie group automorphism can be calculated on the Lie algebra level (using Proposition 55 (ii) in [6, Chapter 3, §3, no. 16]), it follows that $s_G(\alpha) = \Delta_{(V_t)_{++}}(\alpha|_{(V_t)_{++}}) = \Delta_{L(G)_+}(L(\alpha)|_{L(G)_+}) = s_{L(G)}(L(\alpha))$. $\quad\square$

4.7 *p*-adic contraction groups

Exploiting the local linearity of α (see (4.1) in Section 4.6) and the closedness of p-adic contraction groups (see Corollary 4.24), it is possible to reduce the discussion of p-adic contraction groups to contraction groups of linear automorphisms, as in Proposition 4.32. We record the results, due to Wang [43, Theorem 3.5].

Theorem 4.36 *Let G be a p-adic Lie group and $\alpha: G \to G$ be an automorphism. Abbreviate $\beta := L(\alpha)$. Then U_α, M_α and $U_{\alpha^{-1}}$ are closed subgroups* *of G with Lie algebras*

$$L(U_\alpha) = U_\beta, \quad L(M_\alpha) = M_\beta \quad and \quad L(U_{\alpha^{-1}}) = U_{\beta^{-1}},$$

respectively. Furthermore, $U_\alpha M_\alpha U_{\alpha^{-1}}$ is an open identity neighbourhood in G, and the product map

$$U_\alpha \times M_\alpha \times U_{\alpha^{-1}} \to U_\alpha M_\alpha U_{\alpha^{-1}}, \quad (x,y,z) \mapsto xyz$$

is an analytic diffeomorphism. □

In Part (ii) of his Theorem 3.5, Wang also obtained essential information concerning the groups U_α:

Theorem 4.37 *Let G be a p-adic Lie group admitting a contractive automorphism α. Then G is a unipotent linear algebraic group defined over $\mathbf{Q}_p$, and hence nilpotent.* □

We remark that Theorem 4.36 becomes false in general for Lie groups over local fields of positive characteristic (as illustrated in 4.8.1). However, we shall see later that its conclusions remain valid if closedness of U_α is made an extra hypothesis. Assuming closedness of U_α, a certain analogue of Theorem 4.36 can even be obtained in a purely topological setting:

Proposition 4.38 *Let G be a totally disconnected, locally compact group and $\alpha: G \to G$ be an automorphism. If U_α is closed, then $U_\alpha M_\alpha U_{\alpha^{-1}}$ is an open identity neighbourhood in G and the product map*

$$\pi: U_\alpha \times M_\alpha \times U_{\alpha^{-1}} \to U_\alpha M_\alpha U_{\alpha^{-1}}, \quad (x,y,z) \mapsto xyz \qquad (4.1)$$

is a homeomorphism.

Proof If U_α is closed, then every identity neighbourhood in G contains a compact open subgroup tidy for α (see [3] if G is metrizable; the results from [30] can be used to remove the metrizability condition). Thus α is "tidy" in the terminology of [13]. Therefore, the proposition is covered by part (f) of the theorem in [13] (the proof of which heavily uses results from [3]). □

We conclude the section with an example for the calculations in Sections 4.5–4.7.

Example 4.39 We consider the *p*-adic Lie group $G := \mathrm{GL}_2(\mathbf{Q}_p)$ and its inner automorphism $\alpha: A \mapsto gAg^{-1}$ given by

$$g := \begin{pmatrix} 1 & 0 \\ 0 & p \end{pmatrix}.$$

For

$$A = \begin{pmatrix} a & b \\ c & d \end{pmatrix} \in G,$$

we have

$$g^n A g^{-n} = \begin{pmatrix} a & p^{-n}b \\ p^n c & d \end{pmatrix}$$

for all $n \in \mathbf{Z}$, entailing that $M_\alpha \subseteq G$ is the subgroup of all invertible diagonal matrices,

$$U_\alpha = \left\{ \begin{pmatrix} 1 & 0 \\ c & 1 \end{pmatrix} : c \in \mathbf{Q}_p \right\} \quad \text{and} \quad U_{\alpha^{-1}} = \left\{ \begin{pmatrix} 1 & b \\ 0 & 1 \end{pmatrix} : b \in \mathbf{Q}_p \right\}.$$

Since G is an open subset of the space $M_2(\mathbf{Q}_p)$ of p-adic (2×2)-matrices, we can identify the tangent space $L(G)$ at the identity element with $M_2(\mathbf{Q}_p)$ (as usual). Then $L(\alpha)$ corresponds to the linear automorphism

$$\beta: M_2(\mathbf{Q}_p) \to M_2(\mathbf{Q}_p), \quad \begin{pmatrix} a & b \\ c & d \end{pmatrix} \mapsto \begin{pmatrix} a & p^{-1}b \\ pc & d \end{pmatrix}.$$

Note that the matrix units $E_{j,k}$ with only one non-zero-entry, 1, in the jth row and kth column, form a basis of β-eigenvectors for $M_2(\mathbf{Q}_p)$ with eigenvalues p and p^{-1} (of multiplicity one) and 1 as an eigenvalue of multiplicity 2. Thus

$$s_G(\alpha) = s_{L(G)}(\beta) = |1| \cdot |1| \cdot |p^{-1}| = p,$$

by Theorem 4.35 and Proposition 4.33. It is clear from the preceding decomposition in eigenspaces that the maximum-norm on $M_2(\mathbf{Q}_p)$, $\|A\|_\infty := \max\{|a|, |b|, |c|, |d|\}$ for A as above, is adapted to β. Therefore

$$B_r := \{A \in M_2(\mathbf{Q}_p): \|A\|_\infty < r\}$$

is tidy for β for each $r > 0$ (see Proposition 4.33). As the matrix exponential function

$$\exp: U \to \mathrm{GL}_2(\mathbf{Q}_p)$$

(defined on some open 0-neighbourhood $U \subseteq M_2(\mathbf{Q}_p)$) is an exponential function for G (see [6]), we deduce with Theorem 4.35 that

$$V_r := \exp(B_r)$$

is a tidy subgroup for α for all sufficiently small $r > 0$. We mention that $V_r = 1 + B_r$ coincides with the ball of radius r around the identity matrix for small r, by Proposition 4.11, using that the derivative $\exp'(0) = \mathrm{id}_{M_2(\mathbf{Q}_p)}$ is an isometry of $(M_2(\mathbf{Q}_p), \|.\|_\infty)$.

4.8 Pathologies in positive characteristic

Most of our discussion of automorphisms of p-adic Lie groups becomes false for Lie groups over local fields of positive characteristic (without extra assumptions). Suitable extra assumptions will be described later. For the moment, let us have a look at some of the possible pathologies.

4.8.1 Non-closed contraction groups

We describe an analytic automorphism of an analytic Lie group over a local field of positive characteristic, with a non-closed contraction group.

Given a finite field $\mathbf{F}$ with prime order $|\mathbf{F}| = p$, consider the compact group $G := \mathbf{F}^{\mathbf{Z}}$ and the right shift

$$\alpha: G \to G, \quad \alpha(f)(n) := f(n-1).$$

Since the sets $\{f \in \mathbf{F}^{\mathbf{Z}}: f(-n) = f(-n+1) = \cdots = f(n-1) = f(n) = 0\}$ (for $n \in \mathbf{N}$) form a basis of 0-neighbourhoods in G, it is easy to see that the contraction group U_α consists exactly of the functions with support bounded below, i.e.,

$$U_\alpha = \mathbf{F}^{(-\mathbf{N})} \times \mathbf{F}^{\mathbf{N}_0}.$$

This is a dense, non-closed subgroup of G. Also note that G does not have arbitrarily small subgroups tidy for α: in fact, G is the only tidy subgroup.

We now observe that G can be considered as a 2-dimensional Lie group over $\mathbf{K} := \mathbf{F}((X))$: The map

$$G \to \mathbf{F}[[X]] \times \mathbf{F}[[X]], \quad f \mapsto \left(\sum_{n=1}^{\infty} f(-n)X^{n-1}, \sum_{n=0}^{\infty} f(n)X^n \right)$$

is a global chart. The automorphism of $\mathbf{F}[[X]]^2$ corresponding to α coincides on the open 0-neighbourhood $X\mathbf{F}[[X]] \times \mathbf{F}[[X]]$ with the linear map

$$\beta: \mathbf{K}^2 \to \mathbf{K}^2, \quad \beta(v,w) = (X^{-1}v, Xw).$$

Hence α is an analytic automorphism.

4.8.2 An automorphism whose scale cannot be calculated on the Lie algebra level

We retain G, α and β from the preceding example.

Since G is compact, we have $s_G(\alpha) = 1$. However, $L(\alpha) = \beta$ and therefore $s_{L(G)}(L(\alpha)) = s_{\mathbf{K}((X))^2}(\beta) = |X^{-1}| = p$. Thus

$$s_G(\alpha) \neq s_{L(G)}(L(\alpha)),$$

in stark contrast to the p-adic case, where equality did always hold!

4.8.3 An automorphism which is not locally linear in any chart

For $\mathbf{F}$, $\mathbf{K}$ and p as before, consider the 1-dimensional $\mathbf{K}$-analytic Lie group $G := \mathbf{F}[[X]]$. Then

$$\alpha: G \to G, \quad z \mapsto z + X z^p$$

is an analytic automorphism of G with $\alpha'(0) = \mathrm{id}$. If α could be locally linearized, we could find a diffeomorphism $\phi: U \to V$ between open 0-neighbourhoods $U, V \subseteq G$ such that $\phi(0) = 0$ and

$$\alpha \circ \phi|_W = \phi \circ \beta|_W \tag{4.1}$$

for some linear map $\beta: \mathbf{K} \to \mathbf{K}$ and 0-neighbourhood $W \subseteq G$. Then $\phi'(0) = \alpha'(0) \circ \phi'(0) = \phi'(0) \circ \beta$, and hence $\beta = \mathrm{id}$. Substituting this into (4.1), we deduce that $\alpha|_{\phi(W)} = \mathrm{id}_{\phi(W)}$, which is a contradiction.

4.9 Tools from non-linear analysis: invariant manifolds

As an automorphism α of a Lie group G over a local field of positive characteristic need not be locally linear (see 4.8.3), one has to resort to techniques from non-linear analysis to understand the iteration of such an automorphism. The essential tools are stable manifolds and centre manifolds around a fixed point, which are well known tools in the classical case of dynamical systems on real manifolds. The analogy is clear: We are dealing with the (time-) discrete dynamical system (G, α), and are interested in its behaviour around the fixed point 1.

We shall use the following terminology (cf. [6] and [40]): A subset N of an m-dimensional analytic manifold M is called an n-dimensional **submanifold** of M if for each $x \in N$, there is an analytic diffeomorphism $\phi = (\phi_1, \ldots, \phi_m): U \to V$ from an open x-neighbourhood $U \subseteq M$ onto an open subset $V \subseteq \mathbf{K}^m$ such that $\phi(U \cap N) = V \cap (\mathbf{K}^n \times \{0\}) \subseteq \mathbf{K}^n \times \mathbf{K}^{m-n}$. Then the maps $(\phi_1, \ldots, \phi_n)|_{U \cap N}$ form an atlas of charts for an analytic manifold structure on N. An **immersed submanifold** of M is a subset $N \subseteq M$, together with an analytic manifold structure on N for which the inclusion map $\iota: N \to M$ is an analytic immersion (i.e., an analytic map whose tangent maps $T_x(\iota)\colon T_xN \to T_xM$ are injective for all $x \in N$). If a subgroup H of an analytic Lie group G is a submanifold, then the inherited analytic manifold structure turns it into an analytic Lie group. A subgroup H of G, together with an analytic Lie group structure on H, is called an **immersed Lie subgroup** of G, if the latter turns the inclusion map $H \to G$ into an immersion.

To discuss stable manifolds in the ultrametric case, we consider the following setting: $(\mathbf{K}, |.|)$ is a complete ultrametric field and M a finite-dimensional analytic manifold over $\mathbf{K}$, of dimension m. Also, $\alpha\colon M \to M$ is a given analytic diffeomorphism and $z \in M$ a fixed point of α. We let $a \in {]0,1]} \setminus R(T_z(\alpha))$, using notation as in Section 4.5.

Now define $W_a^s(\alpha, z)$ as the set of all $x \in M$ with the following property:[7] For some (and hence each) chart $\phi\colon U \to V \subseteq \mathbf{K}^m$ of M around z such that $\phi(z) = 0$, and some (and hence each) norm $\|.\|$ on $\mathbf{K}^m$, there exists $n_0 \in \mathbf{N}$ such that $\alpha^n(x) \in U$ for all integers $n \geq n_0$ and

$$\lim_{n\to\infty} \frac{\|\phi(\alpha^n(x))\|}{a^n} = 0.$$

It is clear from the definition that $W_a^s(\alpha, z)$ is an α-stable subset of M. The following fact is obtained if we combine [17, Theorem 1.3] and [19, Theorem A] (along with its proof):

Theorem 4.40 *For each $a \in {]0,1]} \setminus R(T_z(\alpha))$, the set $W_a^s(\alpha, z)$ is an immersed submanifold of M. Its tangent space at z is $\bigoplus_{\rho < a} T_z(M)_\rho$, using notation as in Lemma 4.27 with $\beta := T_z(\alpha)\colon T_z(M) \to T_z(M)$ in place of α and $E := T_z(M)$.*

As in the real case, we call $W_a^s(\alpha, z)$ the *a*-**stable manifold** around z.

Some ideas of the proof. The main point is to construct a local a-stable manifold, i.e., an α-invariant submanifold N of M tangent to $\bigoplus_{\rho < a} T_z(M)_\rho$ such that $\alpha|_N\colon N \to N$ is an analytic map.[8] In the real case, M. C. Irwin [26] showed how the construction of a (local) a-stable manifold can be reduced to the implicit function theorem, applied to a Banach space of sequences (cf. also [45]).[9] Since implicit function theorems are available also for analytic mappings between ultrametric Banach spaces (see [5]), it is possible to adapt

---

[7] The letter "s" in $W_a^s(\alpha, z)$ stands for "stable."
[8] Then $W_a^s(\alpha, z) = \bigcup_{n \in \mathbf{N}_0} \alpha^{-n}(N)$, which is easily made an analytic manifold in such a way that N becomes an open subset and α restricts to an analytic diffeomorphism of $W_a^s(\alpha, z)$.
[9] The idea is to construct not the points x of the manifold, but their orbits $(\alpha^n(x))_{n \in \mathbf{N}_0}$.

Irwin's method to the ultrametric case (see [17]).   □

Other classical types of invariant manifolds are also available in the ultrametric case. In particular:

4.41   *For M and α as before, there always exists a so-called **centre manifold**, i.e., an immersed analytic submanifold C of M which is tangent to $M_\beta \subseteq T_z(M)$ at z, α-stable (i.e., $\alpha(C) = C$), and such that the restriction $f|_C \colon C \to C$ is analytic [17, Theorem 1.10]. By Proposition 4.11, after shrinking C we may assume that C is diffeomorphic to a ball and hence compact (if **K** is a local field).*

The neutral element 1 of a Lie group G is a fixed point of each automorphism α of G, but it need not be a hyperbolic fixed point, i.e., it may very well happen that $1 \in R(T_1(\alpha))$. Nonetheless, it is always possible to make $U_\alpha = W_1^s(\alpha, 1)$ a manifold (see [19, Theorem D]):

Proposition 4.42   *Let G be an analytic Lie group over a complete valued field, $\alpha \colon G \to G$ be an analytic automorphism and $\beta := L(\alpha)$ the associated Lie algebra automorphism of $L(G)$. Then U_α can be made an immersed Lie subgroup modelled on $U_\beta = W_1^s(\beta, 0) \subseteq L(G)$, such that $\alpha|_{U_\alpha} \colon U_\alpha \to U_\alpha$ is an analytic automorphism and contractive.*   □

Proof   The idea is to show that $U_\alpha = W_a^s(\alpha, 1)$ for $a \in \,]0, 1[\,\backslash R(\beta)$ close to 1. Details can be found in [19].   □

Although it always is an immersed Lie subgroup, U_α need not be a Lie subgroup (as the example in Section 4.8 shows).

4.10 The scale, tidy subgroups and contraction groups in positive characteristic

We begin with a discussion of the contraction group for a well-behaved automorphism α of a Lie group G over a local field, and the associated local decomposition of G adapted to α. Using the tools from non-linear analysis just described, one obtains the following analogue of Theorem 4.36 (see [20]):

Theorem 4.43   *Let G be an analytic Lie group over a local field and $\alpha \colon G \to G$ be an analytic automorphism. Abbreviate $\beta := L(\alpha)$. If U_α is closed, then U_α, M_α and $U_{\alpha^{-1}}$ are closed Lie subgroups of G with Lie algebras*

$$L(U_\alpha) = U_\beta, \quad L(M_\alpha) = M_\beta \quad and \quad L(U_{\alpha^{-1}}) = U_{\beta^{-1}},$$

respectively. Moreover, $U_\alpha M_\alpha U_{\alpha^{-1}}$ *is an open identity neighbourhood in* G, *and the product map*

$$\pi \colon U_\alpha \times M_\alpha \times U_{\alpha^{-1}} \to U_\alpha M_\alpha U_{\alpha^{-1}}, \quad (x, y, z) \mapsto xyz \qquad (4.1)$$

is an analytic diffeomorphism.

Some ideas of the proof. By Proposition 4.42, U_α and $U_{\alpha^{-1}}$ are immersed Lie subgroups of G with Lie algebras U_β and $U_{\beta^{-1}}$, respectively. By 4.41, there exists a compact centre manifold C for α around 1, modelled on M_β. Since $\alpha^n(C) = C$ for each $n \in \mathbf{Z}$, we have $C \subseteq M_\alpha$. Now $L(G) = U_\beta \oplus M_\beta \oplus M_{\beta^{-1}}$ (see Proposition 4.32). By the inverse function theorem, the product map

$$U_\alpha \times C \times U_{\alpha^{-1}} \to G$$

is a local diffeomorphism at $(1, 1, 1)$. Using Proposition 4.38, we deduce that C is open in M_α, and now a standard argument (Proposition 18 in [6, Chapter 3, §1, no. 9]) can be used to make M_α a Lie group. Then π from (4.1) is a local diffeomorphism at $(1, 1, 1)$, and one deduces as in the proof of Part (f) in the theorem in [13] that π is an analytic diffeomorphism onto its image. $\quad\square$

Remark 4.44 Even if U_α is not closed, its stable manifold structure makes it an immersed Lie subgroup of G, and $\alpha|_{U_\alpha}$ is a contractive automorphism of this Lie group (see Proposition 4.42). Therefore Section 4.11 below provides structural information on U_α (no matter whether it is closed or not).

Next, we discuss tidy subgroups and the scale for well-behaved automorphisms of Lie groups over local fields.

Let G be an analytic Lie group over a local field, and α be an analytic automorphism of G. Let $\|.\|$ be a norm on $L(G)$ adapted to $\beta := L(\alpha)$ and $\phi \colon G \supseteq U \to V \subseteq L(G)$ be an analytic diffeomorphism such that $1 \in U$, $\phi(1) = 0$ and $T_1\phi = \mathrm{id}_{L(G)}$. We know from Lemma 4.14 that $V_r := \phi^{-1}(B_r(0))$ is a subgroup of G if r is small (where $B_r(0) \subseteq L(G)$). Then the following holds:

Theorem 4.45 *If* U_α *is closed, then the subgroup* V_r *of* G *(as just defined) is tidy for* α, *for each sufficiently small* $r > 0$. *Moreover,*

$$s_G(\alpha) = s_{L(G)}(L(\alpha)), \qquad (4.2)$$

where $s_{L(G)}(L(\alpha))$ *can be calculated as in Proposition 4.33. If* U_α *is not closed, then* $s_G(\alpha) \neq s_{L(G)}(L(\alpha))$; *more precisely, the number* $s_G(\alpha)$ *is a proper divisor of* $s_{L(G)}(L(\alpha))$.

Sketch of proof. We explain why V_r is tidy. Combining Theorem 4.43 with the

Ultrametric Inverse Function Theorem, we obtain a diffeomorphic decomposition

$$V_r = (V_r \cap U_\alpha)(V_r \cap M_\alpha)(V_r \cap U_{\alpha^{-1}}) \tag{4.3}$$

for small r. Since β takes $B_r(0) \cap U_\beta$ inside $B_{r\|\beta|_{U_\beta}\|_{\mathrm{op}}}(0) \cap U_\beta$ for each r (where $\|\beta|_{U_\beta}\|_{\mathrm{op}} < 1$), the Ultrametric Inverse Function Theorem implies that

$$\bigcap_{n \in \mathbf{N}_0} \alpha^n(V_r \cap U_\alpha) = \{1\}$$

for all small $r > 0$. Similarly, $\beta(B_r(0) \cap M_\beta) = B_r(0) \cap M_\beta$ and $\beta(B_r(0) \cap U_{\beta^{-1}}) \supseteq B_r(0) \cap U_{\beta^{-1}}$ imply that

$$\bigcap_{n \in \mathbf{N}_0} \alpha^n(V_r \cap M_\alpha) = V_r \cap M_\alpha \quad \text{and} \quad \bigcap_{n \in \mathbf{N}_0} \alpha^n(V_r \cap U_{\alpha^{-1}}) = V_r \cap U_{\alpha^{-1}}$$

for small r. Combining this information with (4.3) and (4.1), we see that

$$(V_r)_+ := \bigcap_{n \in \mathbf{N}_0} \alpha^n(V_r) = (V_r \cap M_\alpha)(V_r \cap U_{\alpha^{-1}}).$$

As a consequence, $(V_r)_+$ has the Lie algebra $M_\beta \oplus U_{\beta^{-1}}$. Likewise, $(V_r)_- = (V_r \cap U_\alpha)(V_r \cap M_\alpha)$ and $L((V_r)_-) = U_\beta \oplus M_\beta$. Hence $V_r = (V_r)_+(V_r)_-$, by (4.3). Thus V_r is tidy above (as in 4.21). Since $(V_r)_{++} = (V_r \cap M_\alpha)U_{\alpha^{-1}}$ as a consequence of (4.1) and the α-stability of $V_r \cap M_\alpha$, we see that $(V_r)_{++}$ is closed (whence V_r is also tidy below and hence tidy). For the remaining assertions, see [20].   □

Remark 4.46 At a first glance, the discussion of the scale and tidy subgroups for automorphisms of Lie groups are easier in the p-adic case than in positive characteristic because any automorphism is locally linear in the p-adic case. But a more profound difference is the automatic validity of the tidiness property **TB** for automorphisms of p-adic Lie groups, which becomes false in positive characteristic.[10]

As observed in [20], results from [3] imply the following closedness criterion for U_α, which is frequently easy to check.

Proposition 4.47 *Let G be a totally disconnected, locally compact group and α be an automorphism of G. If there exists an injective continuous homomorphism $\phi \colon G \to H$ to a totally disconnected, locally compact group H and an automorphism β of H such that U_β is closed and $\beta \circ \phi = \phi \circ \alpha$, then also U_α is closed.*

---

[10] Note that also the pathological automorphism α in Section 4.8 was locally linear.

Proof It is known that $U_\beta \cap M_\beta = \{1\}$ if and only if U_β is closed (see [3, Theorem 3.32] if H is metrizable; the metrizability condition can be removed using techniques from [30]), which holds by hypothesis. Since $\phi(U_\alpha) \subseteq U_\beta$ and $\phi(M_\alpha) \subseteq M_\beta$, the injectivity of ϕ entails that $U_\alpha \cap M_\alpha = \{1\}$. Hence U_α is closed, using [3, Theorem 3.32] again. $\qquad\square$

Since $U_\beta \cap M_\beta = \{1\}$ holds for each inner automorphism of $\mathrm{GL}_n(\mathbf{K})$ (compare Proposition 4.32), the preceding proposition immediately entails:

Proposition 4.48 *If a totally disconnected, locally compact group G admits an injective, continuous homomorphism into $\mathrm{GL}_n(\mathbf{K})$ for some $n \in \mathbf{N}$ and some local field $\mathbf{K}$, then U_α is closed in G for each inner automorphism α of G.* $\qquad\square$

Remark 4.49 In particular, U_α is closed for each inner automorphism of a closed subgroup G of the general linear group $\mathrm{GL}_n(\mathbf{K})$ over a local field $\mathbf{K}$, as already observed in [3, Remark 3.33 (3)]. Also an analogue of (4.2) can already be found in [3, Proposition 3.23], in the special case of Zariski connected reductive linear algebraic groups defined over $\mathbf{K}$. While our approach is analytic, the special case is discussed in [3] using methods from the theory of linear algebraic groups.

Remark 4.50 Let $\mathbf{K} = \mathbf{F}((X))$, G and its automorphism α be as in 4.8.1, and $H := G \rtimes \langle \alpha \rangle$. Then H is a 2-dimensional $\mathbf{K}$-analytic Lie group, and conjugation with α restricts to α on G. In view of the considerations in 4.8.1, we have found an inner automorphism of H with a non-closed contraction group. By Proposition 4.48, H is not a linear Lie group, and more generally it does not admit a faithful continuous linear representation over a local field.

For recent studies of linearity questions on the level of pro-p-groups, the reader is referred to [1], [2], [7], [28], [29], [33] and the references therein.

4.11 The structure of contraction groups in the case of positive characteristic

Wang's structural result concerning p-adic contraction groups (as recalled in Theorem 4.37) can partially be adapted to Lie groups over local fields of positive characteristic (see Theorems A and B in [16]):

Theorem 4.51 *Let G be an analytic Lie group over a local field $\mathbf{K}$ of positive characteristic. If G admits an analytic automorphism $\alpha \colon G \to G$ which is contractive, then G is a torsion group of finite exponent. Moreover, G is nilpotent* *and admits a central series*

$$1 = G_0 \lhd G_1 \lhd \cdots \lhd G_n = G$$

where each G_j is a Lie subgroup of G.

Some ideas of the proof. Let $p := \mathrm{char}(\mathbf{K})$. It is known from [22] that $G = \mathrm{tor}(G) \times H_{p_1} \times \cdots \times H_{p_m}$ with certain α-stable p_j-adic Lie groups H_{p_j}, for suitable primes $p_1, \ldots, p_m$. Since G is locally pro-p, we must have $G = \mathrm{tor}(G) \times H_p$. If H_p was non-trivial, then the size of the sets of p^k-th powers of balls in the p-adic Lie group H_p and such in G would differ too much as k tends to ∞, and one can reach a contradiction (if one makes this idea more precise, as in the proof of [16, Theorem A]). Hence $G = \mathrm{tor}(G)$ is a torsion group.

To see that G is nilpotent, we exploit that the a-stable submanifolds $W_a^s(\alpha, 1)$ are Lie subgroups for all $a \in \,]0,1[\,\backslash R(\alpha)$ (see [19, Proposition 6.2]). Since

$$[W_a^s(\alpha, 1), W_b^s(\alpha, 1)] \subseteq W_{ab}^s(\alpha, 1)$$

holds for the commutator subgroups whenever $a, b, ab \in \,]0,1[\,\backslash R(\alpha)$ (as follows from the second order Taylor expansion of the commutator map), one can easily pick numbers $a_1 < \cdots < a_n$ in $\,]0,1[\,\backslash R(L(\alpha))$ for some n, such that the Lie subgroups $G_j := W_{a_j}^s(\alpha, 1)$ (for $j \in \{1, \ldots, n\}$) form the desired central series. $\square$

4.12 Further results that can be adapted to positive characteristic

We mention that a variety of further results can be generalized from p-adic Lie groups to Lie groups over arbitrary local fields, if one assumes that the relevant automorphisms have closed contraction groups. The ultrametric inverse function theorems (Propositions 4.11 and 4.12) usually suffice as a replacement for the naturality of exp (although we cannot linearize, at least balls are only deformed as by a linear map). And Theorem 4.43 gives control over the eigenvalues of $L(\alpha)$.

We now state two results which can be generalized to positive characteristic following this pattern.

4.52 *To discuss these results, introduce more terminology. First, we recall from [35] that a totally disconnected, locally compact group G is called **uniscalar** if $s_G(x) = 1$ for each $x \in G$. This holds if and only if each group element $x \in G$ normalises some compact, open subgroup V_x (which may depend on x).*

It is natural to ask whether this condition implies that V_x can be chosen independently of x, i.e., whether G has a compact, open, normal subgroup. A suitable p-adic Lie group (see [21, §6]) shows that a positive answer can only be expected if G is compactly generated. But even in the compactly generated case, the answer is negative in general (see [4] and preparatory work in [31]), and one has to restrict attention to particular classes of groups to obtain a positive answer (like compactly generated p-adic Lie groups). If G has the (even stronger) property that every identity neighbourhood contains an open, compact, normal subgroup of G, then G is called **pro-discrete**. Finally, a Lie group G over a local field is **of type R** if the eigenvalues of $L(\alpha)$ have absolute value 1, for each inner automorphism α (see [37]).

Using the above strategy, one can establish generalizations of the results from [37] and [21] (see [20]):

Proposition 4.53 *Let G be an analytic Lie group over a local field* **K**, *and α be an analytic automorphism of G. Then the following properties are equivalent:*

(a) *U_α is closed, and $s_G(\alpha) = s_G(\alpha^{-1}) = 1$;*
(b) *All eigenvalues of $L(\alpha)$ in $\overline{\mathbf{K}}$ have absolute value 1;*
(c) *Every identity neighbourhood of G contains a compact, open subgroup U which is α-stable, i.e., $\alpha(U) = U$.*

In particular, G is of type R if and only if G is uniscalar and U_α is closed for each inner automorphism α of G (in which case $U_\alpha = \{1\}$). $\square$

Using a fixed point theorem for group actions on buildings by A. Parreau, it was shown in [36] (and [21]) that every compactly generated, uniscalar p-adic Lie group is pro-discrete. More generally, one can prove (see [20]):

Proposition 4.54 *Every compactly generated analytic Lie group of type R over a local field is pro-discrete.* $\square$

As mentioned in 4.52, results of the preceding form cannot be expected for general totally disconnected, locally compact groups. The author is grateful to the referee for the following additional example (cf. also [4] and [31]). Consider the Grigorchuk group H. Recall that H is a certain infinite (discrete) group which is a 2-group (in the sense that every element has order a power of 2) and finitely generated (by four elements). It is known that H admits a transitive left action

$$H \times \mathbf{Z} \to \mathbf{Z}, \quad (h,n) \mapsto h.n$$

on the set of integers such that each $h \in H$ commensurates **N** (in the sense that $h.\mathbf{N}$ and $\mathbf{N}$ have finite symmetric difference); such an action exists since H admits Schreier graphs with several ends. The preceding action induces a left action[11]

$$\mathbf{F}_2((X)) \to \mathbf{F}_2((X)), \quad \sum_{n \in \mathbf{Z}} a_n X^n \mapsto \sum_{n \in \mathbf{Z}} a_n X^{h.n}$$

by automorphisms of the topological group $(\mathbf{F}_2((X)), +)$, endowed with its usual topology (such that $\mathbf{F}_2[[X]]$ is a compact open subgroup). We use this action to form the semi-direct product $G := \mathbf{F}_2((X)) \rtimes H$.

Proposition 4.55 $G := \mathbf{F}_2((X)) \rtimes H$ *is a compactly generated, totally discon-nected locally compact torsion group. For every $g \in G$ and identity neighbour-hood $W \subseteq G$, there exists a compact open subgroup V of G such that $V \subseteq W$ and V is normalised by g; moreover, the contraction group U_{α_g} of the inner automorphism $\alpha_g \colon G \to G, x \mapsto gxg^{-1}$ is trivial. However, G does not have a compact, open, normal subgroup; in particular, G is not pro-discrete.*

Proof After shrinking W, we may assume that W is a compact open subgroup of G. Since $\mathbf{F}_2((X))$ and $G/\mathbf{F}_2((X)) \cong H$ are torsion groups, also G is a torsion group. If $m \in \mathbf{N}_0$ is the order of $g \in G$, then $V := \bigcap_{n=0}^{m} g^n(W)$ is a compact open subgroup of G which is normalised by g and contained in W. If $e \neq x \in G$, after shrinking W we may assume that $x \notin W$ and hence $x \notin V$. As the complement of V in G is stable under conjugation with g, we deduce that $g^k x g^{-k} \notin V$ for all $k \in \mathbf{N}$ and hence $g \notin U_{\alpha_g}$, whence $U_{\alpha_g} = \{e\}$ is trivial. Suppose we could find a compact, open, normal subgroup $V \subseteq G$. Then $X^n \mathbf{F}_2[[X]] \subseteq V$ for some $n \in \mathbf{Z}$ and thus $X^n \in V$. Since H acts transitively on $\mathbf{Z}$, we deduce that $X^m \in V$ for each $m \in \mathbf{Z}$ and thus $\mathbf{F}_2((X)) \subseteq V$, as V is a closed subgroup of G. But then $\mathbf{F}_2((X))$ would be a closed subgroup of V and hence $\mathbf{F}_2((X))$ would be compact, a contradiction. $\square$

Acknowledgements. The author was supported by the German Research Foun-dation (DFG), projects GL 357/2-1, GL 357/5-1 and GL 357/6-1.

References for this chapter

[1] M. Abért, N. Nikolov and B. Szegedy, Congruence subgroup growth of arith-metic groups in positive characteristic. *Duke Math. J.* **117** (2003), 367–383.

[2] Y. Barnea and M. Larsen, A non-abelian free pro-p group is not linear over a local field. *J. Algebra* **214** (1999), 338–341.

---

[11] for $(a_n)_{n \in \mathbf{Z}} \in \mathbf{F}_2^{\mathbf{Z}}$ with support bounded below.

[3] U. Baumgartner and G. A. Willis, Contraction groups and scales of automorphisms of totally disconnected locally compact groups. *Israel J. Math.* **142** (2004), 221–248.

[4] M. Bhattacharjee and D. MacPherson, Strange permutation representations of free groups. *J. Aust. Math. Soc.* **74** (2003), 267–285.

[5] N. Bourbaki, *Variétés différentielles et analytiques. Fascicule de résultats.* Hermann, Paris, 1967.

[6] N. Bourbaki, *Lie groups and Lie algebras (Chapters 1–3).* Springer-Verlag, Berlin, 1989.

[7] R. Camina and M. du Sautoy, Linearity of $\mathbf{Z}_p[[t]]$-perfect groups. *Geom. Dedicata* **107** (2004), 1–16.

[8] S. G. Dani and R. Shah, Contraction subgroups and semistable measures on p-adic Lie groups. *Math. Proc. Cambridge Philos. Soc.* **110** (1991), 299–306.

[9] S. G. Dani, N. A. Shah and G. A. Willis, Locally compact groups with dense orbits under $\mathbf{Z}^d$-actions by automorphisms. *Ergodic Theory Dyn. Syst.* **26** (2006), 1443–1465.

[10] J. D. Dixon, M. P. F. du Sautoy, A. Mann and D. Segal, *Analytic pro-p groups.* Cambridge Univ. Press, Cambridge, 1991.

[11] M. du Sautoy, A. Mann and D. Segal, *New horizons in pro-p groups.* Birkhäuser, Boston, 2000.

[12] H. Glöckner, Scale functions on p-adic Lie groups. *Manuscr. Math.* **97** (1998), 205–215.

[13] H. Glöckner, Contraction groups for tidy automorphisms of totally disconnected groups. *Glasgow Math. J.* **47** (2005), 329–333.

[14] H. Glöckner, Implicit functions from topological vector spaces to Banach spaces. *Israel J. Math.* **155** (2006), 205–252.

[15] H. Glöckner, Locally compact groups built up from p-adic Lie groups, for p in a given set of primes. *J. Group Theory* **9** (2006), 427–454.

[16] H. Glöckner, Contractible Lie groups over local fields. *Math. Z.* **260** (2008), 889–904.

[17] H. Glöckner, Invariant manifolds for analytic dynamical systems over ultrametric fields. *Expo. Math.* **31** (2013), 116–150.

[18] H. Glöckner, Finite order differentiability properties, fixed points and implicit functions over valued fields. Preprint, arXiv:math/0511218.

[19] H. Glöckner, Invariant manifolds for finite-dimensional non-archimedean dynamical systems. In: *Advances in Non-Archimedean Analysis* (ed. by H. Glöckner, A. Escassut and K. Shamseddine). Contemp. Math. **665** (2016), 73–90.

[20] H. Glöckner, Endomorphisms of Lie groups over local fields. To appear in: The 2016 MATRIX annals. Springer-Verlag, 2018.

[21] H. Glöckner and G. A. Willis, Uniscalar p-adic Lie groups. *Forum Math.* **13** (2001), 413–421.

[22] H. Glöckner and G. A. Willis, Classification of the simple factors appearing in composition series of totally disconnected contraction groups. *J. Reine Angew. Math.* **634** (2010), 141–169.

[23] W. Hazod and S. Siebert, *Stable probability measures on Euclidean spaces and on locally compact groups.* Kluwer, Dordrecht, 2001.

[24] E. Hewitt and K. A. Ross, *Abstract harmonic analysis, Vol. I*. Springer, Berlin, 1994.

[25] K. H. Hofmann and S. A. Morris, *The Lie theory of connected pro-Lie groups*. EMS, Zürich, 2007.

[26] M. C. Irwin, On the stable manifold theorem. *Bull. London Math. Soc.* **2** (1970), 196–198.

[27] N. Jacobson, *Basic algebra II*. W. H. Freeman and Company, New York, 1989.

[28] A. Jaikin-Zapirain, On linearity of finitely generated R-analytic groups. *Math. Z.* **253** (2006), 333–345.

[29] A. Jaikin-Zapirain and B. Klopsch, Analytic groups over general pro-p domains. *J. London Math. Soc.* **76** (2007), 365–383.

[30] W. Jaworski, On contraction groups of automorphisms of totally disconnected locally compact groups. *Israel J. Math.* **172** (2009), 1–8.

[31] A. Kepert and G. A. Willis, Scale functions and tree ends. *J. Aust. Math. Soc.* **70** (2001), 273–292.

[32] M. Lazard, Groupes analytiques p-adiques. *Publ. Math. IHES* **26** (1965), 5–219.

[33] A. Lubotzky and A. Shalev, On some Λ-analytic groups. *Israel J. Math.* **85** (1994), 307–337.

[34] G. A. Margulis, *Discrete subgroups of semisimple Lie groups*. Springer-Verlag, Berlin, 1991.

[35] T. W. Palmer, *Banach algebras and the general theory of $*$-algebras, Vol. 2*. Cambridge University Press, Cambridge, 2001.

[36] A. Parreau, Sous-groupes elliptiques de groupes linéaires sur un corps valué. *J. Lie Theory* **13** (2003), 271–278.

[37] C. R. E. Raja, On classes of p-adic Lie groups. *New York J. Math.* **5** (1999), 101–105.

[38] A. M. Robert, *A course in p-adic analysis*. Springer-Verlag, New York, 2000.

[39] W. H. Schikhof, *Ultrametric calculus*. Cambridge University Press, Cambridge, 1984.

[40] J.-P. Serre, *Lie algebras and Lie groups*. Springer-Verlag, Berlin, 1992.

[41] E. Siebert, Contractive automorphisms on locally compact groups. *Math. Z.* **191** (1986), 73–90.

[42] E. Siebert, Semistable convolution semigroups and the topology of contraction groups. In: *Probability measures on groups IX. Proceedings of a conference held in Oberwolfach, January 17-23, 1988* (ed. by H. Heyer). Lecture Notes in Mathematics 1379, Springer-Verlag, Berlin, 1989, 325–343.

[43] J. S. P. Wang, The Mautner phenomenon for p-adic Lie groups. *Math. Z.* **185** (1984), 403–412.

[44] A. Weil, *Basic number theory*. Springer-Verlag, New York, 1967.

[45] J. C. Wells, Invariant manifolds of non-linear operators. *Pacific J. Math.* **62** (1976), 285–293.

[46] G. A. Willis, The structure of totally disconnected, locally compact groups. *Math. Ann.* **300** (1994), 341–363.

[47] G. A. Willis, Further properties of the scale function on a totally disconnected group. *J. Algebra* **237** (2001), 142–164.

[48] J. S. Wilson, *Profinite groups*. Oxford University Press, Oxford, 1998.

5

Abstract quotients of profinite groups, after Nikolov and Segal

Benjamin Klopsch

Abstract

We discuss results of Nikolay Nikolov and Dan Segal on abstract quotients of compact Hausdorff topological groups, paying special attention to the class of finitely generated profinite groups. Our primary source is [17]. Sidestepping all difficult and technical proofs, we present a selection of accessible arguments to illuminate key ideas in the subject.

5.1	Introduction	73
5.2	Serre's problem on finite abstract quotients of profinite groups	76
5.3	Nikolov and Segal's results on finite and profinite groups	82
5.4	Abstract versus continuous cohomology	86
	References for this chapter	90

5.1 Introduction

§**5.1.1.** Many concepts and techniques in the theory of finite groups depend intrinsically on the assumption that the groups considered are a priori finite. The theoretical framework based on such methods has led to marvellous achievements, including — as a particular highlight — the classification of all finite simple groups. Notwithstanding, the same methods are only of limited use in the study of infinite groups: it remains mysterious how one could possibly pin down the structure of a general infinite group in a systematic way. Significantly more can be said if such a group comes equipped with additional information, such as a structure-preserving action on a notable geometric object. A coherent approach to studying restricted classes of infinite groups is found by imposing suitable 'finiteness conditions', i.e., conditions that generalize the notion of being finite but are significantly more flexible, such as the group being finitely generated or compact with respect to a natural topology.

One rather fruitful theme, fusing methods from finite and infinite group theory, consists in studying the interplay between an infinite group Γ and the collection of all its finite quotients. In passing, we note that the latter, subject to surjections, naturally form a lattice that is anti-isomorphic to the lattice of finite-index normal subgroups of Γ, subject to inclusions. In this context it is reasonable to focus on Γ being **residually finite**, i.e., the intersection $\bigcap_{N \trianglelefteq_f \Gamma} N$ of all finite-index normal subgroups of Γ being trivial. Every residually finite group Γ embeds as a dense subgroup into its profinite completion $\widehat{\Gamma} = \varprojlim_{N \trianglelefteq_f \Gamma} \Gamma/N$. The latter can be constructed as a closed subgroup of the direct product $\prod_{N \trianglelefteq_f \Gamma} \Gamma/N$ of finite discrete groups and thus inherits the structure of a **profinite group**, i.e., a topological group that is compact, Hausdorff and totally disconnected. The finite quotients of Γ are in one-to-one correspondence with the continuous finite quotients of $\widehat{\Gamma}$.

Profinite groups also arise naturally in a range of other contexts, e.g., as Galois groups of infinite field extensions or as open compact subgroups of Lie groups over non-archimedean local fields, and it is beneficial to develop the theory of profinite groups in some generality. At an advanced stage, one is naturally led to examine more closely the relationship between the underlying abstract group structure of a profinite group G and its topology. In the following we employ the adjective 'abstract' to emphasise that a subgroup, respectively a quotient, of G is not required to be closed, respectively continuous. For instance, as G is compact, an abstract subgroup H is open in G if and only if it is closed and of finite index in G. A profinite group may or may not have (normal) abstract subgroups of finite index that fail to be closed. Put in another way, G may or may not have non-continuous finite quotients. What can be said about abstract quotients of a profinite group G in general? When do they exist and how 'unexpected' can their features possibly be?

To the newcomer, even rather basic groups offer some initial surprises. For instance, consider for a prime p the procyclic group $\mathbf{Z}_p = \varprojlim_{k \in \mathbf{N}} \mathbf{Z}/p^k \mathbf{Z}$ of p-adic integers under addition. Of course, its proper continuous quotients are just the finite cyclic groups $\mathbf{Z}_p/p^k \mathbf{Z}_p$. As we will see below, the abelian group $\mathbf{Z}_p$ fails to map abstractly onto $\mathbf{Z}$, but does have abstract quotients isomorphic to $\mathbf{Q}$.

§**5.1.2.** In [17], Nikolov and Segal streamline and generalize their earlier results that led, in particular, to the solution of the following problem raised by Jean-Pierre Serre in the 1970s: are finite-index abstract subgroups of an arbitrary finitely generated profinite group G always open in G? Recall that a profinite group is said to be (topologically) **finitely generated** if it contains a dense finitely generated abstract subgroup. The problem, which can be found in later editions of Serre's book on Galois cohomology [22, I.§4.2], was solved by Nikolov and Segal in 2003.

Theorem 5.1 (Nikolov, Segal [14]) *Let G be a finitely generated profinite group. Then every finite-index abstract subgroup H of G is open in G.*

Serre had proved this assertion in the special case, where G is a finitely generated pro-p group for some prime p, by a neat and essentially self-contained argument; compare Section 5.2. The proof of the general theorem is considerably more involved and makes substantial use of the classification of finite simple groups; the same is true for several of the main results stated below.

The key theorem in [17] concerns normal subgroups in finite groups and establishes results about products of commutators of the following type. There exists a function $f \colon \mathbf{N} \to \mathbf{N}$ such that, if $H \trianglelefteq \Gamma = \langle y_1, \ldots, y_r \rangle$ for a finite group Γ, then every element of the subgroup $[H, \Gamma] = \langle [h, g] \mid h \in H, g \in \Gamma \rangle$ is a product of $f(r)$ factors of the form $[h_1, y_1][h_2, y_1^{-1}] \cdots [h_{2r-1}, y_r][h_{2r}, y_r^{-1}]$ with $h_1, \ldots, h_{2r} \in H$. Under certain additional conditions on H, a similar conclusion holds based on the significantly weaker hypothesis that $\Gamma = H \langle y_1, \ldots, y_r \rangle = \Gamma' \langle y_1, \ldots, y_r \rangle$, where $\Gamma' = [\Gamma, \Gamma]$ denotes the commutator subgroup. A more precise version of the result is stated as Theorem 5.18 in Section 5.3. By standard compactness arguments, one obtains corresponding assertions for normal subgroups of finitely generated profinite groups; this is also explained in Section 5.3.

Every compact Hausdorff topological group G is an extension of a compact connected group G°, its identity component, by a profinite group G/G°. By the Levi–Mal'cev Theorem, e.g., see [10, Theorem 9.24], the connected component G° is essentially a product of compact Lie groups and thus relatively tractable. We state two results whose proofs require the new machinery developed in [17] that goes beyond the methods used in [14, 15, 16].

Theorem 5.2 (Nikolov, Segal [17]) *Let G be a compact Hausdorff topological group. Then every finitely generated abstract quotient of G is finite.*

Theorem 5.3 (Nikolov, Segal [17]) *Let G be a compact Hausdorff topological group such that G/G° is topologically finitely generated. Then G has a countably infinite abstract quotient if and only if G has an infinite virtually abelian continuous quotient.*

In a recent paper, Nikolov and Segal generalized their results in [17] to obtain the following structural theorem.

Theorem 5.4 (Nikolov, Segal [18])   *Let G be a compact Hausdorff group such that $G/G°$ is topologically finitely generated. Then for every closed normal subgroup $H \trianglelefteq_c G$ the abstract subgroup $[H,G] = \langle [h,g] \mid h \in H, g \in G \rangle$ is closed in G.*

There are many interesting open questions regarding the algebraic properties of profinite groups; the introductory survey [12] provides more information as well as a range of ideas and suggestions for further investigation. In Section 5.4 we look at one possible direction for applying and generalizing the results of Nikolov and Segal, namely in comparing the abstract and the continuous cohomology of a finitely generated profinite group.

5.2 Serre's problem on finite abstract quotients of profinite groups

§**5.2.1.** A profinite group G is said to be **strongly complete** if every finite-index abstract subgroup is open in G. Equivalently, G is strongly complete if every finite quotient of G is a continuous quotient. Clearly, such a group is uniquely determined by its underlying abstract group structure as it is its own profinite completion.

It is easy to see that there are non-finitely generated profinite groups that fail to be strongly complete. For instance, $G = \varprojlim_{n \in \mathbf{N}} C_p^n \cong C_p^{\aleph_0}$, the direct product of a countably infinite number of copies of a cyclic group C_p of prime order p, has $2^{2^{\aleph_0}}$ subgroups of index p, but only countably many of these are open. This can be seen by regarding the abstract group G as a vector space of dimension $2^{\aleph_0}$ over a field with p elements. Implicitly, this simple example uses the axiom of choice.

Without taking a stand on the generalized continuum hypothesis, it is slightly more tricky to produce an example of two profinite groups G_1 and G_2 that are non-isomorphic as topological groups, but nevertheless abstractly isomorphic. As an illustration, we discuss a concrete instance. In [8], Jonathan A. Kiehlmann classifies more generally countably based abelian profinite groups, up to continuous isomorphism and up to abstract isomorphism.

Proposition 5.5   *Let p be a prime. The pro-p group $G = \prod_{i=1}^{\infty} C_{p^i}$ is abstractly isomorphic to $G \times \mathbf{Z}_p$, but there is no continuous isomorphism between the groups G and $G \times \mathbf{Z}_p$.*

Proof The torsion elements form a dense subset in the group G, but they do not in $G \times \mathbf{Z}_p$. Hence the two groups cannot be isomorphic as topological groups.

It remains to show that G is abstractly isomorphic to $G \times \mathbf{Z}_p$. Let $\mathfrak{U} \subseteq \mathscr{P}(\mathbf{N}) = \{T \mid T \subseteq \mathbf{N}\}$ be a non-principal ultrafilter. This means: $\mathfrak{U}$ is closed under finite intersections and under taking supersets; moreover, whenever a disjoint union $T_1 \sqcup T_2$ of two sets belongs to $\mathfrak{U}$, precisely one of T_1 and T_2 belongs to $\mathfrak{U}$; finally, $\mathfrak{U}$ does not contain any one-element sets. Then $\mathfrak{U}$ contains all co-finite subsets of $\mathbf{N}$, and, in fact, one justifies the existence of $\mathfrak{U}$ by enlarging the filter consisting of all co-finite subsets to an ultrafilter; this process relies on the axiom of choice.

Informally speaking, we use the ultrafilter $\mathfrak{U}$ as a tool to form a consistent limit. Indeed, with the aid of $\mathfrak{U}$, we define

$$\psi: G \cong \prod_{i=1}^{\infty} \mathbf{Z}/p^i\mathbf{Z} \longrightarrow \mathbf{Z}_p \cong \varprojlim_j \mathbf{Z}/p^j\mathbf{Z},$$

$$x = (x_1 + p\mathbf{Z}, x_2 + p^2\mathbf{Z}, \dots) \mapsto \varprojlim_j (x\psi_j),$$

where

$$x\psi_j = a + p^j\mathbf{Z} \in \mathbf{Z}/p^j\mathbf{Z} \qquad \text{if } U_j(a) = \{k \geq j \mid x_k \equiv_{p^j} a\} \in \mathfrak{U}.$$

Observe that $\mathfrak{U} \ni \{k \mid k \geq j\} = U_j(0) \sqcup \cdots \sqcup U_j(p^j - 1)$ to see that the definition of ψ_j is valid. It is a routine matter to check that ψ is a (non-continuous) homomorphism from G onto $\mathbf{Z}_p$. Furthermore, G decomposes as an abstract group into a direct product $G = \ker \psi \times H$, where $H = \overline{\langle (1,1,\dots) \rangle} \cong \mathbf{Z}_p$ is the closed subgroup generated by $(1,1,\dots)$. Consequently, $\ker \psi \cong G/H$ as an abstract group. Finally, we observe that

$$\vartheta: G \cong \prod_{i=1}^{\infty} \mathbf{Z}/p^i\mathbf{Z} \longrightarrow G \cong \prod_{i=1}^{\infty} \mathbf{Z}/p^i\mathbf{Z}, \quad (x_i + p^i\mathbf{Z})_i \mapsto (x_i - x_{i+1} + p^i\mathbf{Z})_i$$

is a continuous, surjective homomorphism with $\ker \vartheta = H$. This shows that

$$G \cong G/H \times H \cong G \times \mathbf{Z}_p$$

as abstract groups. $\qquad\qquad\qquad\qquad\qquad\qquad\qquad\qquad\qquad\qquad\qquad\qquad\qquad\square$

Theorem 5.1 states that finitely generated profinite groups are strongly complete: there are no unexpected finite quotients of such groups. What about infinite abstract quotients of finitely generated profinite groups? In view of Theorem 5.1, the following proposition applies to all finitely generated profinite groups.

Proposition 5.6   *Let G be a strongly complete profinite group and $N \trianglelefteq G$ a normal abstract subgroup. If G/N is residually finite, then $N \trianglelefteq_c G$ and G/N is a profinite group with respect to the quotient topology.*

Proof   The group N is the intersection of finite-index abstract subgroups of G. Since G is strongly complete, each of these is open and hence closed in G.   □

Corollary 5.7   *A strongly complete profinite group does not admit any countably infinite residually finite abstract quotients.*

We remark that a strongly complete profinite group can have countably infinite abstract quotients. Indeed, we conclude this section with a simple argument, showing that the procyclic group $\mathbf{Z}_p$ has abstract quotients isomorphic to $\mathbf{Q}$. Again, this makes implicit use of the axiom of choice.

Proposition 5.8   *The procyclic group $\mathbf{Z}_p$ maps non-continuously onto $\mathbf{Q}$.*

Proof   Clearly, $\mathbf{Z}_p$ is a spanning set for the $\mathbf{Q}$-vector space $\mathbf{Q}_p$. Thus there exists an ordered $\mathbf{Q}$-basis $x_0 = 1, x_1, \ldots, x_\omega, \ldots$ for $\mathbf{Q}_p$ consisting of p-adic integers; here ω denotes the first infinite ordinal number. Set

$$y_0 = x_0 = 1, \quad y_i = x_i - p^{-i} \text{ for } 1 \leq i < \omega \quad \text{and} \quad y_\lambda = x_\lambda \text{ for } \lambda \geq \omega.$$

The $\mathbf{Q}$-vector space $\mathbf{Q}_p$ decomposes into a direct sum $\mathbf{Q} \oplus W = \mathbf{Q}_p$, where W has $\mathbf{Q}$-basis $y_1, \ldots, y_\omega, \ldots$, inducing a natural surjection $\eta \colon \mathbf{Q}_p \to \mathbf{Q}$ with kernel W. By construction, $\mathbf{Z}_p + W = \mathbf{Q}_p$ so that we obtain $\mathbf{Z}_p/(\mathbf{Z}_p \cap W) \cong (\mathbf{Z}_p + W)/W \cong \mathbf{Q}$.   □

To some extent this basic example is rather typical; cf. Theorems 5.2 and 5.3.

§5.2.2. We now turn our attention to finitely generated profinite groups. Around 1970 Serre discovered that finitely generated pro-p groups are strongly complete, and he asked whether this is true for arbitrary finitely generated profinite groups. Here and in the following, p denotes a prime.

Theorem 5.9 (Serre)   *Finitely generated pro-p groups are strongly complete.*

It is instructive to recall a proof of this special case of Theorem 5.1, which we base on two auxiliary lemmata.

Lemma 5.10   *Let G be a finitely generated pro-p group and $H \leq_f G$ a finite-index subgroup. Then $|G : H|$ is a power of p.*

Proof   Indeed, replacing H by its core in G, viz., the subgroup $\bigcap_{g \in G} H^g \trianglelefteq_f G$, we may assume that H is normal in G. Thus G/H is a finite group of order m, say, and the set $X = G^{\{m\}} = \{g^m \mid g \in G\}$ of all mth powers in G is contained in H. Being the image of the compact space G under the continuous map $x \mapsto x^m$, the set X is closed in the Hausdorff space G.

Since G is a pro-p group, it has a base of neighbourhoods of 1 consisting of open normal subgroups $N \trianglelefteq_o G$ of p-power index. Hence G/N is a finite p-group for every $N \trianglelefteq_o G$. Let $g \in G$. Writing $m = p^r \tilde{m}$ with $p \nmid \tilde{m}$, we conclude that $g^{p^r} \in XN$ for every $N \trianglelefteq_o G$. Since X is closed in G, this yields

$$g^{p^r} \in \bigcap_{N \trianglelefteq_o G} XN = X \subseteq H.$$

Consequently, every element of G/H has p-power order, and we obtain $|G : H| = m = p^r$. $\qquad\square$

For any subset X of a topological group G we denote by $\overline{X}$ the topological closure of X in G.

Lemma 5.11 *Let G be a finitely generated pro-p group. Then the abstract commutator subgroup $[G,G]$ is closed in G, i.e., $[G,G] = \overline{[G,G]}$.*

Proof We use the following fact about finitely generated nilpotent groups that can easily be proved by induction on the nilpotency class:

(†) if $\Gamma = \langle \gamma_1, \ldots, \gamma_d \rangle$ is a nilpotent group then

$$[\Gamma, \Gamma] = \{ [x_1, \gamma_1] \cdots [x_d, \gamma_d] \mid x_1, \ldots, x_d \in \Gamma \}.$$

Suppose that G is topologically generated by $a_1, \ldots, a_d$, i.e., $G = \overline{\langle a_1, \ldots a_d \rangle}$. Being the image of the compact space $G \times \cdots \times G$ under the continuous map $(x_1, \ldots, x_d) \mapsto [x_1, a_1] \ldots [x_d, a_d]$, the set

$$X = \{ [x_1, a_1] \ldots [x_d, a_d] \mid x_1, \ldots, x_d \in G \} \subseteq [G,G]$$

is closed in the Hausdorff space G. Using (†), this yields

$$\overline{[G,G]} = \bigcap_{N \trianglelefteq_o G} [G,G]N = \bigcap_{N \trianglelefteq_o G} XN = X \subseteq [G,G],$$

and thus $[G,G] = \overline{[G,G]}$. $\qquad\square$

The Frattini subgroup $\Phi(G)$ of a profinite group G is the intersection of all its maximal open subgroups. If G is a finitely generated pro-p group, then $\Phi(G) = \overline{G^p[G,G]}$ is open in G. Here $G^p = \langle g^p \mid g \in G \rangle$ denotes the abstract subgroup generated by the set $G^{\{p\}} = \{ g^p \mid g \in G \}$ of all pth powers.

Corollary 5.12 *Let G be a finitely generated pro-p group. Then $\Phi(G)$ is equal to*

$$G^p[G,G] = G^{\{p\}}[G,G] = \{ x^p y \mid x \in G \text{ and } y \in [G,G] \}.$$

Proof Indeed, $G^p[G,G] = G^{\{p\}}[G,G]$, because $G/[G,G]$ is abelian. By Lemma 5.11, the group $[G,G]$ is closed in G. Being the image of the compact space $G \times [G,G]$ under the continuous map $(x,y) \mapsto x^p y$, the set $G^{\{p\}}[G,G]$ is closed in the Hausdorff space G. Thus $\Phi(G) = \overline{G^p[G,G]} = G^{\{p\}}[G,G]$.   □

Proof of Theorem 5.9 Let $H \leq_f G$. Replacing H by its core in G, we may assume that $H \trianglelefteq_f G$. Using Lemma 5.10, we argue by induction on $|G : H| = p^r$. If $r = 0$ then $H = G$ is open in G. Now suppose that $r \geq 1$.

From Corollary 5.12 we deduce that $M = H\Phi(G)$ is a proper open subgroup of G. Since M is a finitely generated pro-p group and $H \trianglelefteq_f M$ with $|M : H| < |G : H|$, induction yields $H \leq_o M \leq_o G$. Thus H is open in G.   □

For completeness, we record another consequence of the proof of Lemma 5.11 that can be regarded as a special case of Corollary 5.21 below.

Corollary 5.13   *Let G be a finitely generated pro-p group and $N \trianglelefteq G$ a normal abstract subgroup. If $N[G,G] = G$ then $N = G$.*

Proof Suppose that $N[G,G] = G$. Then $N\Phi(G) = G$ and N contains a set of topological generators $a_1, \ldots, a_d$ of G. Arguing as in the proof of Lemma 5.11 we obtain $[G,G] \subseteq N$ and thus $N = N[G,G] = G$.   □

§**5.2.3.** There were several generalizations of Serre's result to other classes of finitely generated profinite groups, most notably by Brian Hartley [6] dealing with poly-pronilpotent groups, by Consuelo Martínez and Efim Zelmanov [9] and independently by Jan Saxl and John S. Wilson [19], each team dealing with direct products of finite simple groups, and finally by Segal [20] dealing with prosoluble groups.

The result of Martínez and Zelmanov, respectively Saxl and Wilson, on powers in non-abelian finite simple groups relies on the classification of finite simple groups and leads to a slightly more general theorem. A profinite group G is called **semisimple** if it is the direct product of non-abelian finite simple groups.

Theorem 5.14 (Martínez and Zelmanov [9]; Saxl and Wilson [19])   *Let $G = \prod_{i \in \mathbf{N}} S_i$ be a semisimple profinite group, where each S_i is non-abelian finite simple. Then G is strongly complete if and only if there are only finitely many groups S_i of each isomorphism type.*

In particular, whenever a semisimple profinite group G as in the theorem is finitely generated, there are only finitely many factors of each isomorphism type and the group is strongly complete. We do not recall the full proof of Theorem 5.14, but we explain how the implication '$\Leftarrow$' can be derived from the following key ingredient proved in [9, 19]:

(‡) for every $n \in \mathbf{N}$ there exists $k \in \mathbf{N}$ such that for every non-abelian finite simple group S whose exponent does not divide n,

$$S = \{x_1^n \cdots x_k^n \mid x_1, \ldots, x_k \in S\}.$$

Suppose that $G = \prod_{i \in \mathbf{N}} S_i$ as in Theorem 5.14, with only finitely many groups S_i of each isomorphism type, and let $H \leq_f G$. Replacing H by its core we may assume that $H \trianglelefteq_f G$. Put $n = |G : H|$ and choose $k \in \mathbf{N}$ as in (‡). Then H contains $X = \{x_1^n \cdots x_k^n \mid x_1, \ldots, x_k \in G\} \supseteq \prod_{i \geq j} S_i$, where $j \in \mathbf{N}$ is such that the exponent of S_i does not divide n for $i \geq j$, and hence H is open in G. The existence of the index j is guaranteed by the classification of finite simple groups.

Observe that implicitly we have established the following corollary.

Corollary 5.15 *Let G be a semisimple profinite group and $q \in \mathbf{N}$. Then $G^q = \langle g^q \mid g \in G \rangle$, the abstract subgroup generated by all qth powers, is closed in G.*

§5.2.4. The groups $[G, G]$ and G^q featuring in the discussion above are examples of verbal subgroups of G. We briefly summarise the approach of Nikolov and Segal that led to the original proof of Theorem 5.1 in [14]; the theorem is derived from a 'uniformity result' concerning verbal subgroups of finite groups.

Let $d \in \mathbf{N}$, and let $w = w(X_1, \ldots, X_r)$ be a group word, i.e., an element of the free group on r generators $X_1, \ldots, X_r$. A w-**value** in a group G is an element of the form $w(x_1, x_2, \ldots, x_r)$ or $w(x_1, x_2, \ldots, x_r)^{-1}$ with $x_1, x_2, \ldots, x_r \in G$. The **verbal subgroup** $w(G)$ is the subgroup generated (algebraically, whether or not G is a topological group) by all w-values in G. The word w is d-**locally finite** if every d-generator group H satisfying $w(H) = 1$ is finite. Finally, a **simple commutator** of length $n \geq 2$ is a word of the form $[X_1, \ldots, X_n]$, where $[X_1, X_2] = X_1^{-1} X_2^{-1} X_1 X_2$ and $[X_1, \ldots, X_n] = [[X_1, \ldots, X_{n-1}], X_n]$ for $n > 2$. It is well known that the verbal subgroup corresponding to the simple commutator of length n is the nth term of the lower central series.

Suppose that w is d-locally finite or that w is a simple commutator. In [14], Nikolov and Segal prove that there exists $f = f(w, d) \in \mathbf{N}$ such that in every d-generator finite group G every element of the verbal subgroup $w(G)$ is equal to a product of f w-values in G. This 'quantitative' statement about (families of) finite groups translates into the following 'qualitative' statement about profinite groups. If w is d-locally finite, then in every d-generator profinite group G, the verbal subgroup $w(G)$ is open in G. From this one easily deduces Theorem 5.1. Similarly, considering simple commutators Nikolov and Segal prove that each term of the lower central series of a finitely generated profinite group G is closed.

By a variation of the same method and by appealing to Zelmanov's celebrated solution to the restricted Burnside problem, one establishes the following result.

Theorem 5.16 (Nikolov, Segal [16])   *Let G be a finitely generated profinite group. Then the subgroup $G^q = \langle g^q \mid g \in G \rangle$, the abstract subgroup generated by qth powers, is open in G for every $q \in \mathbf{N}$.*

The approach in [14] is based on quite technical results concerning products of 'twisted commutators' in finite quasisimple groups that are established in [15]. The proofs in [16] make use of the full machinery in [14]. Ultimately these results all rely on the classification of finite simple groups.

Remark 5.17   An independent justification of Theorem 5.16 would immediately yield a new proof of Theorem 5.1. Furthermore, Andrei Jaikin-Zapirain has shown that the use of the positive solution of the restricted Burnside problem in proving Theorem 5.16 is to some extent inevitable; see [7, Section 5.1].

We refer to the survey article [25] for a more thorough discussion of the background to Serre's problem and further information on finite-index subgroups and verbal subgroups in profinite groups. A comprehensive account of verbal width in groups is given in [21].

5.3 Nikolov and Segal's results on finite and profinite groups

§**5.3.1.** In this section we discuss the new approach in [17]. The key theorem concerns normal subgroups in finite groups. A finite group H is almost-simple if $S \trianglelefteq H \leq \operatorname{Aut}(S)$ for some non-abelian finite simple group S. For a finite group Γ, let $d(\Gamma)$ denote the minimal number of generators of Γ, write $\Gamma' = [\Gamma, \Gamma]$ for the commutator subgroup of Γ and set

$$\Gamma_0 = \bigcap \{T \trianglelefteq \Gamma \mid \Gamma/T \text{ almost-simple}\}$$
$$= \bigcap \{C_\Gamma(M) \mid M \text{ a non-abelian simple chief factor}\},$$

where the intersection over an empty set is naturally interpreted as Γ. To see that the two descriptions of Γ_0 agree, recall that the chief factors of Γ arise as the minimal normal subgroups of arbitrary quotients of Γ. Further, we remark that Γ/Γ_0 is semisimple-by-(soluble of derived length at most 3), because the outer automorphism group of any simple group is soluble of derived length at most 3. This strong form of the Schreier conjecture is a consequence of the classification of finite simple groups.

For $X \subseteq \Gamma$ and $f \in \mathbf{N}$ we write $X^{*f} = \{x_1 \cdots x_f \mid x_1, \ldots, x_f \in X\}$.

Theorem 5.18 (Nikolov, Segal [17]) *Let Γ be a finite group and $\{y_1, \ldots, y_r\} \subseteq \Gamma$ a symmetric subset, i.e., a subset that is closed under taking inverses. Let $H \trianglelefteq \Gamma$.*

(1) If $H \subseteq \Gamma_0$ and $H\langle y_1, \ldots, y_r \rangle = \Gamma'\langle y_1, \ldots, y_r \rangle = \Gamma$ then

$$\langle [h,g] \mid h \in H, g \in \Gamma \rangle = \{[h_1, y_1] \cdots [h_r, y_r] \mid h_1, \ldots, h_r \in H\}^{*f},$$

where $f = f(r, d(\Gamma)) = O(r^6 d(\Gamma)^6)$.

(2) If $\Gamma = \langle y_1, \ldots, y_r \rangle$ then the conclusion in (1) holds without assuming $H \subseteq \Gamma_0$ and with better bounds on f.

While the proof of Theorem 5.18 is rather involved, the basic underlying idea is simple to sketch. Suppose that $\Gamma = \langle g_1, \ldots, g_r \rangle$ is a finite group and M a non-central chief factor. Then the set $[M, g_i] = \{[m, g_i] \mid m \in M\}$ must be 'relatively large' for at least one generator g_i. Hence $\prod_{i=1}^{r} [M, g_i]$ is 'relatively large'. In order to transform this observation into a rigorous proof one employs a combinatorial principle, discovered by Timothy Gowers in the context of product-free sets of quasirandom groups and adapted by Nikolay Nikolov and László Pyber to obtain product decompositions in finite simple groups; cf. [5, 13]. Informally speaking, to show that a finite group is equal to a product of some of its subsets, it suffices to know that the cardinalities of these subsets are 'sufficiently large'. A precise statement of the result used in the proof of Theorem 5.18 is the following.

Theorem 5.19 ([1, Corollary 2.6]) *Let Γ be a finite group, and let $\ell(\Gamma)$ denote the minimal dimension of a non-trivial $\mathbf{R}$-linear representation of Γ.*

If $X_1, \ldots, X_t \subseteq \Gamma$, for $t \geq 3$, satisfy

$$\prod_{i=1}^{t} |X_i| \geq |\Gamma|^t \, \ell(\Gamma)^{2-t},$$

then $X_1 \cdots X_t = \{x_1 \cdots x_t \mid x_i \in X_i \text{ for } 1 \leq i \leq t\} = \Gamma$.

A short, but informative summary of the proof of Theorem 5.18, based on product decompositions, can be found in [12, §10].

§**5.3.2.** By standard compactness arguments, Theorem 5.18 yields a corresponding result for normal subgroups of finitely generated profinite groups. For a profinite group G, let $d(G)$ denote the minimal number of topological generators of G, write $G' = [G, G]$ for the abstract commutator subgroup of G and set

$$G_0 = \bigcap \{T \trianglelefteq_o G \mid G/T \text{ almost-simple}\},$$

where the intersection over an empty set is naturally interpreted as G. As in the finite case, G/G_0 is semisimple-by-(soluble of derived length at most 3). For $X \subseteq G$ and $f \in \mathbf{N}$ we write $X^{*f} = \{x_1 \cdots x_f \mid x_1, \ldots, x_f \in X\}$ as before, and the topological closure of X in G is denoted by $\overline{X}$.

Theorem 5.20 (Nikolov, Segal [17]) *Let G be a profinite group and $H \trianglelefteq_c G$ be a closed normal subgroup. Let also $\{y_1, \ldots, y_r\} \subseteq G$ be a symmetric subset.*
(1) If $H \subseteq G_0$ and $H\langle y_1, \ldots, y_r \rangle = \overline{G'\langle y_1, \ldots, y_r \rangle} = G$ then

$$\langle [h, g] \mid h \in H, g \in G \rangle = \{[h_1, y_1] \cdots [h_r, y_r] \mid h_1, \ldots, h_r \in H\}^{*f},$$

where $f = f(r, d(G)) = O(r^6 d(G)^6)$.
(2) If $y_1, \ldots, y_r$ topologically generate G then the conclusion in (1) holds without assuming $H \subseteq G_0$ and better bounds on f.

Proof We indicate how to prove (1). The inclusion '⊇' is clear. The inclusion '⊆' holds modulo every open normal subgroup $N \trianglelefteq_o G$, by Theorem 5.18. Consider the set on the right-hand side, call it Y. Being the image of the compact space $H \times \cdots \times H$, with rf factors, under a continuous map, the set Y is closed in the Hausdorff space G. Hence

$$\langle [h, g] \mid h \in H, g \in G \rangle \subseteq \bigcap_{N \trianglelefteq_o G} YN = Y. \qquad \square$$

In particular, the theorem shows that, if G is a finitely generated profinite group and $H \trianglelefteq_c G$ a closed normal subgroup, then the abstract subgroup

$$[H, G] = \langle [h, g] \mid h \in H, g \in G \rangle$$

is closed. Thus G' and more generally all terms $\gamma_i(G)$ of the abstract lower central series of G are closed; these consequences were already established in [14]. Theorem 5.4, stated in the introduction, generalizes these results to more general compact Hausdorff groups.

Furthermore, one obtains from Theorem 5.20 the following tool for studying abstract normal subgroups of a finitely generated profinite group G, reducing certain problems more or less to the abelian profinite group G/G' or the profinite group G/G_0 which is semisimple-by-(soluble of derived length at most 3).

Corollary 5.21 (Nikolov, Segal [17]) *Let G be a finitely generated profinite group and $N \trianglelefteq G$ a normal abstract subgroup. If $NG' = NG_0 = G$ then $N = G$.*

Proof Suppose that $NG' = NG_0 = G$, and let $d = d(G)$ be the minimal number of topological generators of G. Then there exist $y_1, \ldots, y_{2d} \in N$ such that

$$G_0 \overline{\langle y_1, \ldots, y_{2d} \rangle} = \overline{G' \langle y_1, \ldots, y_{2d} \rangle} = G.$$

Applying Theorem 5.20, with $H = G_0$, we obtain

$$[G_0, G] \subseteq \langle [G_0, y_i] \cup [G_0, y_i^{-1}] \mid 1 \leq i \leq 2d \rangle \subseteq N,$$

hence $G = NG' = N[NG_0, G] = N$. $\qquad\square$

Using Corollaries 5.21 and 5.15, it is not difficult to derive Theorem 5.1.

Proof of Theorem 5.1 Let $H \leq_f G$ be a finite-index subgroup of the finitely generated profinite group G. Then its core $N = \bigcap_{g \in G} H^g \trianglelefteq_f G$ is contained in H, and it suffices to prove that N is open in G. The topological closure $\overline{N}$ is open in G; in particular, $\overline{N}$ is a finitely generated profinite group and without loss of generality we may assume that $\overline{N} = G$.

Assume for a contradiction that $N \not\geq G$. Using Corollary 5.21, we deduce that

$$NG' \not\geq G \qquad \text{or} \qquad NG_0 \not\geq G. \tag{5.1}$$

Setting $q = |G : N|$, we know that the abstract subgroup $G^q = \langle g^q \mid g \in G \rangle$ generated by all qth powers is contained in N. By Theorem 5.20, the abstract commutator subgroup $G' = [G, G]$ is closed, thus the subgroup $G'G^q$ is closed in G. Hence $G/G'G^q$, being a finitely generated abelian profinite group of finite exponent, is finite and discrete. As $NG'/G'G^q$ is dense in $G/G'G^q$, we deduce that $NG' = NG'G^q = G$.

It suffices to prove that $NG_0 = G$ in order to obtain a contradiction to (5.1). Factoring out by G_0, we may assume without loss of generality that $G_0 = 1$. Then G has a semisimple subgroup $T \trianglelefteq_c G$ such that G/T is soluble. From $G = NG'$ we see that G/NT is perfect and soluble, and we deduce that

$$NT = G. \tag{5.2}$$

Corollary 5.15 shows that $T^q \leq_c T$. Factoring out by T^q (which is contained in N) we may assume without loss of generality that $T^q = 1$. The definition of G_0 shows that T is a product of non-abelian finite simple groups, each normal in G, of exponent dividing q. Using the classification of finite simple groups, one sees that the finite simple factors of T have uniformly bounded order. Thus $G/C_G(T)$ is finite. As $T \cap C_G(T) = 1$, we conclude that T is finite, thus $T \cap N \leq_c G$. This shows that

$$T = [T, G] = [T, \overline{N}] \leq \overline{[T, N]} \leq T \cap N,$$

and (5.2) gives $N = NT = G$.   □

§**5.3.3.** The methods developed in [17] lead to new consequences for abstract quotients of finitely generated profinite groups and, more generally, compact Hausdorff topological groups. In the introduction we stated three such results: Theorems 5.2, 5.3 and 5.4.

We finish by indicating how Corollary 5.21 can be used to see that, for profinite groups, the assertion in Theorem 5.2 reduces to the following special case.

Theorem 5.22   *A finitely generated semisimple profinite group does not have countably infinite abstract images.*

Nikolov and Segal prove Theorem 5.22 via a complete description of the maximal normal abstract subgroups of a strongly complete semisimple profinite group; see [17, Theorem 5.12]. Their argument relies, among other things, on work of Martin Liebeck and Aner Shalev [11] on the diameters of non-abelian finite simple groups.

Proof of 'Theorem 5.22 implies Theorem 5.2 for profinite groups'   For a contradiction, assume that $\Gamma = G/N$ is an infinite finitely generated abstract image of a profinite group G. Replacing G by the closed subgroup generated by any finite set mapping onto a generating set of Γ, we may assume that G is topologically finitely generated, hence strongly complete by Theorem 5.1. Let $\Gamma_1 = \bigcap_{\Delta \trianglelefteq_f \Gamma} \Delta$. Then Γ/Γ_1 is a countable residually finite image of G, and Corollary 5.7 shows that $|\Gamma : \Gamma_1| < \infty$. Replacing Γ by Γ_1 and G by the preimage of Γ_1 in G, we may assume that Γ has no non-trivial finite images.

We apply Corollary 5.21 to G and the normal abstract subgroup $N \trianglelefteq G$ with $G/N = \Gamma$. Since Γ is finitely generated and has no non-trivial finite images, we conclude that $\Gamma' = \Gamma$ and hence $NG' = G$. Thus $NG_0 \lneqq G$. Consequently, G/G_0 maps onto the infinite finitely generated perfect quotient $\Gamma_0 = G/NG_0$ of Γ. We may assume that $G_0 = 1$ so that G is semisimple-by-soluble. Since soluble groups do not map onto perfect groups, we may assume that G is semisimple.   □

5.4 Abstract versus continuous cohomology

In this section we interpret Theorem 5.1 in terms of cohomology groups and indicate some work in preparation; for more details we refer to a forthcoming joint paper [2] with Yiftach Barnea and Jaikin-Zapirain.

§**5.4.1.** Let Γ be a dense abstract subgroup of a profinite group G, and let M be a continuous finite G-module. For $i \in \mathbf{N}_0$, the i-dimensional continuous cohomology group can be defined as the quotient

$$\mathrm{H}^i_{\mathrm{cont}}(G,M) = \mathrm{Z}^i_{\mathrm{cont}}(G,M) / \mathrm{B}^i_{\mathrm{cont}}(G,M)$$

of continuous i-cocycles modulo continuous i-coboundaries with values in M. Alternatively, the continuous cohomology can be described via the direct limit

$$\mathrm{H}^i_{\mathrm{cont}}(G,M) = \varinjlim_{U \trianglelefteq_o G} \mathrm{H}^i(G/U, M^U),$$

where U runs over all open normal subgroups, M^U denotes the submodule of invariant under U, and $\mathrm{H}^i(G/U, M^U)$ denotes the ordinary cohomology of the finite group G/U. Restriction provides natural maps

$$\mathrm{H}^i_{\mathrm{cont}}(G,M) \to \mathrm{H}^i(\Gamma, M), \quad i \in \mathbf{N}_0. \tag{5.1}$$

Here $\mathrm{H}^i(\Gamma, M)$ is the ordinary cohomology, or equivalently the continuous cohomology of Γ equipped with the discrete topology.

In fact, we concentrate on the case $\Gamma = G$, regarded as an abstract group, and we write $\mathrm{H}^i_{\mathrm{disc}}(G,M) = \mathrm{H}^i(\Gamma, M)$ for the cohomology of the abstract group G. Nikolov and Segal's solution to Serre's problem can be reformulated as follows.

Theorem 5.23 (Nikolov, Segal)   *Let G be a finitely generated profinite group and M a continuous finite G-module. Then $\mathrm{H}^1_{\mathrm{cont}}(G,M) \to \mathrm{H}^1_{\mathrm{disc}}(G,M)$ is a bijection, and $\mathrm{H}^2_{\mathrm{cont}}(G,M) \to \mathrm{H}^2_{\mathrm{disc}}(G,M)$ is injective.*

It is natural to ask for analogous results in higher dimensions, but the situation seems to become rather complicated already in dimensions 2 and 3. Currently, very little is known, even at a conjectural level. Some basic results, regarding pro-p groups, can be found in [4]; see also the references therein.

§**5.4.2.** In [22, I.2.6], Serre introduced a series of equivalent conditions to investigate the maps (5.1), in a slightly more general setting. For our purpose the following formulation is convenient: for $n \in \mathbf{N}_0$ and p a prime, we define the property

$\mathsf{E}_n(p)$ For every continuous finite G-module M of cardinality a p-power, the natural map $\mathrm{H}^i_{\mathrm{cont}}(G,M) \to \mathrm{H}^i_{\mathrm{disc}}(G,M)$ is bijective for $0 \le i \le n$.

We say that G satisfies E_n if $\mathsf{E}_n(p)$ holds for all primes p. Since cohomology 'commutes' with taking direct sums in the coefficients, G satisfies E_n if and only if, for every continuous finite G-module M, the maps (5.1) are bijections for $0 \leq i \leq n$. We say that G is **cohomologically p-good** if G satisfies $\mathsf{E}_n(p)$ for all $n \in \mathbf{N}_0$. The group G is called **cohomologically good** if it is cohomologically p-good for all primes p; cf. [22, I.2.6].

The following proposition from [2] provides a useful tool for showing that a group has property $\mathsf{E}_n(p)$.

Proposition 5.24 *A profinite group H satisfies $\mathsf{E}_n(p)$ if there is an abstract short exact sequence*

$$1 \longrightarrow N \longrightarrow H \longrightarrow G \longrightarrow 1,$$

where N and G are finitely generated profinite groups satisfying $\mathsf{E}_n(p)$.

We discuss a natural application. Recall that the **rank** of a profinite group G is

$$\mathrm{rk}(G) = \sup\{d(H) \mid H \leq_c G\},$$

where $d(H)$ is the minimal number of topological generators of H. Using the classification of finite simple groups, it can be shown that if G is a profinite group all of whose Sylow subgroups have rank at most r then $\mathrm{rk}(G) \leq r + 1$. Furthermore, a pro-$p$ group G has finite rank if and only if it is p-adic analytic; in this case the dimension of G as a p-adic manifold is bounded by the rank: $\dim(G) \leq \mathrm{rk}(G)$.

The following result from [2] generalizes [4, Theorem 2.10].

Theorem 5.25 *Soluble profinite groups of finite rank are cohomologically good.*

It is well known that the second cohomology groups, and hence properties $\mathsf{E}_2(p)$ and E_2, are intimately linked to the theory of group extensions. Theorem 5.25 can be used to prove that every abstract extension of soluble profinite groups of finite rank carries again, in a unique way, the structure of a soluble profinite group of finite rank. We refer to [24, Chapter 8] for a general discussion of profinite groups of finite rank. An example of a soluble profinite group of finite rank that is not virtually pronilpotent can be obtained as follows: consider $G = \overline{\langle x \rangle} \ltimes A \cong \widehat{\mathbf{Z}} \ltimes \widehat{\mathbf{Z}}$, where conjugation by x on a Sylow pro-p subgroup $A_p \cong \mathbf{Z}_p$ of $A = \prod_p A_p$ is achieved via multiplication by a primitive $(p-1)$th root or unity.

§**5.4.3** Fix a prime p and consider the questions raised above for a finitely generated pro-p group G. By means of a reduction argument, one is led to focus on cohomology groups with coefficients in the trivial module $\mathbf{F}_p$, i.e., the groups

$$H^i_{\text{cont}}(G) = H^i_{\text{cont}}(G, \mathbf{F}_p) \quad \text{and} \quad H^i_{\text{disc}}(G) = H^i_{\text{disc}}(G, \mathbf{F}_p) \quad \text{for } i \in \mathbf{N}_0.$$

Indeed, G satisfies $\mathsf{E}_n(p)$ if and only if, for $1 \leq i \leq n$, every cohomology $\alpha \in H^i_{\text{disc}}(U)$, where $U \leq_o G$, vanishes when restricted to a suitable smaller open subgroup. It is natural to start with the cohomology groups in dimension 2, and by virtue of Theorem 5.23, we may regard $H^2_{\text{cont}}(G)$ as a subgroup of $H^2_{\text{disc}}(G)$. It is well known that these groups are intimately linked with continuous, respectively abstract, central extensions of a cyclic group C_p by G.

First suppose that the finitely generated pro-p group G is not finitely presented as a pro-p group. For instance, the profinite wreath product $C_p \hat{\wr} \mathbf{Z}_p = \varprojlim_i C_p \wr C_{p^i}$ has this property. Then it can be shown that there exists a central extension $1 \to C_p \to H \to G \to 1$ such that H is not residually finite, hence cannot be a profinite group; compare [4, Theorem A]. It follows that $H^2_{\text{cont}}(G)$ is strictly smaller than $H^2_{\text{disc}}(G)$, and G does not satisfy $\mathsf{E}_2(p)$. This raises a natural question.

Problem   *Characterise finitely presented pro-p groups that satisfy $\mathsf{E}_2(p)$.*

Groups of particular interest, for which the situation is not yet fully understood, include on the one hand non-abelian free pro-p groups and on the other hand p-adic analytic pro-p groups. Indeed, a rather intriguing theorem of Aldridge K. Bousfield [3, Theorem 11.1] in algebraic topology shows that, for a non-abelian finitely generated free pro-p group F, the direct sum $H^2_{\text{disc}}(F) \oplus H^3_{\text{disc}}(F)$ is infinite[1]. This stands in sharp contrast to the well known fact that $H^2_{\text{cont}}(F) = H^3_{\text{cont}}(F) = 0$.

In [2] we establish the following result.

Theorem 5.26   *Let F be a finitely presented pro-p group satisfying $\mathsf{E}_2(p)$. Then every finitely presented continuous quotient of F also satisfies $\mathsf{E}_2(p)$.*

In particular, this shows that, if every finitely generated free pro-p group F has $H^2_{\text{disc}}(F) = H^2_{\text{cont}}(F) = 0$, then all finitely presented pro-p groups satisfy $\mathsf{E}_2(p)$.

We close by stating a theorem from [2] regarding the cohomology of compact p-adic analytic groups that improves results of Balasubramanian Sury [23] on central extensions of Chevalley groups over non-archimedean local fields of characteristic 0.

---

[1] Note added in proof: For recent advances, see the arXiv-preprint arXiv:1705.09131 by Sergei O. Ivanov and Roman Mikhailov.

Theorem 5.27   *Let Φ be an irreducible root system, and let K be a non-archimedean local field of characteristic 0 and residue characteristic p.*
Then every open compact subgroup H of the Chevalley group $\mathrm{Ch}_\Phi(K)$ satisfies $E_2(p)$. Consequently, every central extension of C_p by H is residually finite and carries, in a unique way, the structure of a profinite group.

Acknowledgements. The author is grateful to the referee for valuable suggestions that led, for instance, to the inclusion of Corollary 5.13.

References for this chapter

[1] L. Babai, N. Nikolov and L. Pyber, *Product growth and mixing in finite groups*, Proceedings of the Nineteenth Annual ACM-SIAM Symposium on Discrete Algorithms, 248–257, ACM, New York, 2008.

[2] Y. Barnea, A. Jaikin-Zapirain and B. Klopsch, *Abstract versus topological extensions of profinite groups*, preprint, 2015.

[3] A. K. Bousfield, *On the p-adic completions of nonnilpotent spaces*, Trans. Amer. Math. Soc. **331** (1992), 335–359.

[4] G. A. Fernández-Alcober, I. V. Kazachkov, V. N. Remeslennikov and P. Symonds, *Comparison of the discrete and continuous cohomology groups of a pro-p group*, Algebra i Analiz **19** (2007), 126–142; translation in: St. Petersburg Math. J. **19** (2008), 961–973.

[5] W. T. Gowers, *Quasirandom groups*, Combin. Probab. Comput. **17** (2008), 363–387.

[6] B. Hartley, *Subgroups of finite index in profinite groups*, Math. Z. **168** (1979), 71–76.

[7] A. Jaikin-Zapirain, *On the verbal width of finitely generated pro-p groups*, Rev. Mat. Iberoamericana **24** (2008), 617–630.

[8] J. A. Kiehlmann, *Classifications of countably-based Abelian profinite groups*, J. Group Theory **16** (2013), 141–157.

[9] C. Martínez and E. Zelmanov, *Products of powers in finite simple groups*, Israel J. Math. **96** (1996), 469–479.

[10] K. H. Hofmann and S. A. Morris, The structure of compact groups, de Gruyter, Berlin, 2013.

[11] M. W. Liebeck and A. Shalev, *Diameters of finite simple groups: sharp bounds and applications*, Ann. Math. **154** (2001), 383–406.

[12] N. Nikolov, *Algebraic properties of profinite groups*, preprint, arXiv:1108.5130, 2012.

[13] N. Nikolov and L. Pyber, *Product decompositions of quasirandom groups and a Jordan type theorem*, J. Eur. Math. Soc. **13** (2011), 1063–1077.

[14] N. Nikolov and D. Segal, *On finitely generated profinite groups I. Strong completeness and uniform bounds*, Ann. of Math. **165** (2007), 171–238.

[15] N. Nikolov and D. Segal, *On finitely generated profinite groups II. Products in quasisimple groups*, Ann. of Math. **165** (2007), 239–273.

[16] N. Nikolov and D. Segal, *Powers in finite groups*, Groups Geom. Dyn. **5** (2011), 501–507.

[17] N. Nikolov and D. Segal, *Generators and commutators in finite groups; abstract quotients of compact groups*, Invent. Math. **190** (2012), 513–602.

[18] N. Nikolov and D. Segal, *On normal subgroups of compact groups*, J. Eur. Math. Soc. (JEMS) **16** (2014), 597–618.

[19] J. Saxl and J. S. Wilson, *A note on powers in simple groups*, Math. Proc. Camb. Philos. Soc. **122** (1997), 91–94.

[20] D. Segal, *Closed subgroups of profinite groups*, Proc. London Math. Soc. **81** (2000), 29–54.

[21] D. Segal, Words: notes on verbal width in groups, Cambridge University Press, Cambridge, 2009.

[22] J.-P. Serre, Galois cohomology, Springer-Verlag, Berlin, 1997.

[23] B. Sury, *Central extensions of p-adic groups; a theorem of Tate*, Comm. Algebra **21** (1993), 1203–1213.

[24] J. S. Wilson, Profinite groups, Oxford University Press, New York, 1998.

[25] J. S. Wilson, *Finite index subgroups and verbal subgroups in profinite groups*, Séminaire Bourbaki, Vol. 2009/2010, Exposés 1012–1026. Astérisque No. **339** (2011), Exp. No. 1026.

6

Automorphism groups of trees: generalities and prescribed local actions

Alejandra Garrido, Yair Glasner and Stephan Tornier

Abstract

We recall the basic theory of automorphisms of trees and Tits' simplicity theorem, and present two constructions of tree groups via local actions: the universal group associated to a finite permutation group by Burger and Mozes, and the k-closures of a given group by Banks, Elder and Willis.

6.1	Introduction	92
6.2	Generalities on trees	93
6.3	Independence properties and Tits' simplicity theorem	100
6.4	Universal Groups	102
6.5	Simple totally disconnected locally compact groups with prescribed local actions	109
References for this chapter		116

6.1 Introduction

In the study of totally disconnected locally compact (tdlc) Hausdorff groups, groups of automorphisms of locally finite trees appear naturally and form a significant class of examples. They are also the most basic case of groups acting on buildings or CAT(0) cube complexes. Moreover, they play an important role in the structure theory of compactly generated tdlc groups through Cayley–Abels graph constructions (see [8, §11] and Chapter 1 in this volume).

In Section 2, we present general results and notions about automorphisms of infinite, locally finite trees. Section 6.3 deals with independence properties (the idea that restrictions of the action to subtrees are in some sense independent of each other) and a general form of Tits' simplicity theorem, Theorem 6.14. This theorem extracts abstractly simple subgroups from groups acting on the tree in a sufficiently dense and independent way.

We then present two constructions of closed subgroups of tree automorphisms with prescribed local actions and examine some of the local-to-global phenomena that they exhibit. First, we present in Section 6.4 the universal group construction by Burger and Mozes (see [2]), which to every finite permutation group $F \leq S_d$ associates a subgroup $U(F) \leq \text{Aut}(T_d)$ which locally acts like F (here, T_d denotes the d-regular tree). Properties of the groups $U(F)$ are often determined by those of the finite permutation group F. Universal groups are fundamental in the study of lattices in the product of the automorphism groups of two trees, analogous to the study of lattices in semisimple Lie groups, and are key to the proof of the normal subgroup theorem (see [3] and Chapters 11 and 12 in this volume).

Second, we describe in Section 6.5 a variation of this construction by Banks, Elder and Willis (see [1]): Given $G \leq \text{Aut}(T)$ for an arbitrary tree T it produces a sequence of closed subgroups $G^{(k)} \leq \text{Aut}(T)$ which act like G on balls of radius k. This sequence converges to the topological closure $\overline{G}$ of G from above, in a sense to be made precise. We will use Tits' simplicity theorem to construct infinitely many, pairwise distinct, non-discrete, compactly generated, abstractly simple, tdlc groups. Finding simple tdlc groups (and in particular, compactly generated ones) has become relevant to the structure theory of tdlc groups thanks to results of Caprace and Monod [5], which state that such groups can be decomposed into simple pieces.

6.2 Generalities on trees

6.2.1 Trees

Let T be a simplicial tree with vertex set $V(T)$ and edge set $E(T) \subset V(T) \times V(T)$. We will always assume that $V(T)$ is countable and that T is not isomorphic to a bi-infinite line. Each edge $e \in E(T)$ is determined by its origin $o(e) \in V(T)$ and its terminus $t(e) \in V(T)$, so the edge $\bar{e} = (t(e), o(e))$ is the **inverse** of e. The pair $\{e, \bar{e}\}$ is a **geometric edge**. Given any edge $e = (x, y) \in E(T)$, the subgraph $T \setminus \{e, \bar{e}\}$ has two connected components which we call **half-trees**.

A **path** is given by a sequence of adjacent vertices, and it is a **reduced path** (or **geodesic**) if there is no backtracking in the sequence. Setting adjacent vertices to have distance 1 between them yields the standard metric on T, denoted by d. Given a vertex $x \in V(T)$, we set $E(x) := \{e \in E(T) \mid o(e) = x\}$ to be the set of edges starting at x. We denote the ball of radius n centred at x by $B(x,n)$, and the sphere of radius n around x by $S(x,n) = \{y \in V(T) \mid d(x,y) = n\}$.

An isometric embedding of $\mathbf{R}_{\geq 0}$ into T is called a *ray*. We will say that two rays $\alpha, \beta : \mathbf{R}_{\geq 0} \to T$ are equivalent, denoting it by $\alpha \sim \beta$, if there exists some $R, s \in \mathbf{R}$ such that $\alpha(r+s) = \beta(r), \forall r \geq R$. We call the collection of equivalence classes of such rays the *boundary* of the tree and denote it by

$$\partial T = \{\alpha : \mathbf{R}_{\geq 0} \to T\}/\sim .$$

The rays α, β are **at bounded distance from each other** if the function $f(t) = d(\alpha(t), \beta(t))$ is bounded. This *a priori* weaker equivalence relation is actually the same as the previous one: it is impossible for two rays to be at bounded distance from each other without actually coinciding eventually.

We introduce a topology on ∂T, resp. on $T \cup \partial T$, by taking the sets

$$\{\partial Y \mid Y \text{ is half a tree}\}, \quad \text{resp.}$$

$$\{Y \cup \partial Y \mid Y \text{ is half a tree}\} \cup \{B(x,r) \mid x \in T, r > 0\}$$

as a basis of open sets. This makes T into an open and dense subset of $T \cup \partial T$. When T is locally finite then both ∂T and $T \cup \partial T$ are compact with this topology.

Let $\mathrm{Aut}(T)$ be the automorphism group of T. Suppose that $G \leq \mathrm{Aut}(T)$ and that Y is a subgraph of T. The **stabilizer** $\mathrm{St}_G(Y)$ of Y consists of elements $g \in G$ such that $gY = Y$, while the **fixator** $\mathrm{Fix}_G(Y)$ of Y consists of $g \in G$ such that $gy = y$ for every vertex y of Y. We say that G acts **edge-transitively** (respectively, **vertex-transitively**) if G acts transitively on geometric edges (respectively, vertices). Denote by G^+ the subgroup of G generated by fixators of edges. These are all contained in G^o, the subgroup (of index at most 2) in G consisting of orientation-preserving automorphisms of T, and in the case $G = \mathrm{Aut}(T)$, there is equality $G^+ = G^o$. The group $\mathrm{Aut}(T)^+$ also appears in the literature as the group of **type-preserving** automorphisms of T. Note that, in general, we only have containments $G^+ \leq G^v \leq G^o$ where G^v denotes the subgroup generated by vertex stabilizers in G.

The topology of pointwise convergence on vertices gives $\mathrm{Aut}(T)$ the structure of a Polish (i.e. metrizable, separable and complete) topological group. The basic open sets for this topology are of the form

$$\mathscr{U}(g, \mathscr{F}) := \{h \in \mathrm{Aut}(T) \mid hx = gx \text{ for all } x \in \mathscr{F}\},$$

where $g \in \mathrm{Aut}(T)$ and $\mathscr{F}$ is a finite subset of $V(T)$. In particular, $\mathrm{Fix}(\mathscr{F})$ is open for any finite $\mathscr{F}$. We will sometimes assume that the tree is locally finite (i.e. every vertex has only finitely many neighbours). In this case the balls $(B(x,n))_{n \in \mathbf{N}}$ around any vertex x are finite. Since these balls are preserved by the stabilizer $\mathrm{St}(x)$, this open subgroup is the inverse limit of its restriction to balls of finite radius and is therefore compact. Thus when T is locally finite, $\mathrm{Aut}(T)$ is a tdlc group.

6.2.2 Classification of automorphisms

We will say that two directed edges are *co-oriented* if their orientations agree along the unique geodesic connecting them (see Figure 6.1). For an automorphism $\phi \in \mathrm{Aut}(T)$, define $\ell(\phi) = \inf\{d(x, \phi x) \mid x \in T\}$ and $X(\phi) = \{x \in T \mid d(x, \phi(x)) = \ell(\phi)\}$. The infimum above is taken over all points x in the geometric realization of the tree. Due to the simplicial nature of the action $X(\phi)$ is never empty and hence the infimum above is actually equal to the minimum. Indeed since $d(v, \phi(v)) \in \mathbf{N} \cup \{0\}$ whenever v is a vertex, the collection of vertices in $X^\varepsilon(\phi) = \{x \in T \mid d(x, \phi(x)) \leq \ell(\phi) + \varepsilon\}$ stabilizes as $\varepsilon \to 0$. If for some $\varepsilon > 0$ the set $X^\varepsilon(\phi)$ contains no vertex this convex ϕ-invariant set must be contained within one geometric edge. In this case ϕ must flip the edge and $X(\phi)$ is the middle of the geometric edge. The same argument shows that $\ell : \mathrm{Aut}(T) \to \mathbf{N} \cup \{0\}$ assumes only non-negative integer values. If $x \in X(\phi)$ then $\ell(\psi) \leq \ell(\phi)$ for every $\psi \in \mathrm{Aut}(T)$ such that $\psi(x) = \phi(x)$, which shows that ℓ is upper semicontinuous. We will see a little later that ℓ is actually continuous on $\mathrm{Aut}(T)$.

Definition 6.1 An automorphism ϕ is called *hyperbolic* if $\ell(\phi) > 0$, an *inversion* if it inverts an edge and *elliptic* if it fixes some vertex. We denote by $\mathrm{Hyp}, \mathrm{Inv}, \mathrm{Ell} \subset \mathrm{Aut}(T)$ the subsets of hyperbolic elements, inversions and elliptic elements respectively. It is clear that $\mathrm{Aut}(T) = \mathrm{Hyp} \sqcup \mathrm{Ell} \sqcup \mathrm{Inv}$ is the disjoint union of these three sets.

Let $\phi \in \mathrm{Aut}(T)$ and assume that there exists an edge e that is co-oriented with its image ϕe. Applying ϕ iteratively, it follows that $\{\phi^n e \mid n \in \mathbf{Z}\}$ are all co-oriented along a bi-infinite geodesic X and ϕ restricts to a translation of length $\ell = d(e, \phi e) + 1$ along this axis. Here $d(e, \phi e)$ stands for the distance between the two geometric edges as closed sets. Once we have such an invariant axis, then for every vertex $x \in T$ we have $d(x, \phi x) = \ell + 2d(x, X) \geq \ell$ (see Figure 6.1). In particular, as our notation already suggests, $X = X(\phi), \ell = \ell(\phi)$.

Thus, whenever there exists an edge that is co-oriented with its image, ϕ is hyperbolic. Conversely, if $\ell(\phi) > 0$, let $x \in X(\phi)$, and let $x = x_0, x_1, \ldots, x_{\ell(\phi)}$

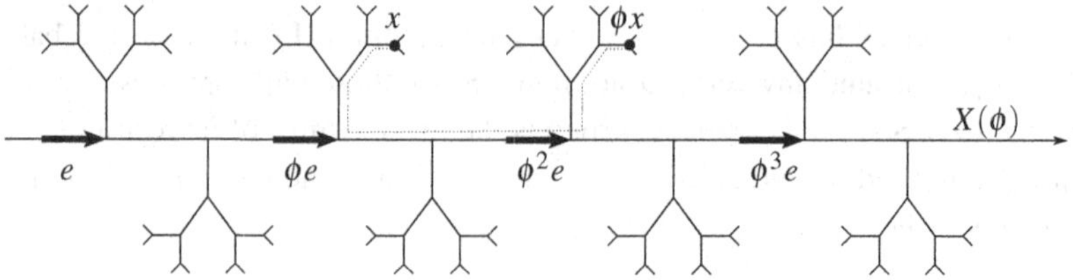

Figure 6.1 A hyperbolic element.

be the geodesic connecting x to ϕx. The directed edge $e = (x_0, x_1)$ must be co-oriented with its image since otherwise we would have $d(x_1, \phi x_1) < d(x, \phi x)$, contradicting our choice of x. A similar picture shows that the general formula $d(x, \phi x) = \ell(\phi) + 2d(x, X(\phi))$ holds also for elliptic elements and inversions. To summarize:

Proposition 6.2   *The map $\ell : \mathrm{Aut}(T) \to \mathbf{N} \cup \{0\}$ is continuous. Each of the sets Hyp, Inv and Ell is clopen and $\mathrm{Aut}(T) = \mathrm{Hyp} \sqcup \mathrm{Inv} \sqcup \mathrm{Ell}$ is a disjoint union of these three classes. Moreover for $\phi \in \mathrm{Aut}(T)$, $x \in T$ we have $d(x, \phi x) = \ell(\phi) + 2d(x, X(\phi))$.*

Proof   For $n \in \mathbf{N}$ set $\ell^{-1}(\{n\}) = \{\phi \in \mathrm{Aut}(T) \mid \ell(\phi) = n\}$ is open. Indeed if $\phi \in \ell^{-1}(n)$ then by the above discussion we can find some e such that $e, \phi(e)$ are co-oriented edges with $d(e, \phi(e)) = n - 1$ and then $U = \{\psi \in \mathrm{Aut}(T) \mid \psi(e) = \phi(e)\}$ is an open neighbourhood such that $\phi \in U \subset \ell^{-1}(n)$. Quite similarly, Ell and Inv are also open. For example, if $\phi \in \mathrm{Ell}$ and $x \in X(\phi)$ then $\phi \in W \subset \mathrm{Ell}$ where $W = \{\psi \in \mathrm{Aut}(T) \mid \psi x = x\} \subset \mathrm{Ell}$. This directly implies all the statements of the proposition, except for the final formula that was already discussed before. $\square$

When $\phi \in \mathrm{Hyp}$ we denote by $a_\phi, r_\phi \in \partial T$ the two points of ∂T corresponding to the two extremes of the axis $X(\phi)$. The point a_ϕ, represented by the ray in the direction of translation, is called *the attracting point of ϕ* and r_ϕ is called its *repelling point*. Note that if $a_\phi \in A, r_\phi \in R$ are open neighbourhoods, then for every large enough $n \in \mathbf{N}$ we have $\phi^n(\partial T \setminus R) \subset A$. Since ϕ^{-1} and ϕ share the same axis, but translate in different directions, we have $a_{\phi^{-1}} = r_\phi, r_{\phi^{-1}} = a_\phi$.

If we replace the tree by its barycentric subdivision (adding one vertex in the middle of every geometric edge), we retain the same automorphism group, but every inversion becomes an elliptic element on the new tree[1]. Using this trick we will assume from now on that there are no inversions in $\mathrm{Aut}(T)$.

---

[1] Recall that we have ruled out the case of a tree consisting of only one bi-infinite geodesic.

Definition 6.3 A subgroup $G \le \mathrm{Aut}(T)$ is *purely elliptic* (respectively, **purely hyperbolic**) if all of its (non-trivial) elements are elliptic (respectively, hyperbolic).

6.2.3 Purely elliptic subgroups

Lemma 6.4 (J. Tits, [9, Proposition 26]) *Let* $\phi, \psi \in \mathrm{Aut}(T)$ *be two elliptic elements with* $X(\phi) \cap X(\psi) = \varnothing$. *Then* $\phi\psi, \psi\phi \in \mathrm{Hyp}$.

Proof Let $x_0, x_1, x_2, \ldots, x_m$ be the shortest geodesic from $X(\phi)$ to $X(\psi)$. Let $e = (x_0, x_1)$ be the first directed edge along this path. It is clear from Figure 6.2 that e and $\psi\phi e$ are co-oriented which proves the lemma.

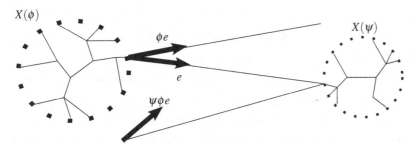

Figure 6.2 Tits' lemma.

☐

The following is a Helly type lemma, capturing the fact that a tree is essentially a one-dimensional object.

Lemma 6.5 *If* X_1, X_2, X_3 *are convex subsets of* T *with pairwise non-empty intersections then* $X_1 \cap X_2 \cap X_3 \ne \varnothing$.

Proof We refer to the indices modulo 3. Pick a point $y_{i,j} \in X_i \cap X_j$. By convexity, the geodesic from $y_{i,i+1}$ to $y_{i,i+2}$ is contained in X_i. Now, in any tree, every triangle has at least one point common to all of its edges (a tree is 0-hyperbolic). For the triangle with vertices $y_{i,j}$ this point will be in the desired intersection $X_1 \cap X_2 \cap X_3$. ☐

Corollary 6.6 (Classification of purely elliptic subgroups) *Let* $G \le \mathrm{Aut}(T)$ *be a purely elliptic subgroup. Then* G *fixes a point in* $T \cup \partial T$. *If* G *is finitely generated then it fixes a point in* T.

Proof Assume first that G is generated by a finite set S. By Lemma 6.4, for any pair s_i, s_j of generators, $X(s_i) \cap X(s_j) \neq \varnothing$. Now, applying Lemma 6.5 successively, we find that $\bigcap_{s \in S} X(s) \neq \varnothing$, which proves the finitely generated case.

Assume that G is purely elliptic but does not fix any point within the tree. Given $x \in V(T)$ and $g \in G$ such that $gx \neq x$ we let $\sigma(x)$ be the neighbour of x in the direction of $X(g)$. The point is that σx does not depend on g as, by the finitely generated case, $X(g), X(h)$ are convex subtrees with a non-trivial intersection. Consider the geodesic ray $\alpha_x := (x, \sigma x, \sigma^2 x, \dots)$. For two different vertices x, y we have $d(\sigma(x), \sigma(y)) \leq d(x, y)$ since we can choose some g that fixes neither x nor y and σ would then just be one step towards the convex set $X(g)$. Thus $\alpha_x \sim \alpha_y$ represent the same point in the boundary. Moreover this boundary point is fixed by G as $\alpha_x \sim \alpha_{gx} = g\alpha_x \in \partial T$. $\qquad\square$

6.2.4 Geometric Density

Definition 6.7   A group $G \leq \mathrm{Aut}(T)$ is called *geometrically dense* if it does not fix any end of T and does not stabilize (as a set) any proper subtree of T.

This notion appears as a condition in Tits' simplicity theorem (our Theorem 6.14). It can be generalized to much more general CAT(0) spaces, and in this capacity plays an important role in the work of Caprace and Monod [6, 7] who argue that it should be thought of as a geometric analogue of Zariski density. For example, given a local field k and considering the action of $\mathrm{PGL}_2(k)$ on its Bruhat–Tits tree, a subgroup $\Gamma < \mathrm{PGL}_2(k)$ is geometrically dense if and only if it is Zariski dense as a subgroup of $\mathrm{PGL}_2(K)$. Here K is the algebraic closure of k.

Lemma 6.8   *If $G \leq \mathrm{Aut}(T)$ contains at least one hyperbolic element then there is a unique minimal G-invariant subtree of T.*

Proof   If $g \in G \cap \mathrm{Hyp}$ then its axis $X(g)$ is contained in every non-empty convex G-invariant set (refer to Figure 6.1). Let $Y \subset T$ be the smallest convex set containing the axes of all hyperbolic elements of T. Since $X(hgh^{-1}) = hX(g)$ the collection $\{X(g) \mid g \in G \cap \mathrm{Hyp}\}$ is invariant under G and so is its convex hull Y. Thus Y is an invariant subtree which is contained in any other G-invariant tree. $\qquad\square$

Remark 6.9   In the setting of the above lemma, let $L = \{a_g \mid g \in G \cap \mathrm{Hyp}\} \subset \partial T$ be the collection of attracting points of all the hyperbolic elements in G. As L is G-invariant, so is its convex hull $\mathrm{conv}(L) \subset T \cup \partial T$. Thus $Y = \mathrm{conv}(L) \cap T$ is a G-invariant tree. Since G contains a hyperbolic element g, the set L contains at least two points $\{a_g, r_g = a_{g^{-1}}\}$ and Y is non-empty. If we further assume that G is geometrically dense then $Y = T$ and consequently L must be dense in ∂T.

Lemma 6.10 *Every geometrically dense subgroup $G \leq \mathrm{Aut}(T)$ contains a hyperbolic element. Furthermore, given any half-tree $Y \subset T$, the group G contains a hyperbolic element whose axis is contained in Y.*

Proof Recalling that we have assumed away the existence of inversions, the existence of hyperbolic elements follows directly from Corollary 6.6. By Remark 6.9 above for every half-tree Y, ∂Y contains an attracting point a_h for some $h \in \mathrm{Hyp} \cap G$.

Let $g \in \mathrm{Hyp} \cap G$ be another hyperbolic element with attracting and repelling fixed points a_g, r_g. The conjugations $g_n := h^n g h^{-n}$ are again hyperbolic elements with $X(g_n) = h^n X(g), a_{g_n} = h^n a_g, r_{g_n} = h^n r_g$. If $a_g \neq r_h, r_g \neq r_h$ then, due to the proximal action of h on ∂T, all of these can be "pushed" arbitrarily close to a_h, and hence deep into the half-tree Y, by applying high powers of h. Thus it is enough to find such a g. We choose an element of the form $g = f h f^{-1}$ where $f \in G$. This is hyperbolic with attracting and repelling points $a_g = f a_h, r_g = f r_h$. Assume by way of contradiction that $f^{-1} r_h \in \{r_h, a_h\}$ for all $f \in G$. Since G is a group this means that either the point $\{r_h\}$ or the set $\{r_h, a_h\}$ is G-invariant, contradicting our assumptions that G is geometrically dense and that the tree is not an infinite line. □

Lemma 6.11 *Let $G \leq \mathrm{Aut}(T)$ be geometrically dense and assume that $N \leq \mathrm{Aut}(T)$ is a non-trivial group normalised by G. Then N is geometrically dense too.*

Proof We start with the boundary. Let $A := \{\xi \in \partial T \mid n\xi = \xi \ \forall n \in N\} = X(N) \cap \partial T$. We claim that A is empty. Clearly A is closed and, since N is normalised by G, the collection of its fixed points is stabilized by G. If A consists of one point then this point is fixed by G, contradicting geometric density. If $|A| > 1$ let Y be the convex hull of A inside the tree T. This is a non-empty G-invariant subtree of T and, by the geometric density of G, we have $Y = T$. This readily implies that A is dense and hence $A = \partial T$, contradicting our assumption that N is non-trivial. Thus N has no fixed point on the boundary. A very similar argument shows that N does not fix any point inside the tree.

Since N does not admit global fixed points in $T \cup \partial T$, Corollary 6.6 implies that N contains hyperbolic elements. Thus, by Lemma 6.8, there is a unique minimal N-invariant subtree $Y \subset T$. This unique tree is stabilized by G. Since the latter is geometrically dense we actually have $Y = T$, which proves the theorem. □

6.3 Independence properties and Tits' simplicity theorem

For any finite or (bi)infinite path C in T and any $k \in \mathbf{N}$ let C^k denote the k-neighbourhood of C (i.e. the subtree of T spanned by all vertices at distance at most k from C). Denote by $\pi_C : T \to C$ the nearest point projection on C and let

$$T_x := \pi_C^{-1}(x) = \{z \in T \,|\, \forall y \in C: \ d(x,z) \le d(y,z)\}.$$

We think of T_x as a subtree rooted at x. If $G \le \mathrm{Aut}(T)$, then, for each vertex x of C, the pointwise stabilizer $\mathrm{Fix}_G(C^{k-1})$ acts on T_x. Denoting by F_x the permutation group induced by restricting $\mathrm{Fix}_G(C^{k-1})$ to T_x, we obtain a map

$$\Phi : \mathrm{Fix}_G(C^{k-1}) \to \prod_{x \in C} F_x$$

which is clearly an injective homomorphism.

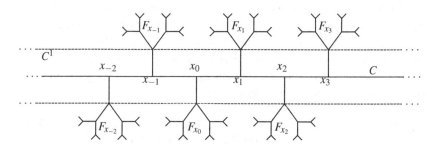

Figure 6.3   The restrictions F_{x_i} of $\mathrm{Fix}_G(C^1)$ to the subtrees T_{x_i}.

Definition 6.12   We say that G satisfies **Property** P_k if for every finite or (bi)infinite path C in T the map $\Phi : \mathrm{Fix}_G(C^{k-1}) \to \prod_{x \in C} F_x$ defined above is an isomorphism.

Since $\mathrm{Fix}_G(C^k) \le \mathrm{Fix}_G(C^{k-1})$ and restriction to T_x is a homomorphism, Property P_{k-1} implies Property P_k for every $k \ge 1$. Notice that when $k = 1$ we recover the original Property P defined by Tits ([11]) so we sometimes omit the subscript when referring to Property P_1. We remark that Property P is also known as Tits' Independence Property in the literature, because it ensures that the actions on subtrees rooted at a path can be chosen independently from each other.

To find simple subgroups of $\mathrm{Aut}(T)$ we will use a generalization of Tits' theorem ([11, Théorème 4.5]).

Definition 6.13 Let

$$G^{+k} := \langle \mathrm{Fix}_G(e^{k-1}) \mid e \in E \rangle$$

denote the subgroup of G generated by pointwise stabilizers of "$(k-1)$-thick" edges. In particular, $G^+ := G^{+1}$ is generated by pointwise stabilizers of edges.

Theorem 6.14 *Suppose that $G \leq \mathrm{Aut}(T)$ is geometrically dense and satisfies Property P_k. Then G^{+k} is either simple or trivial.*

Proof Figure 6.4 illustrates the proof.

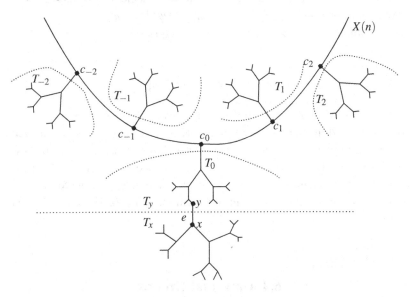

Figure 6.4 Illustration of Tits' simplicity proof.

Write $H := G^{+k}$. We assume that H is non-trivial as otherwise there is nothing to prove. Let $N \neq 1$ be normalised by H. We will show that $N \geq H$, thus proving the theorem. Since H is generated by the pointwise stabilizers $G_e(k) := \mathrm{Fix}_G(e^{k-1})$, it is enough to fix an edge $e \in E(T)$ and show that $N \geq G_e(k)$. Let x, y be the vertices incident with e. Writing $F := G_e(k)$, by Property P_k, there is a decomposition $F = F_x \times F_y$. By the symmetry of the situation it is enough to show that $F_x \leq N$.

By Lemma 6.11, $H \lhd G$ is geometrically dense. A second application of the same lemma shows that N is geometrically dense too. Lemma 6.10 shows that there exists a hyperbolic element $n \in N$ with an axis whose k-neighbourhood is completely contained in the half-tree $T_y (= \pi_e^{-1}(y))$. We denote the axis by $C =$

$X(n) = (\ldots c_{-1}, c_0, c_1, c_2 \ldots)$ with $nc_i = c_{i+\ell}$ where $\ell = \ell(n)$ is the translation length. Assume that $c_0 = \pi_C(e)$ is the projection of the edge e on this axis.

We have to show that a given element $\phi \in F_x$ is contained in N. Applying Property P_k to the axis C we obtain $\phi = \ldots \phi_{-1} \phi_0 \phi_1 \phi_2 \ldots = \phi_0$. In the last equality we used the fact that $T_i = \pi_C^{-1}(c_i) \subset T_y$ for every $i \neq 0$ which implies that $\phi_i = \mathrm{id}$. Since N is normalised by H, we have $[n, \psi] = n\psi n^{-1} \psi^{-1} \in N$ for all $\psi \in H$. To prove that $\phi \in N$ it is enough to exhibit some $\psi \in H$ such that $\phi = [n, \psi]$. We will in fact find such an element in $\mathrm{Fix}_G(C^{k-1})$. If $\psi \in \mathrm{Fix}_G(C^{k-1})$ is an element we denote by $\psi_i = \psi_{T_i}$ its restriction to T_i. By property P_k we have the freedom to construct ψ by prescribing each element ψ_i separately.

With this notation we have $[n, \psi]_i = (n\psi n^{-1} \psi^{-1})_i = n|_{i-\ell} \circ \psi_{i-\ell} \circ (n^{-1})|_i \circ (\psi^{-1})_i$. Solving for $[n, \psi] = \phi$ we obtain two equations

$$\psi_i = (\phi_i)^{-1} n|_{i-\ell} \circ \psi_{i-\ell} \circ (n|_{i-\ell})^{-1}$$
$$\psi_i = (n|_i)^{-1} \circ \phi_{i+\ell} \circ \psi_{i+\ell} \circ n|_i$$

where for the second one we shifted all indices by ℓ. Now assuming we have arbitrarily fixed the values of $\psi_0, \psi_1, \ldots, \psi_{\ell-1}$ we can now solve recursively for all other values of ψ_i using the first of the above equations for the positive values of i and the second one for negative values of i. This method enables us to realize any element $\phi \in \mathrm{Fix}_G(C^{k-1})$ as a commutator of the form $[\psi, n]$. Note that in our specific case we have $\phi_i = \mathrm{id}$ for all $i \neq 0$. This completes the proof. $\qquad\square$

6.4 Universal Groups

In this section we introduce the universal group construction that was developed and studied by Burger and Mozes in [2], and later on played an important role in the study of lattices in the product of the automorphism groups of two regular trees, see [3].

Let $T_d = (V, E)$ be the d-regular tree ($d \in \mathbf{N}$, $d \geq 3$). Recall that $E(x)$ denotes the set of edges with origin $x \in V$. Further, let $l : E \to \{1, \ldots, d\}$ be a legal labelling of T_d, i.e. for every $x \in V$ the map

$$l_x : E(x) \to \{1, \ldots, d\}, \quad y \mapsto l(y)$$

is a bijection, and $l(y) = l(\bar{y})$ for all $y \in E$. A ball of radius 2 in T_3 looks as in Figure 6.5.

Now, given $x \in V$, every automorphism $g \in \mathrm{Aut}(T_d)$ induces a permutation

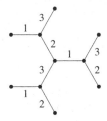

Figure 6.5 A legally labelled ball of radius 2 in T_3.

at x via the following map:

$$c : \mathrm{Aut}(T_d) \times V \to S_d, \ (g,x) \mapsto l_{gx} \circ g \circ l_x^{-1}.$$

Definition 6.15 Let $F \leq S_d$. Define

$$\mathrm{U}^{(l)}(F) := \{g \in \mathrm{Aut}(T_d) \mid \forall x \in V : c(g,x) \in F\}.$$

It is immediate from the following cocycle property of the map c that the sets introduced in Definition 6.15 are in fact groups.

Lemma 6.16 *Let $x \in V$ and $g,h \in \mathrm{Aut}(T_d)$. Then $c(gh,x) = c(g,hx)c(h,x)$.*

$\square$

The groups of Definition 6.15 are termed **universal groups** because of Proposition 6.23 below. In words, they consist of those automorphisms of the regular tree which around every vertex act like one of the allowed permutations from F. The permutation group F is called the associated **local action**.

To determine the dependence of $\mathrm{U}^{(l)}(F)$ on the labelling l, we record the following.

Lemma 6.17 *Given a quadruple (l,l',b,b') consisting of legal labellings l,l' of T_d and vertices $b,b' \in V$, there is a unique automorphism $g \in \mathrm{Aut}(T_d)$ with $gb = b'$ and $l' = l \circ g$.*

Proof By assumption, $gb = b'$. Now assume inductively that g is uniquely determined on $B(b,n)$ $(n \in \mathbf{N}_0)$ and let $x \in V$ be at distance n from b. Then g is also uniquely determined on $E(x)$ by the requirement that $l' = l \circ g$, namely $g|_{E(x)} := l|_{E(gx)}^{-1} \circ l'|_{E(x)}$. $\square$

Corollary 6.18 *Let l and l' be legal labellings of T_d. Further, let $F \leq S_d$. Then $\mathrm{U}^{(l)}(F)$ and $\mathrm{U}^{(l')}(F)$ are conjugate in $\mathrm{Aut}(T_d)$.*

Proof   Choose $b \in V$. If $\tau \in \mathrm{Aut}(T_d)$ is the automorphism of T_d associated to (l,l',b,b) by Lemma 6.17, then $\mathrm{U}^{(l)}(F) = \tau \mathrm{U}^{(l')}(F) \tau^{-1}$. □

With Corollary 6.18 in mind, we henceforth omit the reference to an explicit labelling.

Example 6.19   Clearly, $\mathrm{U}(S_d) = \mathrm{Aut}(T_d)$. On the other hand, $\mathrm{U}(\{\mathrm{id}\}) \cong \mathbf{Z}/2\mathbf{Z} * \cdots * \mathbf{Z}/2\mathbf{Z}$ where the number of copies of $\mathbf{Z}/2\mathbf{Z}$ is d. To see this, fix $b \in V$ and for $i \in \{1,\ldots,d\}$ denote by $e_i \in E$ the edge with $o(e_i) = b$ and $l(e_i) = i$. Further, let $\sigma_i \in \mathrm{U}(\{\mathrm{id}\})$ denote the unique label-respecting inversion of the edge e_i, which is associated to $(l,l,o(e_i),t(e_i))$ by Lemma 6.17. Then the subgroups $\langle \sigma_1 \rangle, \ldots, \langle \sigma_d \rangle$ generate the asserted free product within $\mathrm{U}(\{\mathrm{id}\})$ by an application of the ping-pong lemma. Finally, every $\alpha \in \mathrm{U}(\{\mathrm{id}\})$ is the unique automorphism of T_d associated to $(l,l,b,\alpha(b))$ by Lemma 6.17 which can be realized as an element of $\langle \sigma_1 \rangle * \cdots * \langle \sigma_d \rangle$ by composing the inversions along the edges that occur in the unique reduced path from b to $\alpha(b)$.

Lemma 6.17 also plays an important role in proving the following list of basic properties of the groups $\mathrm{U}(F)$. We say that $H \leq \mathrm{Aut}(T_d)$ is **locally permutation isomorphic** to $F \leq S_d$ if for every $x \in V$ the actions $\mathrm{St}_H(x) \curvearrowright E(x)$ and $F \curvearrowright \{1,\ldots,d\}$ are isomorphic. Also, recall that $\mathrm{U}(F)^+ := \langle g \in \mathrm{U}(F) \mid \exists e \in E : ge = e \rangle$.

Proposition 6.20   *Let $F \leq S_d$. Then the following statements hold.*

(i) $\mathrm{U}(F)$ *is closed in* $\mathrm{Aut}(T_d)$.

(ii) $\mathrm{U}(F)$ *is locally permutation isomorphic to F.*

(iii) $\mathrm{U}(F)$ *is vertex-transitive.*

(iv) $\mathrm{U}(F)$ *is edge-transitive if and only if the action $F \curvearrowright \{1,\ldots,d\}$ is transitive.*

(v) $\mathrm{U}(F)$ *is discrete in $\mathrm{Aut}(T_d)$ if and only if the action $F \curvearrowright \{1,\ldots,d\}$ is free.*

(vi) $\mathrm{U}(F)^+$ *has all local permutations in $F^+ = \langle \mathrm{St}_F(i) \mid i \in \{1,\ldots,d\} \rangle$.*

Proposition 6.20 illustrates the principle that properties of $\mathrm{U}(F)$ correspond to properties of F, which constitutes part of the beauty of the universal group construction.

Proof (Proposition 6.20). For (i), suppose that $g \in \mathrm{Aut}(T_d) \setminus \mathrm{U}(F)$. Then $c(g,x) \notin F$ for some $x \in V$ and hence the open neighbourhood $\{h \in \mathrm{Aut}(T_d) \mid h|_{B(x,1)} = g|_{B(x,1)}\}$ of g is also contained in the complement of $\mathrm{U}(F)$ in $\mathrm{Aut}(T_d)$.

For (ii), let $b \in V$ and $a \in F$. Further, let $\alpha \in \mathrm{Aut}(T_d)$ be the automorphism associated to $(l, a \circ l, b, b)$. by Lemma 6.17. Then $c(\alpha, x) = a$ for every $x \in V$ and hence $\alpha \in \mathrm{St}_{\mathrm{U}(F)}(b)$ realizes the permutation a at the vertex b.

For part (iii), let $b, b' \in V$ and let $g \in \mathrm{Aut}(T_d)$ be the automorphism of T_d associated to (l, l, b, b') by Lemma 6.17. Then $g \in \mathrm{U}_k(F)$ as $c(g, x) = \mathrm{id} \in F$ for all $x \in V$.

As to (iv), suppose that F is transitive. Given $e, e' \in E$, choose $\alpha' \in \mathrm{U}(F)$ such that $\alpha' o(e) = o(e')$ by (iii). Then pick $\alpha'' \in \mathrm{U}(F)_{o(e')}$ such that $\alpha''(\alpha' e) = e'$, by (ii) and transitivity of F, and set $\alpha := \alpha'' \circ \alpha'$.

Conversely, if $\mathrm{U}(F)$ is edge-transitive then $\mathrm{St}_{\mathrm{U}(F)}(x)$ acts transitively on $E(x)$ for a given vertex $x \in V$ and hence F is transitive by (ii).

For part (v), fix $b \in V$ and suppose that the action $F \curvearrowright \{1, \ldots, d\}$ is not free, say $a \in F - \{e\}$ fixes $i \in \{1, \ldots, d\}$. For every $n \in \mathbf{N}$ we define $\alpha \in \mathrm{U}(F)$ such that $\alpha|_{B(b,n)} = \mathrm{id}$ but $\alpha|_{B(b,n+1)} \neq \mathrm{id}$ as follows: Set $\alpha|_{B(b,n)} := \mathrm{id}$ and let $e \in E$ be an edge with $o(e) \in S(b, n)$, $t(e) \in S(b, n-1)$ and $l(e) = i$. Then we may extend α to $B(b, n+1)$ as desired by setting $\alpha|_{E(x)} := l_x^{-1} \circ a \circ l_x$ and $\alpha|_{E(x')} := \mathrm{id}$ for all $x' \in S(b, n) - x$. Now, inductively extend α to T_d such that $c(\alpha, x') = a$ whenever $x' \in V$ maps to x under the projection onto $B(x, n)$ and $c(\alpha, x') = \mathrm{id}$ otherwise.

Conversely, assume that $\mathrm{U}(F)$ is non-discrete. Then there are $\alpha \in \mathrm{U}(F)$ and $n \in \mathbf{N}$ such that $\alpha|_{B(b,n)} = \mathrm{id}$ but $\alpha|_{B(b,n+1)} \neq \mathrm{id}$. Hence there is $x \in S(b, n)$ such that $c(\alpha, x) \in F$ is non-trivial and fixes a point.

Concerning (vi), it suffices to prove the statement for generators of $\mathrm{U}(F)^+$. Let $g \in \mathrm{U}(F)^+$ fix an edge $e = (x, y) \in E$. Then both $c(g, x)$ and $c(g, y)$ are in F^+ by definition. Now assume inductively that all local permutations at vertices up to distance $n \in \mathbf{N}$ from the edge e are in F^+ and consider a vertex v at distance $n + 1$. Let e' be the edge with origin v and terminus $v' \in V$ where $d(v, e) = d(v', e) + 1$. Then $c(g, v) \in c(g, v) \mathrm{St}_F(l(e'))$ where $l(e')$ is the label of e'. Hence $c(g, v)$ is an element of F^+, which is generated by point stabilizers of F, by the induction hypothesis. $\square$

For the sake of clarity we extract the following statement from Proposition 6.20.

Proposition 6.21 *Let $F \leq S_d$. Then $\mathrm{U}(F)$ is compactly generated, totally disconnected, locally compact Hausdorff. It is discrete if and only if the action $F \curvearrowright \{1, \ldots, d\}$ is free.*

Proof The group $\mathrm{U}(F)$ is totally disconnected, locally compact Hausdorff as a closed subgroup of $\mathrm{Aut}(T_d)$. Furthermore, $\mathrm{U}(F)$ is generated by the compact set $\mathrm{U}(F)_b \cup \{\sigma_1, \ldots, \sigma_d\}$ where $b \in V$ is a fixed vertex and σ_i is the edge-inversion of Example 6.19. This follows from vertex-transitivity of $\mathrm{U}(\{\mathrm{id}\}) =$

$\langle \sigma_1 \rangle * \cdots * \langle \sigma_d \rangle$: For $\alpha \in U(F)$ pick $\beta \in U(\{id\})$ such that $\beta(\alpha b) = b$. Then $\beta \alpha \in U(F)_b$, hence the assertion. $\qquad \square$

6.4.1 Simplicity

Since $U(F)$ is vertex-transitive it cannot stabilize any subtree. Since it is transitive on the (directed) edges of any given colour it is easy to see that it cannot fix any end of the tree. Thus $U(F)$ is geometrically dense. We claim that the group satisfies Property P. The following argument for when $C = e = (x,y) \in E$ is a single edge carries over to arbitrary finite or infinite paths. We have

$$\mathrm{Fix}_{U(F)}(e) \xrightarrow{\cong} \mathrm{St}_{U(F)}(x) \times \mathrm{St}_{U(F)}(y), \quad \alpha \mapsto (\alpha_x, \alpha_y)$$

where α_x is given by α on T_x and the identity elsewhere, and similarly for α_y. Then for $x' \in T_x$ we have $c(\alpha_x, x') = c(\alpha, x') \in F$ and $c(\alpha_x, -) = \mathrm{id}$ otherwise.

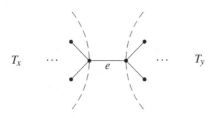

Figure 6.6 Illustration of Tits' simplicity criterion.

As a consequence, the subgroup $U(F)^+ := \langle g \in U(F) \mid \exists e \in E : ge = e \rangle$ of $U(F)$ is simple in many cases. More precisely we have the following theorem, which, among other things, characterizes when $U(F)^+ = U(F)^o$, the subgroup of orientation-preserving automorphisms of $U(F)$. Part of its proof is inspired by [4, Proposition 4.6].

Theorem 6.22 *Let $F \leq S_d$. Then $U(F)^+$ is either trivial or simple. Furthermore, $U(F)^+$ has finite index in $U(F)$ if and only if F is transitive and generated by point stabilizers. In this case, $U(F)^+ = U(F)^o (:= U(F) \cap \mathrm{Aut}(T_d)^+)$ is of index two in $U(F)$.*

Proof The simplicity assertion follows from Theorem 6.14. If F is transitive and generated by point stabilizers then $U(F)^+$ acts transitively on geometric edges: Indeed, the action of every $g \in \mathrm{St}_{U(F)}(x)$ on $E(x)$ ($x \in V$) is realized by a composition of elements in $\mathrm{St}_{U(F)^+}(x)$. This implies the non-trivial inclusion in the equality $U(F)^+ = U(F) \cap \text{Aut}(T_d)^+$: Namely, let $g \in U(F) \cap \text{Aut}(T_d)^+$ and $e \in E$. Since g is orientation-preserving and $U(F)^+$ acts transitively on geometric edges there is $h \in U(F)^+$ such that $h(ge) = e$. Then $hg \in U(F)^+$ and hence $g \in U(F)^+$. Since $[\text{Aut}(T_d) : \text{Aut}(T_d)^+] = 2$, the non-trivial coset being given by an edge-inversion, this implies $[U(F) : U(F)^+] \le 2$ and equality follows from, say, the existence of edge-inversions in $U(F)$.

Conversely, if F is intransitive or not generated by its point stabilizers, then the subgroup F^+ of F generated by point stabilizers is intransitive. As a consequence, the group $U(F)^+$, whose local permutations are in F^+ by Proposition 6.20, has infinitely many vertex orbits which prevents it from being of finite index in $U(F)$: Set $\Omega := \{1, \dots, d\}$ and let $\Omega/F^+ = \Omega_1 \sqcup \dots \sqcup \Omega_r$ be the decomposition of Ω into F^+-orbits. Furthermore, given $i \in \Omega$ let $\omega(i)$ be the unique integer such that $i \in \Omega_{\omega(i)}$. Identify T_d with the Cayley graph Γ of the group $\langle \{x_i \mid i \in \Omega\} \mid \forall i \in \Omega : x_i^2 = 1 \rangle$. Adding the relations $x_i = x_j$ if and only if $\omega(i) = \omega(j)$ yields a quotient graph $p_{F^+} : \Gamma \to \Gamma_{F^+}$ which is a regular tree of degree r. We argue that $U(F)^+$ preserves the vertex fibres of p_{F^+} and therefore has infinitely many vertex orbits given that $r \ge 2$. Let $h \in U(F)^+$ fix an edge and suppose that $gv = v'$ for some vertices $v, v' \in V$. Then $d(v, v')$ is even and the labels $i_1, \dots, i_{2n}$ of the unique geodesic path between v and v' yield a palindromic word $\omega(i_1) \cdots \omega(i_{2n})$ because all local permutations of g are in F^+ by Proposition 6.20. As a consequence, $p_{F^+}(v) = p_{F^+}(v')$. $\square$

6.4.2 Universality

The groups $U(F)$ are universal in the following sense.

Proposition 6.23 *Let $H \le \text{Aut}(T_d)$ be vertex-transitive and locally permutation isomorphic to $F \le S_d$. If F is transitive then there is a legal labelling l of T_d such that $H \le U^{(l)}(F)$.*

Proof Fix $b \in V$. Since H is locally permutation isomorphic to F, there is a bijection $l_b : E(b) \to \{1, \dots, d\}$ such that $H_b|_{E(b)} = l_b^{-1} \circ F \circ l_b$. We now inductively define a legal labelling $l : E \to \{1, \dots, d\}$ such that $H \le U^{(l)}(F)$. Set $l|_{E(b)} := l_b$. We proceed inductively and suppose that l is defined on

$$E(b, n) := \bigcup_{x \in B(b, n-1)} E(x).$$

To extend l to $E(b, n+1)$, let $x \in S(b, n)$ and let $e_x \in E$ be the unique edge with $o(e_x) = x$ and $d(b, t(e_x)) + 1 = d(b, x)$. Since H is vertex-transitive and locally permutation isomorphic to the transitive group F, there is an element

$\sigma_{e_x} \in H$ which inverts the edge e_x. We may thus legally extend l to $E(x)$ by setting $l|_{E(x)} := l \circ \sigma_{e_x}$.

To check the inclusion $H \leq \mathrm{U}^{(l)}(F)$, let $x \in V$ and $h \in H$.

If $(b, b_1, \ldots, b_n, x)$ and $(b, b'_1, \ldots, b'_m, h(x))$ denote the unique reduced paths from b to x and $h(x)$, then

$$s := \sigma_{e_{b'_1}} \cdots \sigma_{e_{b'_m}} \sigma_{e_{h(x)}} \circ h \circ \sigma_x \sigma_{e_{b_n}} \cdots \sigma_{e_{b_2}} \sigma_{e_{b_1}} \in H_b$$

and we have $c(h, x) = c(s, b) \in F$ by Lemma 6.16. $\qquad\square$

6.4.3 Structure of a point stabilizer

In this section, we exhibit a point stabilizer in $\mathrm{U}(F)$ as a profinite group in terms of F for transitive $F \leq S_d$. To this end, let $b \in V$, $\Delta := \{1, \ldots, d\}$, $D := \{1, \ldots, d - 1\}$ and set $\Delta_n := \Delta \times D^{n-1}$. We fix bijections $b_n : S(b, n) \to \Delta_n$ as follows: Given that F is transitive we may for every $i \in \{1, \ldots, d\}$ fix an element $a_i \in F$ with $a_i(i) = d$. Define inductively

(i) $b_1 : S(b, 1) \to \Delta_1$, $x \mapsto l((b, x))$, and
(ii) $b_{n+1} : S(b, n+1) \to \Delta_{n+1} = \Delta_n \times D$, $x \mapsto (b_n p_n x, a_{l(p_{n-1}x, p_n x)}(l(p_n x), x))$

for $n \in \mathbf{N}$, where $p_n : \bigcup_{k \geq n} S(b, k) \to S(b, n)$ is the canonical projection. We now capture the action of $\mathrm{U}(F)_b$ on $S(b, n)$ by inductively defining $F(n) \leq \mathrm{Sym}(\Delta_n)$ as follows: Let $F_d := \mathrm{stab}_F(d)$, set

(i) $F(1) := F \leq \mathrm{Sym}(\Delta_1)$, and define
(ii) $F(n+1) := F(n) \ltimes F_d^{\Delta_{n+1}} \leq \mathrm{Sym}(\Delta_n)$

to be the wreath product for the action of $F(n)$ on Δ_n. Further, let $\pi_n : F(n) \to F(n-1)$ denote the canonical projection. The bijection b_n induces the surjective homomorphism

$$\varphi_n : \mathrm{U}(F)_b \to F(n) \leq \mathrm{Sym}(\Delta_n), \quad g \mapsto b_n \circ g \circ b_n^{-1}$$

with kernel $\{g \in \mathrm{U}(F)_b \mid g|_{S(b,n)} = \mathrm{id}\} = \{g \in \mathrm{U}(F)_b \mid g|_{B(b,n)} = \mathrm{id}\}$ and one readily checks that the map

$$\varphi := (\varphi_n)_{n \in \mathbf{N}} : \mathrm{U}(F)_b \to \varprojlim F(n) =$$

$$\left\{ (f_n)_{n=1}^{\infty} \in \prod_{n=1}^{\infty} F(n) \;\middle|\; \forall n \in \mathbf{N} : \pi_n f_n = f_{n-1} \right\}$$

is an isomorphism of topological groups: Abbreviate $G := \varprojlim F(n)$. Clearly, φ is a bijective homomorphism. To prove that it is a homeomorphism, note that

$U(F)_b$ is compact and $\varprojlim F(n)$ is Hausdorff; therefore φ is closed and it suffices to show continuity. Let $U := G \cap \prod_{n=1}^{\infty} U_n$ be a basic open neighbourhood of $f \in \varprojlim F(n)$. Then there is $N \in \mathbf{N}$ such that $U_n = F(n)$ for all $n \geq N$ and hence for every $g \in \varphi^{-1}(U)$ the open neighbourhood $\{h \in U(F)_b \mid h|_{B(b,N)} = g|_{B(b,N)}\}$ of g is contained in $\varphi^{-1}(U)$.

6.5 Simple totally disconnected locally compact groups with prescribed local actions

The purpose of this section is to find new examples of tdlc groups which are (abstractly) simple, compactly generated and non-discrete. The motivation for this comes from classification results of locally compact groups by Caprace and Monod [5] which yield cases in which compactly generated tdlc groups decompose into (topologically) simple compactly generated non-discrete pieces. We are still at the stage of collecting examples of such simple groups, aiming in the long term for some sort of classification. The examples collected so far can be classified into the following types:

- simple Lie groups
- simple algebraic groups over local fields
- complete Kac–Moody groups over finite fields
- automorphism groups of trees, some CAT(0) cube-complexes, and right-angled buildings
- variations on the above (e.g. almost automorphisms of trees, see Chapter 8 in this volume).

To obtain these new examples of compactly generated, simple, non-discrete tdlc groups, we shall introduce some "k-thickened" (for $k \in \mathbf{N}$) variations of the universal group construction: the k-**closures** of a given $G \leq \mathrm{Aut}(T)$. These prescribe the action on all balls of radius k by elements of G. We will then see that the k-closure of a group of tree automorphisms satisfies Property P_k and use Theorem 6.14 to obtain abstractly simple tdlc groups which are compactly generated. The last step will be ensuring that these are non-discrete and different.

6.5.1 k-closures and Property P_k

Definition 6.24 Let $G \leq \mathrm{Aut}(T)$ and $k \in \mathbf{N}$. The k-**closure** of G is

$$G^{(k)} := \{h \in \mathrm{Aut}(T) \mid \forall x \in V(T) : \exists g \in G : h|_{B(x,k)} = g|_{B(x,k)}\},$$

all automorphisms of T that agree on each ball of radius k with some element of G.

In this setting, G is the analogue of F in the definition of $U(F)$, providing a list of "allowed" actions. However, here we do not require that the tree be regular. Note that a given $h \in G^{(k)}$ need not agree with the same element of G on every ball; the point is that for each ball there is *some* element of G agreeing with h, and they may all be different for each ball. The k-closure of G has the following basic properties, which justify the term "closure".

Proposition 6.25   *Let $G \leq \mathrm{Aut}(T)$ and $k \in \mathbf{N}$.*

(i)   *$G^{(k)}$ is a closed subgroup of $\mathrm{Aut}(T)$.*
(ii) *For every $l \in \mathbf{N}$ with $l > k$ we have $G \leq G^{(l)} \leq G^{(k)}$.*
(iii) *We have $\bigcap_{k \in \mathbf{N}} G^{(k)} = \overline{G}$, where $\overline{G}$ denotes the topological closure of G in $\mathrm{Aut}(T)$.*

Proof   (i) To see that $G^{(k)}$ is indeed a subgroup of $\mathrm{Aut}(T)$, notice that if $a, b \in G^{(k)}$ and $x \in V(T)$ then there exist $g, h \in G$ such that $a|_{B(x,k)} = g|_{B(x,k)}$ and $b|_{B(ax,k)} = h|_{B(ax,k)}$, so $b \circ a|_{B(x,k)} = h \circ g|_{B(x,k)}$ and $b \circ a \in G^{(k)}$. Also, there exists some $f \in G$ such that $a|_{B(a^{-1}x,k)} = f|_{B(a^{-1}x,k)}$, so $a^{-1}|_{B(x,k)} = f^{-1}|_{B(x,k)}$ and $a^{-1} \in G^{(k)}$.

For the closure part, note that for each $a \notin G^{(k)}$ there is some vertex x_a such that no element of G agrees with a on $B(x_a, k)$. Thus $\mathrm{Aut}(T) \setminus G^{(k)} = \bigcup_{a \notin G^{(k)}} \mathscr{U}(a, B(x_a, k))$ is a union of basic open sets.

(ii) The group G agrees with itself on balls of all radii so $G \leq G^{(l)}$ for all l and if $l > k$ then $G^{(l)}$ certainly agrees with G on balls of smaller radius k, so $G^{(l)} \leq G^{(k)}$.

(iii) Since $G \leq G^{(k)}$ for all k and $G^{(k)}$ is closed, the closure of G must be contained in every $G^{(k)}$ and therefore in the intersection of all of them. For the other direction, we show that every element $a \in \bigcap_{k \in \mathbf{N}} G^{(k)}$ is a point of closure of G. For this, note that $\mathscr{U}(a, B(x,k))$ with $a \in \bigcap_{k \in \mathbf{N}} G^{(k)}$, $x \in V$, $k \in \mathbf{N}$ are basic open sets of the induced topology on $\bigcap_{k \in \mathbf{N}} G^{(k)}$ and that each of them contains some $g \in G$ because $a \in G^{(k)}$. This proves the claim. $\square$

Just as $U(F)$ satisfies Property P the k-closure of G satisfies Property P_k.

Proposition 6.26   *Let $G \leq \mathrm{Aut}(T)$ and $k \in \mathbf{N}$. Then $G^{(k)}$ satisfies Property P_k.*

Proof   Let $C = (\dots, x_0, x_1, \dots, x_n, \dots)$ be any finite or (bi)infinite path and suppose that $f := (\dots, f_0, f_1, \dots, f_n, \dots) \in \prod_{x_i \in C} F_{x_i}$. We claim that $f \in G^{(k)}$.

Pick a vertex v, which must be in T_{x_i} for some $x_i \in C$. By definition, f_i is the restriction to T_{x_i} of some $h \in \text{Fix}_{G^{(k)}}(C^{k-1})$. Thus, if $B(v,k)$ is entirely contained in T_{x_i} then $f|_{B(v,k)} = h|_{B(v,k)} = g|_{B(v,k)}$ for some $g \in G$, since $h \in G^{(k)}$. If there is some part of $B(v,k)$ outside T_{x_i} then it is contained in C^{k-1} so both f and h act trivially on it and we still obtain that $f|_{B(v,k)} = h|_{B(v,k)} = g|_{B(v,k)}$ for some $g \in G$. Thus f is indeed in $G^{(k)}$. □

Satisfying Property P_k characterizes when the process of taking k-closures stabilizes.

Theorem 6.27 *If $G \leq \text{Aut}(T)$ satisfies Property P_k then $G^{(k)} = \overline{G}$ (and if $G^{(k)} = \overline{G}$ then $\overline{G}$ satisfies P_k by Proposition 6.26).*

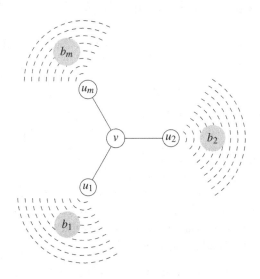

Figure 6.7 Each of the automorphisms b_i acts on the subtree rooted at u_i.

Proof We know from Proposition 6.25 that $\overline{G} = \bigcap_{k \in \mathbf{N}} G^{(k)}$, so it suffices to show that $G^{(k)} = G^{(n)}$ for all $n \geq k$. To illustrate the proof, we only show the case $G^{(1)} = G^{(2)}$ (see [1, Theorem 5.4] for the full proof). Let $x \in G^{(1)}$. For each vertex v there is some $g \in G$ such that $x|_{B(v,1)} = g|_{B(v,1)}$; thus $xg^{-1} \in \text{Fix}_{G^{(1)}}(B(v,1))$. Suppose that $u_1, \dots, u_m$ are the neighbours of v. Since $xg^{-1} \in G^{(1)}$, for each i there exist $a_i \in G$ such that $xg^{-1}|_{B(u_i,1)} = a_i|_{B(u_i,1)}$. So a_i fixes the edge (v, u_i). Because G satisfies property P_1, there exist unique $b_i, c_i \in G$ such that $a_i = b_i c_i$, where b_i only acts non-trivially on T_{u_i} (and c_i fixes T_{u_i}, see Figure 6.5.1). Then the product $b_1 \dots b_m$ fixes all neighbours of v and v itself;

that is, it fixes $B(v,1)$. Furthermore, c_i fixes T_{u_i}, so $b_i|_{B(u_i,1)} = xg^{-1}|_{B(u_i,1)}$ and hence $b_1 \dots b_m|_{B(u_i,1)} = xg^{-1}|_{B(u_i,1)}$ for each i. Thus $b_1 \dots b_m|_{B(v,2)} = xg^{-1}|_{B(v,2)}$ and $b_1 \dots b_m g|_{B(v,2)} = x-1|_{B(v,2)}$. Since $b_1 \dots b_m g \in G$ we conclude that $x \in G^{(2)}$, as required.   □

More importantly, we deduce the following statement which will be used to find infinitely many distinct simple subgroups.

Corollary 6.28   *There are infinitely many distinct k-closures of G if and only if $\overline{G}$ does not satisfy Property P_k for any k.*

Proof   If $\overline{G}$ does not satisfy Property P_k for any k, then $G^{(k)} \neq \overline{G}$ (by Proposition 6.26). Hence $G^{(k)} \neq \bigcap_n G^{(n)}$ for all k and therefore the sequence $(G^{(k)})_k$ never stabilizes; in particular there are infinitely many distinct k-closures $G^{(k)}$ of G. For the converse, we have that $(G^{(k)})_k$ never stabilizes, therefore there is no k such that $G^{(k)} = \bigcap_n G^{(n)} = \overline{G}$ and so G does not satisfy P_k for any k.   □

6.5.2 Local rigidity for k-closures

We digress a moment from our objective in this section of finding infinitely many simple groups, to point out a local-global result of Burger and Mozes that is relevant to k-closures.

Theorem 6.29 (Proposition 3.3.1 in [2])   *Let $F \leq \mathrm{Sym}(d)$ be a finite 2-transitive permutation group on the set $\{1,2,\dots,d\}$ and F_1 the stabilizer of a point under this action. Assume that F_1 is non-abelian and simple.*

Let T be a d-regular tree and $G \leq \mathrm{Aut}(T)$ a vertex-transitive subgroup. If $x \in V(T)$ is any vertex we have a map

$$\phi : \mathrm{St}_G(x) \to \mathrm{Sym}(d)$$

given by the action of $\mathrm{St}_G(x)$ on $B(x,1)$. Assume that $\phi(\mathrm{St}_G(x)) = F$ for all $x \in V(T)$. Write $K := \ker(\phi) = \mathrm{Fix}_G(B(x,1))$ and consider the action of K on $B(y,1)$ for each $y \in S(x,1)$. Since K fixes x, the image of this action is contained in F_1. Moreover, since the supports of these actions on different neighbours of x are disjoint, we obtain an injective homomorphism

$$\phi_2 : K/\mathrm{Fix}_G(B(x,2)) \to \prod_{y \in S(x,1)} F_1.$$

Then $\phi_2(K/\mathrm{Fix}_G(B(x,2))) = \prod_a F_1$ with $a \in \{0,1,d\}$ and we have the following dichotomy:

- *$a \in \{0,1\}$ if and only if G is discrete.*

- $a = d$ *if and only if* $\overline{G} = U(F)$.

We omit the proof of this theorem, but we quote it to emphasize that in some cases there are *local conditions* on a group G which already imply the stabilization of its k-closures. Indeed, if the action of G on every 1-ball is contained in F, then the case $a = d$ in the above theorem yields that $\overline{G} = G^{(1)}$.

Let k be a local field with integer ring $\mathcal{O}$, maximal ideal $\mathscr{P} \lhd \mathcal{O}$ and residue field $f = \mathcal{O}/\mathscr{P}$. For the action of the group $G = \mathrm{PGL}(2,k)$ on its Bruhat–Tits tree, the local action is given by the group $F = GL(2,f)$ and its action on the projective line $\mathbf{P}^1 f$. Below we discuss the fact that in this case the sequence of k-closures $\{G^{(k)} \mid k \in \mathbf{N}\}$ never stabilizes and hence gives rise to an infinite sequence of simple groups containing G. The theorem of Burger and Mozes above implies in particular that this kind of behaviour would never be possible when the "local group" F is $\mathrm{Sym}(6)$, for example.

6.5.3 Finding infinitely many non-discrete simple groups

Returning to the main goal of the section, we have the following recipe to find simple subgroups of $\mathrm{Aut}(T)$:

(i) start off with some geometrically dense $G \leq \mathrm{Aut}(T)$,
(ii) form its k-closures (which all satisfy Property P_k),
(iii) use Theorem 6.14 to obtain the simple subgroups $(G^{(k)})^{+k}$.

We still need to ensure that these subgroups are non-discrete and different from each other, which will follow from the results below.

Lemma 6.30 *If $G \leq \mathrm{Aut}(T)$ does not stabilize a proper subtree of T we have*

(i) $(G^{(k)})^{+k}$ is an open subgroup of $G^{(k)}$.
(ii) $(G^{(k)})^{+k}$ is non-discrete if and only if $G^{(k)}$ is non-discrete.
(iii) $(G^{(k)})^{+k}$ satisfies Property P_k.

Proof (i) The group $(G^{(k)})^{+k}$ is generated by fixators of k-edges (in particular, fixators of finite sets of vertices), which are basic open sets. Hence it is open.
(ii) This follows from the facts that all subgroups of discrete groups are discrete and that all open subgroups of a non-discrete group are non-discrete.
(iii) Let C be some path in T. Since $G^{(k)}$ satisfies P_k we have that

$$\prod_{x \in C} \mathrm{Fix}_{(G^{(k)})^{+k}}(C^{k-1})_x \leq \prod_{x \in C} \mathrm{Fix}_{(G^{(k)})}(C^{k-1})_x = \mathrm{Fix}_{(G^{(k)})}(C^{k-1})$$

and

$$\mathrm{Fix}_{(G^{(k)})}(C^{k-1}) = \bigcap (\mathrm{Fix}_{(G^{(k)})}(e^{k-1}) \mid e \text{ is an edge contained in } C) \leq (G^{(k)})^{+k}.$$

$\square$

Theorem 6.31 *Suppose that $G \leq \mathrm{Aut}(T)$ is geometrically dense. Then we have*

$$(G^{(r)})^{+r} \leq (G^{(k)})^{+k}$$

for every $r \geq k$, with equality if and only if $G^{(r)} = G^{(k)}$.

Proof The first claim follows from the fact that $(G^{(r)}) \leq (G^{(k)})$ for every $r \geq k$.

Suppose that $G^{(r)} = G^{(k)}$ and let g be a generator of $(G^{(k)})^{+k}$ (in particular, it fixes some $(k-1)$-thick edge $(e = (v,w))^{k-1})$. Now, $(G^{(k)})$ satisfies P_k and therefore $g = (g_1, g_2)$ where $g_1 \in (G^{(k)})$ fixes T_w pointwise and $g_2 \in (G^{(k)})$ fixes T_v pointwise. In particular, there exist edges $e_1 \in E(T_w)$ and $e_2 \in E(T_v)$ such that g_1 fixes e_1^{r-1} and g_2 fixes e_2^{r-1}. Hence g_1, g_2 are generators of $(G^{(r)})^{+r}$ and $g \in (G^{(r)})^{+r}$, as required.

Conversely, suppose that $(G^{(r)})^{+r} = (G^{(k)})^{+k}$. For any $x \in G^{(k)}$ and any vertex v there is some $g \in G$ such that xg^{-1} fixes $B(v,k)$. In particular, xg^{-1} fixes e^{k-1} where $e = (v,u)$ is some edge starting at v. Thus $xg^{-1} \in (G^{(k)})^{+k} = (G^{(r)})^{+r}$, that is $xg^{-1} \in G^{(r)}$. Since $g \in G \leq G^{(r)}$ we conclude that $x \in (G^{(r)})$. $\square$

Thus, in order to construct infinitely many distinct tdlc simple non-discrete subgroups of $\mathrm{Aut}(T)$ it suffices to find examples with infinitely many distinct k-closures. By Corollary 6.28, this amounts to finding examples which do not satisfy Property P_k for any k.

Example 6.32 The Baumslag–Solitar group $\mathrm{BS}(m,n) := \langle a, t \mid t^{-1}a^m t = a^n \rangle$ does not satisfy P_k for any k when m,n are coprime and the group acts on its Bass–Serre tree (which is isomorphic to T_{m+n}).

Recall that T, the Bass–Serre tree of $\mathrm{BS}(m,n)$, has vertices the left (say) cosets of $\langle a \rangle$ and (directed) edges labelled by $t^{\pm 1}$ from $u\langle a \rangle$ to $v\langle a \rangle$ if and only if there is some i such that $v\langle a \rangle = ua^i t^{\pm 1} \langle a \rangle$. $\mathrm{BS}(m,n)$ acts on T by left multiplication and, in the non-solvable case (when neither m nor n equals 1), this action is geometrically dense.

We claim that when m,n are coprime $G := \mathrm{BS}(m,n)$ does not satisfy P_k for any k. To see this (Figure 6.8 may be helpful), consider the edge $e = (\langle a \rangle, t^{-1}\langle a \rangle)$. We will show that $\mathrm{Fix}_G(T_{\langle a \rangle}) = \mathrm{Fix}_G(T_{t^{-1}\langle a \rangle})$ while $\mathrm{Fix}_G(e^{k-1}) \neq 1$ (it contains, for instance, a^n). Let $x \in \mathrm{Fix}_G(T_{t^{-1}\langle a \rangle})$, then x must also fix $\langle a \rangle$ as all other neighbours of $t^{-1}\langle a \rangle$ are fixed by x. Thus x must be of the form a^* where $*$ is a multiple of n; say $x = a^{cn^j}$ for some $c, j \in \mathbf{N}$ with $j > 0$ and c not divisible by n. Note that $T_{t^{-1}\langle a \rangle}$ contains vertices of the form $(t^{-1}a)^i t^{-1}\langle a \rangle$ for all $i \in \mathbf{N}$. Pick some $i \geq j$. If $x = a^{cn^j}$ fixes $(t^{-1}a)^i t^{-1} \langle a \rangle$ then, since $ta^{cn^j} t^{-1} = (ta^n t^{-1})^{cn^{j-1}} = a^{mcn^{j-1}}$, we have $a^{cn^j}(t^{-1}a)^i t^{-1} = (t^{-1}a)^j a^{cm^j}(t^{-1}a)^{i-j} t^{-1} = (t^{-1}a)^i t^{-1} a^*$. For the last equality to hold, cm^j must be a multiple of n, which it cannot be by the assumption that m, n are coprime and the choice of c, unless $c = 0$. Thus we must have $x = 1$. A similar argument yields that $\mathrm{Fix}_G(T_{\langle a \rangle}) = 1$.

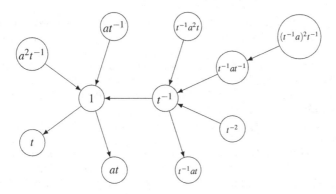

Figure 6.8 Part of the Bass–Serre tree for $\mathrm{BS}(2,3)$. The nodes are labelled by their coset representatives. The arrows on the edges indicate travelling in the t direction.

Example 6.33 The group $G = \mathrm{PSL}(2, \mathbf{Q}_p)$ acting on its Bruhat–Tits tree (which is isomorphic to T_{p+1}) also does not satisfy P_k for any k. Indeed it is well known that the action of G on ∂T is isomorphic to the action of the same group on the projective line $\mathbf{P}^1(\mathbf{Q}_p)$ by fractional linear transformations. In particular the stabilizer of three boundary points is trivial. This means that an element of G is completely determined by its action on three distinct points of ∂T (these elements should be thought of as p-adic Möbius transformations).

Now let $C = (\ldots, c_{-1}, c_0, c_1, c_2, \ldots)$ be an infinite or finite geodesic. And assume that $(\ldots f_0, f_1, f_2, \ldots) \in \prod_{i=-\infty}^{\infty} F_i$ with $f_i \in \mathrm{Fix}_G(C^{k-1})|_{T_i}$. Then each such f_i is defined on ∂T_i which contains many boundary points. Hence each f_i admits a unique extension to the whole tree. This is a strong obstruction to satisfying Property P_k, for any k.

We note that this method finds infinitely many tdlc simple non-discrete groups which are pairwise distinct as subgroups of $\mathrm{Aut}(T)$. It would be desirable to know whether these subgroups are pairwise non-isomorphic. This is stated as work in progress in [1]. Using different methods, Smith ([10]) has found uncountably many tdlc simple non-discrete groups which are pairwise non-isomorphic. This is discussed in Chapter 7 of this volume.

Acknowledgements. The authors owe thanks to Thomas Taylor as well as an anonymous referee for helpful remarks. The second author was supported by ISF grant 2095/15.

References for this chapter

[1] C. Banks, M. Elder and G. A. Willis, *Simple groups of automorphisms of trees determined by their actions on finite subtrees*, Journal of Group Theory **18** (2015), 235261.

[2] M. Burger and S. Mozes, *Groups acting on trees: from local to global structure*, Publications Mathématiques de l'IHÉS **92** (2000), 113–150.

[3] M. Burger and S. Mozes, *Lattices in product of trees*, Publications Mathématiques de l'IHÉS **92** (2000), 151–194.

[4] A. Le Boudec, *Groups acting on trees with almost prescribed local action*, arXiv preprint 1505.01363 (2015).

[5] P.-E. Caprace and N. Monod, *Decomposing locally compact groups into simple pieces*, Math. Proc. Cambridge Philos. Soc. **150** Nr. 1 (2011), 97128.

[6] P.-E. Caprace and N. Monod, *Isometry groups of non-positively curved spaces: structure theory*, J. Topol. **2**(4) (2009), 661–700.

[7] P.-E. Caprace and N. Monod, *Isometry groups of non-positively curved spaces: discrete subgroups*, J. Topol. **2**(4) (2009), 701–746.

[8] N. Monod, *Continuous bounded cohomology of locally compact groups*, LNM vol. 1758 (Springer 2001).

[9] J.-P. Serre, *Trees*, Springer Monographs in Mathematics.

[10] S. M. Smith, *A product for permutation groups and topological groups*, Duke Math. J. **166**(15) (2017), 2965–2999.

[11] J. Tits, *Sur le groupe d'automorphismes d'un arbre*, in Essays on Topology and Related Topics, Springer Berlin Heidelberg (1970), 188–211.

7

Simon Smith's construction of an uncountable family of simple, totally disconnected, locally compact groups

Colin Reid and George Willis

Abstract

We explain Smith's construction of tree automorphism groups and how it leads to an uncountable family as in the title.

7.1	Introduction	117
7.2	Bi-regular trees and legal colourings	118
7.3	The 1-closure and property (P)	121
7.4	The universal group $\mathscr{U}(M,N)$	124
7.5	A continuum of simple groups	128
	References for this chapter	130

7.1 Introduction

Simon M. Smith [7] shows that there are $2^{\aleph_0}$ non-isomorphic compactly generated, abstractly simple, totally disconnected, locally compact groups. Previous to Smith's construction, only countably many such groups were known (see for instance [3]). Smith's construction modifies the method by which Marc Burger and Shahar Mozes [2] associate a totally disconnected, locally compact "universal group" with a finite permutation group. Whereas Burger and Mozes construct a group of automorphisms of a regular tree that has local action given by the finite permutation group, the product constructed by Smith acts on a bi-regular tree with local actions given by two permutation groups and these permutation groups need not be finite. In both cases, the groups constructed satisfy an independence property of Jacques Tits [8] that implies that they contain large simple subgroups.

Smith's construction takes as input two non-trivial (topological) permutation groups M and N and produces a group $\mathscr{U}(M,N)$ acting on a bi-regular tree. For the purposes of this article we will assume that M and N are transitive, so the permutation actions of M and N correspond to the natural actions on coset spaces M/M_* and N/N_* respectively, where M_* and N_* are core-free subgroups of M and N respectively. There is then an associated faithful action of $M \times N$ on $M/M_* \sqcup N/N_*$, where M acts trivially on N/N_* and N acts trivially on M/M_*.

In [7], a **legal colouring** is defined on a bi-regular tree T, which is unique up to the action of $\mathrm{Aut}(T)$, and then $\mathscr{U}(M,N)$ is defined in terms of this colouring. Here we take a different approach to constructing T and $\mathscr{U}(M,N)$, via Bass–Serre theory (as set out in [6]) and the theory of k-closures introduced in [1].

7.2 Bi-regular trees and legal colourings

Let Γ be the graph with vertices $\{m, n\}$, with an arc a from m to n and with the reverse $\bar{a}$ of a. Define a graph of groups A on Γ by setting $A_m = M \times N_*$, $A_n = M_* \times N$ and $A_a = A_{\bar{a}} = M_* \times N_*$, with the natural inclusions from A_a into A_m and A_n. Then the fundamental group of A is the amalgamated free product

$$F = A_m *_{A_a} A_n,$$

which acts on the Bass–Serre covering tree T of A. (See [6, Theorem 7].) The set of vertices of T is the disjoint union $\{fA_m \mid f \in F\} \sqcup \{fA_n \mid f \in F\}$, on which F acts by left multiplication, and there is an arc from gA_m to hA_n whenever $g^{-1}h \in A_m A_n$. In particular, T is a bi-regular tree and F acts transitively without inversions on the set of undirected edges $\{e, \bar{e}\}$ in T. We say a vertex gA_m or hA_n is of type M or N respectively. We regard A_m, A_n and A_a as subgroups of F: note that these are respectively the stabilisers in F of the vertices A_m and A_n, and the directed edge from A_m to A_n.

From now on, we will assume that M, M_*, N and N_* have been specified, and the groups A_m, A_n, A_a and F and the tree T have all been constructed as above.

Let us identify M with the subgroup $M \times \{1\} \leq A_m$, and N with $\{1\} \times N \leq A_n$, so M and N are subgroups of F. Note that with this identification, the subgroups M and N do not commute in F. Indeed, F is isomorphic to the group

$$(M * N)/\langle\langle [M, N_*], [M_*, N] \rangle\rangle,$$

and the product $A_m A_n$ is just MN. Let C be the commutator group $[M, N]$. Then C is normal in F (since it is normalised by both M and N) and C has trivial intersection with M and N, so the quotient F/C is naturally isomorphic to $M \times N$. We thus have a natural action ρ of F on $M/M_* \sqcup N/N_*$ with kernel C. At the same time, $C \cap A_m = C \cap A_n = 1$, and thus (since C is normal) C has trivial intersection with the stabiliser of each vertex in T, in other words C acts freely on VT.

We can now assign a colouring to the arcs of T in a canonical way, such that arcs from gA_m to hA_n (for $g, h \in F$) are coloured by elements of M/M_* and the reverse arcs by elements of N/N_*.

Let e be an arc terminating at the vertex fA_n of type N. Then $f = ckl$ for some $c \in C$, $k \in M$ and $l \in N$. Suppose that $c, c' \in C$, $k, k' \in M$ and $l, l' \in N$ satisfy that $cklA_n = c'k'l'A_n$. Then, by passing to F/C, it may be seen that $kM_* = k'M_*$ and we use the uniquely specified coset kM_* to colour e.

Similarly, given an arc e terminating at the vertex fA_m of type M, then $f = clk$ for some $c \in C$, $k \in M$ and $l \in N$, and we colour e with the uniquely specified coset lN_*.

We have thus defined a map $\chi : ET \to M/M_* \sqcup N/N_*$. This map intertwines the action of F on ET with the action ρ on the set of colours $M/M_* \sqcup N/N_*$.

Lemma 7.1 *Let χ, ρ, F and T be as above. Then $\rho(f)\chi(e) = \chi(fe)$ for all $f \in F$ and $e \in ET$.*

Proof Firstly, observe that $\chi(ce) = \chi(e)$ for all $c \in C$ and $e \in ET$: given an arc e terminating at $c'k'A_n$ for $c' \in C$ and $k' \in M$, then the arc ce terminates at $(cc')k'A_n$, and thus $\chi(e) = \chi(ce) = k'M_*$, and similarly $\chi(ce) = \chi(e)$ if e is an arc terminating at a vertex of type M. As C is also the kernel of ρ, in effect we can freely replace f with any element of the coset fC in F.

Fix $f \in F$ and $e \in ET$. By symmetry, we may suppose that e terminates at a vertex ckA_n of type N, where $c \in C$ and $k \in M$, so that $\chi(e) = kM_*$. By replacing f with an element of fC as needed, we may assume $f = rs$ where $r \in M$ and $s \in N$, and we have $\rho(f)\chi(e) = rkM_*$. On the other hand,

$$fckA_n = rsckA_n = c'rksA_n = c'rkA_n,$$

for some $c' \in C$, using the fact that $C = [M,N]$, so $\chi(fe) = rkM_*$. Hence $\rho(f)\chi(e) = \chi(fe)$ as required. $\square$

We now introduce the definitions of legal colouring and local action, and show that the construction satisfies the specified conditions.

Definition 7.2 Let T be a tree with vertex set $VT = Y_1 \sqcup Y_2$, such that Y_1 and Y_2 are the parts of the natural bipartition of VT (in some specified order), and let A_1 and A_2 be non-empty sets. Given $v \in VT$, write E_v for the set of arcs with initial vertex v. An (A_1, A_2)-**legal colouring** of T is a function $\chi : ET \to A_1 \sqcup A_2$ such that the following conditions hold:

(i) Given $v \in VT$, then χ is injective on E_v, with $\chi(E_v) = A_i$ if $v \in Y_i$.
(ii) Given $e, e' \in ET$ such that e and e' have the same terminal vertex, then
$\chi(e) = \chi(e')$.

Given an (A_1, A_2)-legal colouring, for each $v \in Y_i$ $(i = 1, 2)$, write χ_v^{-1} for the function from A_i to E_v such that $\chi \chi_v^{-1}$ is the identity on A_i. Let $g \in \mathrm{Aut}(T)$ and let $v \in Y_i$ $(i = 1, 2)$. We define the **local action of g at v** to be the permutation $g|_v$ of A_i induced by the composition $\chi g \chi_v^{-1}$. Given subgroups P_i of $\mathrm{Sym}(A_i)$ $(i = 1, 2)$, we say that a subgroup $G \leq \mathrm{Aut}(T)$ is **locally-(P_1, P_2)** or has **local action (P_1, P_2)** if, for $i = 1, 2$ and for every $v \in Y_i$,

$$\{g|_v : g \in G\} = P_i.$$

Proposition 7.3 *Let χ, F and T be as before, with VT partitioned as $VT = Y_1 \sqcup Y_2$, where Y_1 is the set of vertices of type M and Y_2 is the set of vertices of type N. Then χ is an $(M/M_*, N/N_*)$-legal colouring and F is locally-(M, N) (with respect to the natural actions of M and N on M/M_* and N/N_* respectively).*

Proof Let $y \in Y_1$, so that $y = gA_m$ for some $g \in F$. We see that every arc with initial vertex y terminates at a vertex of type N, so $\chi(E_y) \subseteq M/M_*$. Moreover, y is fixed by the subgroup gMg^{-1} of F, which acts transitively via ρ on the set M/M_*. By Lemma 7.1 we see that $\chi(E_y)$ is a $\rho(gMg^{-1})$-invariant set, so $\chi(E_y) = M/M_*$.

We now claim that χ is injective on E_y. Let $z_1 = h_1 A_n$ and $z_2 = h_2 A_n$ be distinct vertices adjacent to y and let e_1 and e_2 be the arcs from y to z_1 and z_2 respectively. Then $g^{-1} h_i \in A_m A_n = MA_n$ for $i = 1, 2$, and by change of representative we may in fact ensure $g^{-1} h_i \in M$ for $i = 1, 2$, so that $h_1^{-1} h_2 \in M$. Write h_1 as $h_1 = ckl$ for $c \in C$, $k \in M$ and $l \in N$, and set $k' = h_1^{-1} h_2$. Then $z_1 = ckA_n$ and $z_2 = cklk'A_n = c'kk'A_n$ for some $c' \in C$. Thus $\chi(e_1) = kM_*$ and $\chi(e_2) = kk'M_*$. If $\chi(e_1) = \chi(e_2)$, then $kM_* = kk'M_*$, so $M_* = k'M_*$, that is, $k' \in M_*$. But then $h_2 A_n = h_1 k' A_n = h_1 A_n$, so $z_1 = z_2$ and hence $e_1 = e_2$. This proves that χ is injective on E_y as claimed.

A similar argument shows that given $y \in Y_2$, then χ is injective on E_y and $\chi(E_y) = N/N_*$.

Let $e, e' \in ET$ such that e and e' have the same terminal vertex. Then $\chi(e) = \chi(e')$ by definition. This completes the proof that χ is an $(M/M_*, N/N_*)$-legal colouring.

The fact that F has local action (M, N) is now clear from Lemma 7.1. $\square$

7.3 The 1-closure and property (P)

Given a set X, we equip $\mathrm{Sym}(X)$ with the permutation topology, that is, the coarsest group topology such that the stabiliser of every $x \in X$ is open. Given a tree T, we give $\mathrm{Aut}(T)$ the subspace topology, regarding $\mathrm{Aut}(T)$ as a subgroup of $\mathrm{Sym}(VT)$. Observe that in fact $\mathrm{Aut}(T)$ corresponds to a closed subgroup of $\mathrm{Sym}(VT)$; if VT is countable, this ensures that $\mathrm{Aut}(T)$ is Polish (that is, separable and completely metrisable) and also totally disconnected, but $\mathrm{Aut}(T)$ is not necessarily locally compact. Assuming T has no leaves, one could equivalently define the topology of $\mathrm{Aut}(T)$ with respect to the permutation topology on arcs or undirected edges, or the permutation topology on one part of the natural bipartition of the vertices of T: this can be seen by noting that two undirected edges suffice to specify a vertex, two vertices to specify an arc, and two vertices in one part of the bipartition to specify a vertex in the other part.

Given $G \leq \mathrm{Aut}(T)$ and $k \geq 1$, the *k*-**closure**[1] $G^{(k)}$ of G is the set of automorphisms $g \in \mathrm{Aut}(T)$ such that for all $v \in VT$, and every finite set of vertices X all of which are at distance at most k from v, there exists $g_X \in G$ such that $gw = g_X w$ for every vertex $w \in X$. We say G is *k*-**closed** if $G = G^{(k)}$.

The notion of the k-closure of a group acting on a tree was introduced in [1]. We will use some basic properties of the k-closure, specifically the 1-closure. We include proofs here, as the approach of [1] implicitly assumes that trees are locally finite.

For the rest of this article, we define $G_{(X)} := \{g \in G \mid \forall x \in X : gx = x\}$, where X is a set of vertices of T.

Proposition 7.4 (See [1] Proposition 3.4) *Let T be a tree, let $G \leq \mathrm{Aut}(T)$ and let $k \in \mathbf{N}$.*

(i) *$G^{(k)}$ is a closed subgroup of $\mathrm{Aut}(T)$.*
(ii) *$G^{(l)} = (G^{(k)})^{(l)}$ whenever $l \leq k$. In particular, $(G^{(k)})^{(k)} = G^{(k)}$, so $G^{(k)}$ is k-closed.*

Proof (i) Write $A := \mathrm{Aut}(T)$. Let $g, h \in G^{(k)}$, let $v \in VT$ and let X be a finite set of vertices all of which are at distance at most k from v. Then there exists $h_X \in G$ such that $h_X w = hw$ for all $w \in X$. In turn, $hX := \{hw \mid w \in X\}$ is a finite set of vertices, all of which are at distance at most k from hv, so there exists $g_{hX} \in G$ such that $g_{hX} w = gw$ for all $w \in hX$. Thus $g_{hX} h_X$ is an element of G such that $g_{hX} h_X w = ghw$ for all $w \in X$. We conclude that $gh \in G^{(k)}$. Similarly, it is clear that $G^{(k)}$ is closed under inverses. Thus $G^{(k)}$ is a subgroup of A.

Let $\mathscr{X}_k$ be the set of all finite sets X of vertices in T, such that there is a

---

[1] This definition is different from the notion of k-closure introduced by Wielandt.

vertex v at distance at most k from every vertex in X. Then $A_{(X)}$ is an open subgroup of A for every $X \in \mathscr{X}_k$. Observe that given $g \in A \setminus G^{(k)}$, then there exists $X_g \in \mathscr{X}_k$ such that no element of G agrees with g on X_g, and hence no element of $G^{(k)}$ agrees with g on X_g, that is, $G^{(k)} \cap g A_{(X_g)} = \varnothing$. We can therefore express the complement of $G^{(k)}$ as the following union of open sets:

$$A \setminus G^{(k)} = \bigcup_{g \in A \setminus G^{(k)}} g A_{(X_g)}.$$

Hence $G^{(k)}$ is closed in $\mathrm{Aut}(T)$.

(ii) Since $G \leq G^{(k)}$ then $G^{(l)} \leq (G^{(k)})^{(l)}$. Let $g \in (G^{(k)})^{(l)}$ and let X be a finite set of vertices of T, all of which are at distance at most l from some vertex v. Then there exists $g_X \in G^{(k)}$ such that $g_X w = gw$ for all $w \in X$. But then all the vertices in X are at distance at most k from v, so there exists $g'_X \in G$ such that $g'_X w = g_X w = gw$ for all $w \in X$. Hence $G^{(l)} = (G^{(k)})^{(l)}$. The remaining conclusions are clear. $\qquad\square$

It is useful to note that the property of being k-closed is inherited by fixators of sets of vertices.

Lemma 7.5 *Let T be a tree, let $G \leq \mathrm{Aut}(T)$, let $k \geq 1$ and let X be a set of vertices of T. If G is k-closed, then so is $G_{(X)}$.*

Proof Suppose G is k-closed, and let $H = (G_{(X)})^{(k)}$. Given $v \in X$ and $g \in H$, we see from the definition of the k-closure that g must fix v. So in fact H is a subgroup of $(G^{(k)})_{(X)}$, which is just $G_{(X)}$. Hence $H = G_{(X)}$. $\qquad\square$

We also recall Tits' property (P), introduced in [8].

Definition 7.6 Let T be a tree and let G be a subgroup of $A = \mathrm{Aut}(T)$. Given a non-empty (finite or infinite) path L in T, let $\pi_L : VT \to VL$ be the closest point projection of the vertices of T onto L; observe that for each $x \in L$, the set $\pi_L^{-1}(x)$ is the set of vertices of a non-empty subtree of T, and also that $A_{(L)}$ preserves setwise each of the fibres $\pi_L^{-1}(x)$ of π_L. Then for each vertex $x \in L$, there is a natural homomorphism $\phi_x : A_{(L)} \to \mathrm{Sym}(\pi_L^{-1}(x))$ induced by the action of $A_{(L)}$ on $\pi_L^{-1}(x)$. We can combine the homomorphisms ϕ_x in the obvious way to obtain a homomorphism

$$\phi_L : A_{(L)} \to \prod_{x \in VL} \phi_x(A_{(L)}),$$

which then restricts to a homomorphism

$$\phi_{L,G} : G_{(L)} \to \prod_{x \in VL} \phi_x(G_{(L)}).$$

For the full automorphism group of the tree, one observes that ϕ_L is actually an isomorphism. In general, $\phi_{L,G}$ is injective, but not necessarily surjective. We say G has **property (P)** (with respect to a collection $\mathscr{L}$ of paths) if $\phi_{L,G}$ is an isomorphism for every possible choice of L (such that $L \in \mathscr{L}$).

A major motivation of [1] was to generalize Tits' property (P), and indeed (P) has a natural interpretation in terms of the 1-closure.

Theorem 7.7 (See [1] Theorem 5.4 and Corollary 6.4) *Let T be a tree and let G be a closed subgroup of $\mathrm{Aut}(T)$. Then $G = G^{(1)}$ if and only if G satisfies Tits' property (P). Indeed, it suffices for G to have property (P) with respect to the edges of T.*

Proof Let L be a non-empty path in T, and let $g \in \mathrm{Aut}(T)_{(L)}$ such that

$$\phi_L(g) \in \prod_{v \in VL} \phi_v(G_{(L)});$$

say $\phi_L(g) = (s_v)_{v \in VL}$. We now claim that $g \in G^{(1)}$ (indeed, $g \in (G_{(L)})^{(1)}$). Let X be a finite set of vertices, all adjacent to some vertex w of T. We may assume that $X \cap VL = \varnothing$, since the vertices of L are all fixed by both g and $G_{(L)}$. Let $x = \pi_L(w)$. We observe that since X is disjoint from VL and the set $X \cup \{w\}$ spans a subtree, any path from X to L must pass through x, in other words $X \subseteq \pi_L^{-1}(x)$. There is then $g_X \in G_{(L)}$ such that $\phi_v(g_X) = s_x$, so that g agrees with g_X on $\pi_L^{-1}(x)$ and in particular on X. Given the freedom of choice of X, we conclude that $g \in G^{(1)}$ as claimed. Thus if $G = G^{(1)}$, then G has property (P).

Conversely, suppose that G is closed and satisfies property (P) (with respect to edges). Suppose that $G \neq G^{(1)}$ and let $g \in G^{(1)} \setminus G$. Since G is closed, the set $G^{(1)} \setminus G$ is a neighbourhood of g in $G^{(1)}$, so there is a finite set X of vertices such that $g(G^{(1)})_{(X)} \cap G = \varnothing$. Let S be the smallest subtree of T containing X; note that $\mathrm{Aut}(T)_{(X)} = \mathrm{Aut}(T)_{(S)}$, since every vertex of S lies on the unique path between a pair of vertices in X. Let us suppose that X has been chosen so that $|S|$ is minimised.

By the definition of $G^{(1)}$, we see that S is not a star (in other words, for every $x \in S$, there is a vertex in S at distance 2 from x), so there exist adjacent vertices x and y of S such that neither x nor y is a leaf of S. Let L be the path consisting of the single arc (x, y). By the minimality of $|S|$, there is some $h \in G$ such that $gx = hx$ and $gy = hy$, so that $h^{-1}g$ fixes L pointwise. Let

$$S_1 = (S \cap \pi_L^{-1}(x)) \cup \{y\} \text{ and } S_2 = (S \cap \pi_L^{-1}(y)) \cup \{x\}.$$

Note that for $i = 1, 2$, then S_i is the set of vertices of a subtree of S that contains L. The condition that neither x nor y is a leaf of S ensures that there is some neighbour of x in S that is not contained in S_2, and similarly there is some neighbour of y in S that is not contained in S_1. Hence S_1 and S_2 are both proper subtrees of S, so by the minimality of $|S|$, there exists $h_1, h_2 \in G$ such that $h_i w_i = h^{-1} g w_i$ for all $w_i \in S_i$ ($i = 1, 2$). Indeed, h_1 and h_2 are elements of $G_{(L)}$, since h_1 and h_2 both agree with $h^{-1} g$ on L. In particular, we see that the action of $h^{-1} g$ induces an element of $\phi_x(G_{(L)}) \times \phi_y(G_{(L)})$. But then by (the restricted) property (P), we have $h^{-1} g \in G_{(L)}$ and hence $g \in G$, a contradiction.   □

One significant consequence of property (P) is that it gives a criterion for simplicity (as an abstract group) of a group acting on a tree.

Theorem 7.8 ([8] Théorème 4.5)   *Let T be a tree and let G be a subgroup of* Aut(T) *with property (P). Suppose that the action of G does not leave invariant any end or proper subtree of T. Then the subgroup G^+ of G generated by the arc stabilisers is either trivial or (abstractly) simple.*

7.4 The universal group $\mathscr{U}(M, N)$

We are now ready to define the group $\mathscr{U}(M, N)$. Specifically, we set $\mathscr{U}(M, N)$ to be the 1-closure of F acting on T, where F and T are as in Section 7.2. Thus $\mathscr{U}(M, N)$ is a (topologically) closed and 1-closed subgroup of Aut(T), by Proposition 7.4, so it has property (P) by Theorem 7.7. In addition, $\mathscr{U}(M, N)$ is totally disconnected, and it is Polish provided that M/M_* and N/N_* are both countable.

Remark 7.9   We note at this point that the earlier construction of Burger–Mozes can be recovered as a special case. Let M be a transitive permutation group, let M_* be a point stabiliser of the action of M, let N be the cyclic group of order 2 and let N_* be trivial. Then $\mathscr{U}(M, N)$ is a closed subgroup of Aut(T), where T is a bi-regular tree in which the vertices of type N have degree 2. Thus T is the barycentric subdivision of a regular tree T' on which $\mathscr{U}(M, N)$ acts vertex-transitively; this action is equivalent to the action of the group $\mathscr{U}(M)$ defined by Burger and Mozes in [2].

Since F already acts transitively on the undirected edges and $\mathscr{U}(M, N)$ acts without inversions, the difference between F and $\mathscr{U}(M, N)$ is captured by the arc stabilisers.

Lemma 7.10   *Let $G = \mathscr{U}(M, N)$. Let e be the arc with initial vertex A_m (of type M) and terminal vertex A_n (of type N). Let T_m and T_n be the preimages* *of A_m and A_n respectively under the nearest point projection map from T to $\{m,n\}$. Then the arc stabiliser G_e is a direct product of two groups*

$$G_e = K_m \times K_n,$$

such that K_m fixes T_n pointwise and K_n fixes T_m pointwise. Moreover, K_m and K_n are both iterated wreath products, with the self-similar structure

$$K_m \cong K_n \wr_X M_*; \quad K_n \cong K_m \wr_Y N_*,$$

where $X = (M/M_*) \setminus \{M_*\}$ and $Y = (N/N_*) \setminus \{N_*\}$.

Proof Define K_m and K_n to be the pointwise stabilisers in G of T_n and T_m respectively. Observe that K_m preserves T_m setwise and acts faithfully on it, and similarly for K_n and T_n. Since e is the unique arc starting in T_m and finishing in T_n, we see that $K_m \le G_e$ and $K_n \le G_e$. We have $\langle K_m, K_n \rangle = K_m \times K_n$ since K_m and K_n have disjoint support on G, and in fact $G_e = K_m \times K_n$ by the fact that G has property (P).

Let B be the set of neighbours of A_m in T_m (that is, the set of all neighbours of A_m in T except for A_n) and let π be the nearest point projection from T_m to $B \cup \{A_m\}$. Then B corresponds (via the colours of edges from A_m to B) to the set X, and K_m induces the permutation group $M_* \curvearrowright X$ with respect to this correspondence. We see that the pointwise stabiliser of B in K_m is a direct product of copies $K_{n,x}$ of K_n, one for each vertex $x \in B$, such that $K_{n,x}$ is the fixator in G of the half-tree complementary to $\pi^{-1}(x)$ (in particular, $K_{n,x}$ is conjugate to K_n in G). Similar observations apply with the roles of m and n reversed. The self-similar structure of K_m and K_n is now evident. $\square$

We can now apply Tits' simplicity criterion to the group $\mathscr{U}(M,N)^+$ generated by the arc stabilisers. Under some mild conditions, we in fact have $\mathscr{U}(M,N)^+ = \mathscr{U}(M,N)$.

Theorem 7.11 (See [7] Theorem 23) *Let $G = \mathscr{U}(M,N)$, acting on the tree T constructed as in Section 7.2. Suppose that $\max\{|M/M_*|, |N/N_*|\} \ge 3$ and that at least one of M_* and N_* is non-trivial. Then G^+ is abstractly simple and non-discrete in the permutation topology. If M and N are generated by the conjugates of M_* and N_* respectively, then $G^+ = G$.*

Proof Suppose that $\max\{|M/M_*|, |N/N_*|\} \ge 3$. Then we see that T has infinitely many ends. From the fact that M acts transitively on the neighbours of the principal vertex A_m of type M, we see that M fixes no end of T, so G fixes no end of T. Since G is transitive on undirected edges, G does not leave any proper subtree invariant. Thus G^+ is either trivial or abstractly simple, by Theorem 7.8. Given Lemma 7.10, we see that any arc stabiliser G_e is non-discrete in the permutation topology (equivalently, the fixator of any finite set of vertices is non-trivial) as soon as either M_* or N_* is non-trivial, so that G^+ is also non-discrete.

Now suppose that M and N are generated by the conjugates of M_* and N_* respectively. We observe that G^+ is a normal subgroup of G that contains both M_* and N_*. Thus $G^+ \cap M$ is a normal subgroup of M containing M_*, so $G^+ \geq M$, and similarly $G^+ \geq N$, so $G^+ \geq \langle M, N \rangle = F$. In particular, G^+ is transitive on the undirected edges of T. Since G acts without inversions, and by definition $G^+ \geq G_e$, we conclude that $G^+ = G$.   □

We recall a standard criterion for a subgroup of the symmetric group to be locally compact:

Lemma 7.12 *Let X be a set and let G be a closed subgroup of $\mathrm{Sym}(X)$. Then G is compact if and only if it has finite orbits on X, and G is locally compact if and only if there exists a finite subset Y of X such that the fixator $G_{(Y)}$ of Y in G is compact.*

Proof The fixators of finite subsets of X form a base of neighbourhoods of the identity in G, so G is locally compact if and only if $G_{(Y)}$ is compact for some finite subset Y of X.

Suppose that G has finite orbits on X. Then there exists a directed set $(Y_i)_{i \in I}$ of finite G-invariant subsets of X, ordered by inclusion, so that $X = \bigcup_{i \in I} Y_i$. We see that the quotients $G/G_{(Y_i)}$ are all finite and discrete; moreover, there is a natural faithful action on X of the inverse limit $\hat{G} = \varprojlim_{i \in I} G/G_{(Y_i)}$, in other words there is a continuous injective map $\rho : \hat{G} \to \mathrm{Sym}(X)$. The group $\hat{G}$ is compact by Tychonov's Theorem, so $\rho(\hat{G})$ is compact (in particular, closed). Now G agrees with $\hat{G}$ on finite subsets, so G is dense in $\rho(\hat{G})$; since G is closed, in fact $G = \rho(\hat{G})$, so G is compact.

Conversely, suppose that G is compact and let $x \in X$. By definition, G_x is an open subgroup of G. Since an open subgroup of a compact group necessarily has finite index, we conclude that $|G : G_x|$ is finite, so by the Orbit-Stabiliser Theorem the orbit Gx is finite.   □

Using this criterion, we can characterise when $\mathscr{U}(M,N)$ is locally compact.

Proposition 7.13 (See [7] Theorem 30)   *Let M and N be non-trivial groups, let M_* be a core-free subgroup of M, let N_* be a core-free subgroup of N and let $G = \mathscr{U}(M,N)$. Then the following are equivalent:*

  (i) *G is locally compact;*
  (ii) *M_* and N_* are relatively compact as subgroups of $\mathrm{Sym}(M/M_*)$ and $\mathrm{Sym}(N/N_*)$ respectively;*

(iii) M_* *is commensurated by* M *and* N_* *is commensurated by* N;

(iv) G_e *is compact, where* e *is some (any) arc in* T.

Proof Suppose that G is locally compact. Then there is some finite set X of vertices in T such that $G_{(X)}$ is compact; it then follows that $G_{(X')}$ is compact for every finite set of vertices X' containing X, so we are free to add finitely many more vertices to X as necessary. We may thus assume that X is the set of vertices of some finite subtree S of T with at least two vertices. Let x be a leaf of S, so one neighbour of x lies in S and the rest in $T \setminus S$. By extending the tree as necessary, we may arrange for x to be a vertex of type N, and the arc of S terminating at x has some colour kM_*. Then $G_{(X)}$ contains a subgroup H that has local action kM_*k^{-1} at x; since $G_{(X)}$ has finite orbits, we conclude that kM_*k^{-1} has finite orbits, so it is relatively compact in $\mathrm{Sym}(M/M_*)$, and hence M_* is relatively compact in $\mathrm{Sym}(M/M_*)$. Similarly, we conclude that N_* is relatively compact in $\mathrm{Sym}(N/N_*)$. Thus (i) implies (ii).

Observe that given a transitive permutation group G with point stabiliser H, then H has finite orbits if and only if $|H : H \cap gHg^{-1}| < \infty$ for all $g \in G$. The equivalence of (ii) and (iii) is now clear from Lemma 7.12.

Now suppose that M_* and N_* are relatively compact as subgroups of the full symmetric groups $\mathrm{Sym}(M/M_*)$ and $\mathrm{Sym}(N/N_*)$ respectively, that is, M_* has finite orbits on M/M_* and N_* has finite orbits on N/N_*. Then we see from the structure of the arc stabiliser G_e in G (as given in Lemma 7.10) that G_e has finite orbits on the vertices of T, and hence G_e is compact, so (ii) implies (iv).

Finally, it is clear that (iv) implies (i), which completes the proof that the statements (i), (ii), (iii) and (iv) are equivalent. $\qquad\square$

Here is a sufficient condition for $\mathscr{U}(M,N)$ to be generated by a compact subset.

Lemma 7.14 (See [7] Theorem 31) *Suppose that* M *and* N *are closed subgroups of* $\mathrm{Sym}(M/M_*)$ *and* $\mathrm{Sym}(N/N_*)$ *respectively, such that both* M *and* N *are compactly generated and both* M_* *and* N_* *are compact. Then* $\mathscr{U}(M,N)$ *is compactly generated.*

Proof Let $G = \mathscr{U}(M,N)$. We see that G_e is compact by Proposition 7.13. The subgroup F of G is generated by M and N, so $F = \langle A, B \rangle$ where A and B are compact generating sets of M and N respectively. Since F is transitive on undirected edges and G acts without inversions, we have $G = FG_e$, so $G = \langle A, B, G_e \rangle$. Hence G is compactly generated. $\qquad\square$

Remark 7.15 Let Y be the set of vertices of T of type N. Then $\mathscr{U}(M,N)$ acts faithfully on Y, so there is an injective map from $\mathscr{U}(M,N)$ to $\mathrm{Sym}(Y)$. The image of this map is denoted $M \boxtimes N$ and is the 'product' of M and N referred to in the title of [7]. The principal motivation for introducing this definition is that in contrast to $\mathscr{U}(M,N)$, as a permutation group $M \boxtimes N$ is transitive, and often primitive. Our focus in this article is on topological groups, and from this perspective $M \boxtimes N$ is isomorphic to $\mathscr{U}(M,N)$, since as noted earlier, there are enough stabilisers of sets of vertices of type N to generate the permutation topology of a group acting on T.

7.5 A continuum of simple groups

We have seen that $\mathscr{U}(M,N)$ is a totally disconnected topological group, and under suitable conditions, $\mathscr{U}(M,N)$ is also locally compact, non-discrete, abstractly simple and compactly generated. It remains to find a continuum of suitable examples and to show that they are pairwise non-isomorphic as groups.

By Ol'shanskii ([4],[5]), for each prime $p > 10^{75}$, there is a continuum of non-isomorphic infinite simple groups M (known as **Tarski–Ol'shanskii monsters**) such that every proper non-trivial subgroup of M has order p. Let M be such a group, and choose a proper non-trivial subgroup M_* of M. Evidently, M is generated by the conjugates of M_*. Let N be the symmetric group of degree 3 and let N_* be a point stabiliser in N: observe that N_* is non-trivial and that N is generated by the conjugates of N_*.

We now produce the group F, the tree T (which is $(\aleph_0,3)$-regular bipartite) and the 1-closure $\mathscr{U}(M,N)$ of F on T as in the previous sections. Then $\mathscr{U}(M,N)$ is totally disconnected since it is a subgroup of $\mathrm{Aut}(T)$; it is locally compact by Proposition 7.13; $\mathscr{U}(M,N)$ is non-discrete and abstractly simple by Theorem 7.11; and $\mathscr{U}(M,N)$ is compactly generated by Lemma 7.14 (note that M is generated by 2 elements).

If we repeat the construction with some other Tarski–Ol'shanskii monster M' that is not isomorphic to M, we obtain another group $\mathscr{U}(M',N)$ acting on a tree T'. To show that $\mathscr{U}(M,N)$ and $\mathscr{U}(M',N)$ are not isomorphic as abstract groups, we need to eliminate the possibility that $\mathscr{U}(M,N)$ has some exotic action on the tree T' used in the construction of $\mathscr{U}(M',N)$.

A group G is said to have **property (FA)** if, whenever G acts on a tree without inversions, then G fixes a vertex. This concept is due to Jean-Pierre Serre and is introduced and used in [6]. We highlight the following:

Proposition 7.16 ([6] Example 6.3.1) *A finitely generated torsion group has property (FA).*

In particular, every Tarski–Ol'shanskii monster has property (FA). Now suppose there is a non-trivial homomorphism $\theta : \mathscr{U}(M,N) \to \mathscr{U}(M',N)$. Then θ is injective since $\mathscr{U}(M,N)$ is abstractly simple, so it induces a faithful action without inversions of $\mathscr{U}(M,N)$ on T'. Then the standard copy of M inside $\mathscr{U}(M,N)$ also acts faithfully without inversions on T', and by property (FA) there is some vertex x of T' fixed by M. Since M acts non-trivially, we can choose x so that M does not fix every neighbour of x. We thus obtain an injective (since M is simple) homomorphism from M to the group P of permutations of the neighbours of x induced by $\mathscr{U}(M',N)_x$. Since $\mathscr{U}(M',N)$ has a locally (M',N)-action on T', we have either $P \cong M'$ or $P \cong N$. If $P \cong N$, then P is finite, so clearly does not contain any copy of M. If $P \cong M'$, then P itself is not isomorphic to M by assumption, and every proper subgroup of P is finite, so again there is no subgroup of P that is isomorphic to M. In either case, there is a contradiction.

Thus we have arrived at a continuum of groups in the class $\mathscr{S}$ of non-discrete, totally disconnected, locally compact, abstractly simple, compactly generated groups. In fact, these examples determine the cardinality of isomorphism types in $\mathscr{S}$ (as abstract groups or topological groups), namely there are exactly $2^{\aleph_0}$ such groups. This is because $\mathscr{S}$ is contained in a much more general class of topological groups that still only has $2^{\aleph_0}$ isomorphism types:

Lemma 7.17 *Up to topological group isomorphism, there are only $2^{\aleph_0}$ Polish groups.*

Proof By definition, a Polish space is separable and completely metrisable, so it has a countable subset S such that every point is the limit of some sequence in S. Thus every Polish space has cardinality at most $2^{\aleph_0}$. Moreover, there is a universal Polish group $\mathscr{U}$ in which all Polish groups appear as closed subgroups (see [9]). Since each closed subgroup of $\mathscr{U}$ can be specified as the closure of a countable set, this means there are only $|\mathscr{U}|^{\aleph_0} = 2^{\aleph_0}$ Polish groups up to topological group isomorphism.   □

We thus arrive at the following theorem.

Theorem 7.18 (See [7] Corollary 39) *There are exactly $2^{\aleph_0}$ (abstract or topological) isomorphism types of topological groups that are totally disconnected, locally compact, compactly generated, abstractly simple and not discrete.*

Remark 7.19 Although the family of groups we are considering in this section are non-isomorphic as abstract groups, for a fixed prime p they are all **locally** isomorphic as topological groups, that is, they have an open subgroup in common. Indeed, it can be seen from Lemma 7.10 that the topological iso-morphism type (indeed, the action on the tree) of an arc stabiliser in $\mathscr{U}(M,N)$ is determined by the embeddings

$$M_* \to \mathrm{Sym}((M/M_*) \setminus \{M_*\}) \text{ and } N_* \to \mathrm{Sym}((N/N_*) \setminus \{N_*\}),$$

taken up to conjugation in the relevant symmetric group. In the present situa-tion, M_* acts freely on $(M/M_*) \setminus \{M_*\}$ (up to conjugation in $\mathrm{Sym}((M/M_*) \setminus \{M_*\})$, there is only one such action of a cyclic group of order p), and the action of N_* on $(N/N_*) \setminus \{N_*\}$ is the regular action of N_*.

References for this chapter

[1] C.C. Banks, M. Elder and G.A Willis, *Simple groups of automorphisms of trees determined by their actions on finite subtrees*, J. Group Theory **18** (2015), 235–261.

[2] M. Burger and S. Mozes, *Groups acting on trees: from local to global structure*, Publ. Math. IHES **92** (2000), 113–150.

[3] P.-E. Caprace and T. de Medts, *Simple locally compact groups acting on trees and their germs of automorphisms*, Transformation Groups **16** (2011) 375–411.

[4] A.Y. Ol'shanskii, *An infinite group with subgroups of prime orders*, Math. USSR Izv. **16** (1981), 279–289; translation of Izvestia Akad. Nauk SSSR Ser. Matem. **44** (1980), 309–321.

[5] A.Y. Ol'shanskii, *Groups of bounded period with subgroups of prime order*, Al-gebra and Logic **21** (1983), 369–418; translation of Algebra i Logika **21** (1982), 553–618.

[6] J.-P. Serre, Trees, Springer-Verlag, 2003.

[7] S.M. Smith, *A product for permutation groups and topological groups*, Duke Math. J. **166**(15) (2017), 2965–2999.

[8] J. Tits, *Sur le groupe des automorphismes d'un arbre*, in Essays on topology and related topics (Memoires dédiés à Georges de Rham), Springer, New York, 1970, pp. 188–211.

[9] V.V. Uspenskii, *A universal topological group with a countable basis*, Funct. Anal. and its Appl. **20** (1986), 86–87.

8

The Neretin groups

Łukasz Garncarek and Nir Lazarovich

Abstract

We propose an introduction to groups of partial automorphisms of locally finite trees. In particular we discuss Neretin's locally compact groups of spheromorphisms.

8.1	Introduction	131
8.2	Preliminaries	132
8.3	The Neretin group of spheromorphisms	134
8.4	Topology on the Neretin group	136
8.5	The Higman–Thompson groups	139
8.6	A convenient generating set for N_q	141
8.7	The simplicity of the Neretin group	142
	References for this chapter	144

8.1 Introduction

The Neretin group N_q was introduced in [6] as an analogue of the diffeomorphism group of the circle. It is a subgroup of the homeomorphism group of the boundary of an infinite q-regular tree T, consisting of elements which locally act by similarities of some (any) visual metric. We define the Neretin group, endow it with a group topology and present the proof of its simplicity, following [5].

In Section 8.2 we discuss the structure of the boundary of a regular tree. Sections 8.3 and 8.4 define the Neretin group, and describe its locally compact totally disconnected group topology. In Sections 8.5 and 8.6 we define the

Higman–Thompson groups $G_{q,r}$, another family of groups related to boundaries of regular trees, and show how they can be embedded into the Neretin
group. Then we prove that the Neretin group N_q is generated by any of the embedded copies of the Higman–Thompson group $G_{q,2}$ together with the canonically embedded group of type-preserving automorphisms of the tree T. Finally,
Section 8.7 presents the proof of simplicity of the Neretin group. Throughout
the text, we assume that all topological groups are Hausdorff.

8.2 Preliminaries

A tree T is a non-empty connected undirected simple graph without non-trivial
cycles. We will interchangeably treat T as a set of vertices endowed with a
binary relation of adjacency, or as a topological space obtained from the set of
vertices by gluing in unit intervals corresponding to edges. A fixed basepoint
$o \in T$ defines a partial order on T, namely $v \leq_o w$ if and only if the unique
shortest path from v to o passes through w. The basepoint o is the greatest
element in this order. The tree structure on T can be recovered from the poset
$(T, \leq_o)$ as follows. Two elements $v, w \in T$ are adjacent if and only if they are
comparable, and there are no other elements between them. It follows that tree
automorphisms of T fixing o are exactly the order automorphisms of $(T, \leq_o)$.
Also, if $v \leq_o w$ are adjacent, we refer to v as a **child** of w.

The distance between vertices $v, w \in T$, i.e. the number of edges on the
unique path joining them, will be denoted by $|vw|$. A vertex $v \in T$ of degree 1
is called a **leaf**. A tree is q-**regular** for some $q \in \mathbf{N}$ if all its vertices have degree $q + 1$. A **finite** q-**regular tree** is a finite tree, whose every vertex is either
a leaf, or has degree $q + 1$, in which case we call it **internal**. A **rooted tree** is a
tree with a fixed choice of base vertex $o \in T$, called its **root**. In case of rooted
trees we slightly modify the definition of q-regularity by requiring the root to
be of degree q instead of $q + 1$. In the subsequent sections we will deal only
with regular trees, so let us assume from now on that T is a rooted or unrooted
q-regular tree with $q \geq 2$. This will relieve us from considering some special
cases, which would otherwise appear in the following discussion.

By a **ray** in T we understand an infinite path, i.e. a sequence $(v_0, v_1, \ldots)$ of
distinct vertices of T such that the consecutive ones are connected by edges.
Two rays are said to be **asymptotic** if, after removing some finite initial sub-
sequences, they become equal. Equivalence classes of rays in T are called the
ends of T. The set of all ends of T is denoted by ∂T and referred to as the
boundary of T. Any end $\xi \in \partial T$ has a unique representative ξ_v with a given
initial vertex $v \in T$. To see it, one has to pick a representative $(v_0, v_1, \ldots)$ of

ξ, find a minimal path from v to one of the vertices v_i and replace the initial segment $(v_0, \ldots, v_i)$ by this path. Thus, if we choose a base vertex $o \in T$, we may identify ∂T with the set of rays emanating from o.

For $o \in T$ denote by $(\xi, \eta)_o$ the length of the common initial segment of the representatives ξ_o and η_o of two ends $\xi, \eta \in \partial T$. It takes values in $\mathbf{N} \cup \{\infty\}$. Together with a choice of $\varepsilon > 0$ this allows to define a **visual metric** $d_{o,\varepsilon}$ on ∂T by

$$d_{o,\varepsilon}(\xi, \eta) = e^{-\varepsilon(\xi, \eta)_o}.$$

The change of the basepoint o leads to a bi-Lipschitz equivalent metric, and changing ε still gives the same topology. It is an exercise to check that this unique natural topology on ∂T is compact and second countable, provided that T is **locally finite**, i.e. every vertex has finite degree.

If $j \colon T_1 \to T_2$ is an embedding of trees, it sends rays to rays, and preserves asymptoticity. Hence, it induces a map $j_* \colon \partial T_1 \to \partial T_2$. If we choose basepoints $o_i \in T_i$ in such a way that $j(o_1) = o_2$, then j_* can be seen to be an isometric embedding, so in particular it is continuous. Taking the boundary in fact gives a functor from the category of trees and tree embeddings into the category of topological spaces and continuous embeddings.

From now on we will fix $\varepsilon = 1$ and $o \in T$, making T a rooted q-regular tree, and suppress them from notation whenever possible. There may exist more natural choices for ε, but they will not be of any use to us. The geometrically obvious inequality $(\xi, \eta) \geq \min\{(\xi, \zeta), (\zeta, \eta)\}$ implies that d is in fact an ultrametric, i.e. it satisfies a stronger variant of the triangle condition,

$$d(\xi, \eta) \leq \max\{d(\xi, \zeta), d(\zeta, \eta)\}.$$

As a consequence, two open balls in $(\partial T, d)$ are either disjoint, or one of them is contained in the other. It follows that the covering of ∂T by open balls of fixed radius is in fact a partition into open—and hence also closed—sets, and ∂T is totally disconnected. Additionally, since the metric d takes values in a discrete set, any closed ball is also an open ball with a slightly larger radius, and vice versa.

Ultrametricity implies that any point of a ball in $(\partial T, d)$ is its centre. It is however still possible to effectively enumerate the balls in a one-to-one manner. To this end, for $v \in T$ define T_v as the subtree of T spanned by the vertices $\{w \in T : w \leq_o v\}$. It is a rooted q-regular tree with root v, and its boundary ∂T_v is a subset of ∂T. It is in fact a closed ball of radius equals $e^{-|ov|}$, and the embedding $(\partial T_v, d_v) \to (\partial T, d_o)$ is a similarity. On the other hand, any ball $B \subseteq \partial T$ is a closed ball of radius e^{-n} for some $n \in \mathbf{N}$, and can be written as ∂T_v where v is the last vertex of the common initial segment of all the rays ξ_o

representing points $\xi \in B$. The family of non-empty balls in ∂T is therefore in a one-to-one correspondence with vertices of T.

As a consequence of the above discussion, the assignment $v \mapsto \partial T_v$ is an order-isomorphism between $(T, \leq_o)$ and the set $\mathscr{B}(\partial T, d_0)$ of all balls in $(\partial T, d_o)$ ordered by inclusion. Moreover, if $\phi : T \to T'$ is a basepoint-preserving iso-morphism of trees, then $\phi(T_v) = T_{\phi(v)}$, and

$$\phi_*(\partial T_v) = \partial(\phi(T_v)) = \partial T_{\phi(v)}.$$

This means that the order-isomorphism between $\mathscr{B}(\partial T, d_o)$ and $\mathscr{B}(\partial T', d_{\phi(o)})$ induced by ϕ is the same as the one induced by $\phi_* : \partial T \to \partial T'$. This correspondence can be reversed, namely if $\Phi : \partial T \to \partial T'$ is a homeomor-phism preserving balls, it necessarily preserves their inclusion, and induces an order-isomorphism of $\mathscr{B}(\partial T, d_o)$ and $\mathscr{B}(\partial T', d_{\phi(o)})$ yielding a basepoint-preserving isomorphism $\phi : T \to T'$, which can be constructed by induction on the distance from the root. It satisfies

$$\phi_*(\partial T_v) = \partial T_{\phi(v)} = \Phi(\partial T_v),$$

but since the balls form a basis of the topology of ∂T, this means that $\Phi = \phi_*$.

Finally, let us introduce the notion of a **forest**. It is what we obtain if we remove the assumption of connectedness from the definition of a tree. In other words, a forest is a graph F which decomposes into a disjoint union of trees. We may define its boundary ∂F as the disjoint union of the boundaries of its constituent trees. It is again functorial. Most of the discussion above extends to forests.

8.3 The Neretin group of spheromorphisms

Let T be a q-regular tree. For a non-empty finite q-regular subtree $F \subseteq T$, by the difference $T \setminus F$ we will understand the rooted q-regular forest obtained by removing from T all the edges and internal vertices of F, and designating the leaves of F as the roots; geometrically, this amounts to removing the interior of F from T. Clearly, $\partial(T \setminus F) = \partial T$, as every ray in T has a subray disjoint from F.

Now, let $F_1, F_2 \subseteq T$ be two finite q-regular subtrees, such that there exists an isomorphism of forests $\phi : T \setminus F_1 \to T \setminus F_2$. It induces a homeomorphism ϕ_* of ∂T, called a **spheromorphism** of ∂T. The isomorphism ϕ will be referred to as a representative of ϕ_*.

Observe that the identity map of ∂T is a spheromorphism. More generally, if $\phi \in \mathrm{Aut}(T)$, then for any subtree $F \subseteq T$ the map ϕ restricts to an isomorphism of forests $T \setminus F \to T \setminus \phi(F)$, and thus the induced homeomorphism ϕ_* is a spheromorphism. Moreover, the inverse of a spheromorphism is also a spheromorphism, and for any pair of spheromorphisms ϕ_* and ψ_* we may find representatives which are composable, showing that $\psi_* \circ \phi_*$ is also a spheromorphism.

Definition 8.1 The **Neretin group** N_q is the group of all spheromorphisms of the boundary of a q-regular tree.

The group N_q has another description, based upon the metric structure of the boundary. We will call a homeomorphism of metric spaces $\Phi \colon X \to Y$ a **local similarity** if for each $x \in X$ there exists an open neighbourhood U of x and a constant $\lambda_U > 0$ such that for every $x_1, x_2 \in U$ we have

$$d_Y(\Phi(x_1), \Phi(x_2)) = \lambda_U d_X(x_1, x_2),$$

i.e. the restriction $\Phi|_U \colon U \to \Phi(U)$ is a similarity [4]. The requirement that ϕ is a homeomorphism allows to choose U to be a ball $B(x, r)$ centred at x, such that $\Phi(B(x, r)) = B(\Phi(x), \lambda_U r)$. It is clear that all local similarities of a metric space form a group.

Proposition 8.2 *For a homeomorphism $\Phi \in \mathrm{Homeo}(\partial T)$ the following conditions are equivalent.*

(i) Φ is a spheromorphism,
(ii) Φ is a local similarity with respect to any visual metric on ∂T,
(iii) Φ is a local similarity with respect to some visual metric on ∂T.

Proof We begin by showing that (1) implies (2). Fix a basepoint $o \in T$ and the corresponding visual metric d. Let $\Phi = \phi_*$ be a spheromorphism represented by $\phi \colon T \setminus F_1 \to T \setminus F_2$. We may assume that both F_1 and F_2 contain o as internal vertex. Let $T \setminus F_1 = T_1 \cup \cdots \cup T_k$ be the decomposition into disjoint trees. Then $T \setminus F_2$ decomposes into $\phi(T_1) \cup \cdots \cup \phi(T_k)$. These decompositions induce partitions of ∂T into open balls ∂T_i and $\partial(\phi(T_i)) = \phi_*(\partial T_i)$.

Let v be the root of T_i. Then the root of $\phi(T_i)$ is $\phi(v)$. For $\xi, \eta \in \partial T_i \subseteq \partial T$ we have

$$(\phi_*(\xi), \phi_*(\eta))_o = (\phi_*(\xi), \phi_*(\eta))_{\phi(v)} + |o\phi(v)|$$
$$= (\xi, \eta)_v + |o\phi(v)| = (\xi, \eta)_o - |ov| + |o\phi(v)|,$$

which implies that

$$d(\phi_*(\xi), \phi_*(\eta)) = e^{|ov| - |o\phi(v)|} d(\xi, \eta),$$

and $\phi_*|_{\partial T_i} : \partial T_i \to \phi_*(\partial T_i)$ is a similarity.

The other non-trivial implication is from (3) to (1). Let $\Phi \in \text{Homeo}(\partial T)$ be a local similarity of $(\partial T, d_o)$. By compactness, we may cover ∂T by finitely many balls B on which Φ is a similarity, and $\Phi(B)$ is also a ball. By ultrametricity, we may assume that this covering is disjoint, and contains at least 2 balls. The balls in the covering are of the form ∂T_v with v in some finite set $L \subseteq T$, and $\Phi(\partial T_v) = \partial T_{v'}$ for some $v' \in T$. The restriction $\Phi|_{\partial T_v} : \partial T_v \to \partial T_{v'}$ preserves balls, and therefore is induced by a root-preserving isomorphism $\phi_v : T_v \to T_{v'}$. It now remains to observe that the forests $\bigcup T_v$ and $\bigcup T_{v'}$ are obtained by removing finite regular subtrees from T, so that the isomorphisms ϕ_v assemble into an isomorphism of these forests, representing a spheromorphism. $\qquad\square$

8.4 Topology on the Neretin group

The Neretin group N_q is a subgroup of the homeomorphism group of ∂T, which is a topological group when endowed with the compact-open topology. Since ∂T is compact, this topology is metrizable: for $\Phi, \Psi \in \text{Homeo}(\partial T)$ we have

$$d(\Phi, \Psi) = \sup_{\xi \in \partial T} d_o(\Phi(\xi), \Psi(\xi)).$$

A first choice for the group topology on N_q would be to restrict the compact-open topology. Unfortunately, this restriction is not locally compact, as we will now observe, using the following lemma.

Lemma 8.3 *If a subgroup H of a topological group G is locally compact, then it is closed.*

Proof First, assume that H is dense in G. Let U be an open neighbourhood of 1 in G such that the closure K of $U \cap H$ in H is compact. Since a compact subset of a Hausdorff space is closed, and G is Hausdorff by definition, K is closed in G. Now, we have $K \cap U = H \cap U$, and this set is both closed and dense in U, hence it is equal to U. Therefore $U \subseteq H$, so the subgroup H is open, and hence closed.

In the general case H is dense, and hence closed, in its closure in G. This means that it is closed in G. $\qquad\square$

Now, we can see that if N_q with the compact-open topology was locally compact, it would be a closed subgroup. We will show this is false by constructing a sequence of spheromorphisms converging to a homeomorphism outside N_q, using the description of spheromorphisms as local similarities. Let B_i be a sequence of pairwise disjoint balls in ∂T. For each i we may construct a spheromorphism $\Phi_i \in N_q$, which is identity outside B_i, and on some ball inside B_i it restricts to a similarity with scale greater that i. The sequence $\Psi_k = \Phi_k \circ \cdots \circ \Phi_1$ is Cauchy, as for $k > l$

$$d(\Psi_k, \Psi_l) = d(\Phi_k \circ \cdots \circ \Phi_{l+1}, \mathrm{id}) \leq \max_{l < i \leq k} \mathrm{diam}\, B_i \xrightarrow[l \to \infty]{} 0.$$

Therefore, Ψ_k converge to a homeomorphism $\Psi \in \mathrm{Homeo}(\partial T)$. It has the same restriction to B_i as Φ_i, and therefore on some ball it restricts to a similarity of scale at least i. But local similarities are Lipschitz, so $\Psi \notin N_q$.

The issue of endowing N_q with a locally compact group topology can be resolved by observing that it already contains a locally compact group as a subgroup. Indeed, $\mathrm{Aut}(T)$ naturally embeds in N_q (and will be identified with its image) and carries the compact-open topology coming from its action on the set of vertices of T. It is locally compact and totally disconnected. We can extend it to N_q by declaring the left cosets of $\mathrm{Aut}(T)$ to be open and homeomorphic to $\mathrm{Aut}(T)$ by the translation maps. If $g\mathrm{Aut}(T) = h\mathrm{Aut}(T)$, then the translation maps induce the same topology, so it is well defined, and clearly locally compact and totally disconnected. What is not so clear is whether this makes N_q a topological group—if it does, then this topology is clearly the unique one making $\mathrm{Aut}(T)$ with its original topology an open subgroup of N_q. This issue is addressed by the following two lemmas.

Lemma 8.4 *Suppose that an abstract group G contains a topological group H as a subgroup. Then G admits a unique group topology such that the inclusion $H \to G$ is continuous and open (in particular, the topology on G restricts to the original topology on H), provided that for all open subsets $U \subseteq H$ and $g, g' \in G$ the intersection $gUg' \cap H$ is open in H.*

Proof For H with its original topology to be an open subgroup of G, it is necessary and sufficient that the family of left translates of open subsets of H forms a basis of the topology of G. It is easy to see that this family satisfies the axioms of a basis; let us therefore endow G with the resulting topology. We need to see that it is a group topology.

First, observe that the right translates of open subsets of H are also open, since for $U \subseteq H$ open and $g' \in G$ the set Ug' can be written as

$$Ug' = \bigcup_{g \in G} (gH \cap Ug') = \bigcup_{g \in G} g(H \cap g^{-1}Ug'),$$

which is a union of left translates of open subsets of H. As a consequence, left and right translations are homeomorphisms of G.

Now we need to ensure that multiplication and inversion are continuous. Let $g_\alpha \to g$ and $g'_\alpha \to g'$ be two convergent nets in G. The cosets gH and Hg' are open neighbourhoods of g and g' respectively, so without loss of generality we may assume that $g_\alpha = gh_\alpha$ and $g'_\alpha = h'_\alpha g'$ with $h_\alpha, h'_\alpha \in H$ converging to 1. Then $h_\alpha h'_\alpha \to 1$ in H, and therefore $g_\alpha g'_\alpha = gh_\alpha h'_\alpha g' \to gg'$. Similarly $g_\alpha^{-1} = h_\alpha^{-1} g^{-1} \to g^{-1}$. $\qquad\qquad\square$

In order to show that the topology we put on N_q is indeed a group topology, it remains to show that for every open $U \subseteq \text{Aut}(T)$ and $g, g' \in N_q$ the subset $\text{Aut}(T) \cap gUg'$ is open in $\text{Aut}(T)$. Observe that this property is preserved under unions and finite intersections, so it is enough to show it for U in a certain subbasis of the topology on $\text{Aut}(T)$.

Let $o \in T$ be a base vertex, and denote by K the stabilizer of o in $\text{Aut}(T)$. Then for $g, h \in \text{Aut}(T)$ the set gKh consists exactly of the automorphisms sending $h^{-1}(o)$ to $g(o)$, so the finite intersections of the sets of the form gKh yield the standard basis for the topology of $\text{Aut}(T)$. We are thus left with proving the following.

Lemma 8.5 *If K is the stabilizer of the base vertex $o \in T$ in the group $\text{Aut}(T)$, then for all $\phi_*, \psi_* \in N_q$ the intersection $\psi_* K \phi_* \cap \text{Aut}(T)$ is open in $\text{Aut}(T)$.*

Proof The spheromorphisms ϕ_* and ψ_* admit representatives $\phi: T \setminus F_1 \to T \setminus B$ and $\psi: T \setminus B \to T \setminus F_2$, where B is a sufficiently large ball in T, centred at o. Let $K_B \subseteq K$ denote the pointwise stabilizer of this ball; it is an open subgroup of $\text{Aut}(T)$. We have

$$\psi_* K \phi_* = \bigcup_{k \in K} \psi_* k K_B \phi_* = \bigcup_{k \in K} \psi_* k \phi_* (\phi_*^{-1} K_B \phi_*),$$

where $\phi_*^{-1} K_B \phi_*$ consists of elements η_* whose representatives $\eta: T \setminus F_1 \to T \setminus F_1$ leave the trees of the forest $T \setminus F_1$ in place, and thus extend to automorphisms of T. Hence, it is an open subgroup of $\text{Aut}(T)$, namely the pointwise stabilizer of F_1, and therefore the intersection $\psi_* K \phi_* \cap \text{Aut}(T)$ is open. $\qquad\square$

This shows that the topology we defined on N_q is indeed a group topology. We may summarize this as follows.

Theorem 8.6 *The Neretin group N_q admits a unique group topology such that the natural embedding $\text{Aut}(T) \to N_q$ is continuous and open. With this topology, N_q is a totally disconnected locally compact group.*

8.5 The Higman–Thompson groups

A tree T is **planar** if it is rooted and for every $v \in T$ there is a fixed linear order on the set of children of v. This corresponds to specifying a way to draw the tree on the plane, so that for every $v \in T$ its children are below it, ordered from left to right. The structure of a planar tree is very rigid—an isomorphism of planar trees, which is required to preserve the roots and orders on the sets of children, is always unique, if it exists.

Let $\mathscr{F}$ be a forest consisting of r planar q-regular trees $T_1, \ldots, T_r$. For every i choose rooted (in particular, this implies that F_i and F_i' have the same root as T_i) finite regular subtrees $F_i, F_i' \subseteq T_i$ in such a way that the forests $\mathscr{F}_1 = \mathscr{F} \setminus \bigcup F_i$ and $\mathscr{F}_2 = \mathscr{F} \setminus \bigcup F_i'$ have the same number of trees. The forests $\mathscr{F}_1$ and $\mathscr{F}_2$ consist of planar q-regular trees, and hence for every bijection of the sets of trees in $\mathscr{F}_1$ and $\mathscr{F}_2$ there exists a unique isomorphism $\phi \colon \mathscr{F}_1 \to \mathscr{F}_2$ of planar forests realizing it. It induces a homeomorphism ϕ_* of $\partial \mathscr{F}$, and the subgroup of $\mathrm{Homeo}(\partial \mathscr{F})$ containing all such homeomorphisms is called the Higman–Thompson group $G_{q,r}$. The group $G_{2,1}$ is known as Thompson group V.

This definition shows some ties between $G_{q,r}$ and the permutation groups S_n, which we will now make more explicit. The order of children on each T_i induces a lexicographic order on paths starting from the root, which correspond to vertices. This defines a linear order on the set of vertices of each T_i. Moreover, the trees themselves can be ordered from T_1 to T_r, so we have a linear order on the set of vertices of $\mathscr{F}$. This allows to order the trees in $\mathscr{F}_1$ and $\mathscr{F}_2$ by looking at the order of their roots. An isomorphism $\phi \colon \mathscr{F}_1 \to \mathscr{F}_2$ of planar forests is now completely determined by a permutation $\sigma \in S_n$, where n is the number of trees in the forests $\mathscr{F}_i$.

We will use this to define a homomorphism $\theta \colon G_{q,r} \to \mathbf{Z}/2\mathbf{Z}$. If q is even, θ is just the zero homomorphism. On the other hand, if q is odd, we claim that the sign of the permutation σ associated to ϕ in the above discussion depends only on the element $\phi_* \in G_{n,r}$, and we put $\theta(\phi_*) = \mathrm{sgn}\,\sigma$. To see this, observe that if $\mathscr{F}_1 = \bigcup_{i=1}^n L_i$ and $\mathscr{F}_2 = \bigcup_{i=1}^n L_i'$ are decompositions into trees, numbered in accordance with the order, we may modify the representative ϕ in an elementary way as follows. Choose one of the trees L_i and remove its root, replacing it with q new trees. Do the same with $\phi(L_i) = L'_{\sigma(i)}$. This gives a new representative ϕ' of ϕ_*, obtained by restricting ϕ. The number of inversions

$I(\sigma')$ in the permutation σ' associated to ϕ' is equal to

$$
\begin{aligned}
I(\sigma') &= \left|\{(j,k) \in \{1,\dots,n\}^2 : j \neq i \neq k,\, j < k,\, \sigma(j) > \sigma(k)\}\right| \\
&\quad + q\left|\{k \in \{i+1,\dots,n\} : \sigma(k) < i\}\right| \\
&\quad + q\left|\{j \in \{1,\dots,i-1\} : \sigma(j) > i\}\right| \\
&= I(\sigma) + (q-1)C,
\end{aligned}
$$

where $(q-1)C$ is even. This means that the sign of the permutation does not change when we apply the described elementary modification to a representative of ϕ_*. It remains to observe that any two representatives of ϕ_* can be transformed by a sequence of elementary modifications into the same third representative.

Using the homomorphism θ defined above, we may now describe the commutator subgroup of $G_{q,r}$. This theorem comes from [3], and the argument below is based on an idea of Mati Rubin [1].

Theorem 8.7 *The commutator subgroup $G'_{q,r}$ of the Higman–Thompson group $G_{q,r}$ is equal to the kernel of the homomorphism $\theta\colon G_{q,r} \to \mathbf{Z}/2\mathbf{Z}$. It is a simple group.*

Proof It is clear that $G'_{q,r} \subseteq \ker\theta$, and we need to prove the opposite inclusion. First, we claim that $G_{q,r}$ is generated by elements with representatives $\phi\colon \mathscr{F}_1 \to \mathscr{F}_2$ such that $\mathscr{F}_1 = \mathscr{F}_2$. Indeed, if $\phi_* \in G_{q,r}$ is represented by $\phi\colon \mathscr{F}_1 \to \mathscr{F}_2$, in both $\mathscr{F}_1$ and $\mathscr{F}_2$ we may find families of q trees whose roots have the same parent in $\mathscr{F}$. If we compose ϕ with a suitable $\psi\colon \mathscr{F}_2 \to \mathscr{F}_2$ such that $\psi \circ \phi$ sends the q fixed trees from $\mathscr{F}_1$ to the q fixed trees in $\mathscr{F}_2$ in an order preserving way, then $(\psi \circ \phi)_*$ can be represented by a map $\chi\colon \mathscr{F}'_1 \to \mathscr{F}'_2$, where $\mathscr{F}'_i$ is obtained from $\mathscr{F}_i$ by adding the common parent of the fixed q trees, and joining them into a single tree. This process stops after finitely many steps, yielding a decomposition of ϕ_* into a product of the claimed generators.

An element of $G_{q,r}$ supported in a proper subset of $\partial\mathscr{F}$, represented by $\phi\colon \mathscr{F}_1 \to \mathscr{F}_1$ which exchanges two trees of $\mathscr{F}_1$ and leaves the rest in place, will be called a **transposition**. It is now clear that $G_{q,r}$ is generated by transpositions, and if q is odd, then $\ker\theta$ is generated by products of pairs of transpositions, which can be further assumed to have disjoint supports not covering the whole boundary $\partial\mathscr{F}$ (we will always assume this when speaking about a product of a pair of transpositions). Moreover, if q is even, any transposition can be decomposed into a product of an even number of transpositions with disjoint supports. Hence, $\ker\theta$ is always generated by products of pairs of disjoint transpositions.

Now, observe that any two products of pairs of transpositions are conjugate in $G_{q,r}$. Thus, in order to complete the proof of the inclusion we need to show that the commutator subgroup $G'_{q,r}$ contains a product of two disjoint transpositions, supported in a proper subset of $\partial \mathscr{F}$. To this end, we just need to take the commutator of two transpositions, one exchanging two balls in $\partial \mathscr{F}$, and the other supported inside one of these balls.

We are now left with observing that the commutator subgroup $G'_{q,r}$ is simple. Let N be a normal subgroup of $G'_{q,r}$ containing a non-trivial element ϕ_*. There exists an open set $U \subseteq \partial \mathscr{F}$ such that $\phi_*(U)$ is disjoint from U, and $U \cup \phi_*(U)$ is a proper subset of $\partial \mathscr{F}$. If ψ_* is a product of two disjoint transpositions supported in U then the commutator $[\phi_*, \psi_*]$ is a product of four disjoint transpositions, is supported in $U \cup \phi_*(U)$ and belongs to N. If χ_* is a transposition supported outside $U \cup \phi_*(U)$, then $[\phi_*, \psi_*]$ is invariant under conjugation by χ_*. Since any product of four disjoint transpositions, supported in a proper subset of $\partial \mathscr{F}$, is conjugate to $[\phi_*, \psi_*]$ by an element of $G_{q,r} = G'_{q,r} \cup \chi_* G'_{q,r}$, the normal subgroup N contains all such products. But from simplicity of the alternating groups it follows that for $n \geq 8$ the alternating group A_n is generated by products of four disjoint transpositions, hence N contains all elements possessing a representative whose associated permutation is even. This means that $N = G'_{q,r}$. □

8.6 A convenient generating set for N_q

Fix a 2-colouring of vertices of T; in this context one usually refers to colours as **types**. An automorphism ϕ of T is **type-preserving** if and only if it preserves this colouring. The subgroup of $\mathrm{Aut}(T)$ consisting of type-preserving automorphisms is denoted by $\mathrm{Aut}^+(T)$.

It is generated by the union of pointwise edge stabilizers in $\mathrm{Aut}(T)$. Indeed, denote by H the subgroup of $\mathrm{Aut}(T)$ generated by the edge stabilizers. It is clearly a subgroup of $\mathrm{Aut}^+(T)$. Any two edges with a common endpoint can be exchanged using an element of the stabilizer of another edge with the same endpoint. It follows that all edges with a common endpoint lie in the same orbit of H, and thus H acts transitively on the edges of T. Hence, if $\phi \in \mathrm{Aut}^+(T)$ and e is an edge of T, then there exists $\psi \in H$ such that $\psi(e) = \phi(e)$. The element $\psi^{-1}\phi \in \mathrm{Aut}^+(T)$ fixes the edge e pointwise, so $\psi^{-1}\phi \in H$.

As an instance of the Tits Simplicity Theorem [7] (see Theorem 6.14 in Chapter 6 of this volume), we obtain the following.

Theorem 8.8 *The group $\mathrm{Aut}^+(T)$ of type-preserving automorphisms is simple.*

Now, consider the Higman–Thompson group $G_{q,2}$ acting on the boundary of the planar forest $\mathscr{F}$. Pick an edge e of T and an embedding $i\colon \mathscr{F} \to T$ sending $\mathscr{F}$ onto $T \setminus e$. It defines an embedding of $G_{q,2}$ into N_q given by $\Phi \mapsto i_* \circ \Phi \circ i_*^{-1}$, whose image we will denote by $G_{q,2}^i$. If $j\colon \mathscr{F} \to T$ is another such embedding, with image $T \setminus e'$, it can be written as $j = \eta \circ i \circ \varepsilon$ where $\eta \in \mathrm{Aut}^+(T)$, and ε is either the identity map, or the unique automorphism of $\mathscr{F}$ preserving its structure of a planar forest, and exchanging its two trees. As a consequence, in N_q the subgroup $G_{q,2}^j$ is conjugate to $G_{q,2}^i$ by an element of $\mathrm{Aut}^+(T)$.

Lemma 8.9 *The Neretin group N_q is generated by $\mathrm{Aut}^+(T)$ and $G_{q,2}^i$, for any embedding i.*

Proof Let $\phi_* \in N_q$ be represented by $\phi\colon T \setminus F_1 \to T \setminus F_2$. We may suppose that the subtrees F_1 and F_2 share a common edge e. Choose an isomorphism $j\colon \mathscr{F} \to T \setminus e$; it defines the subgroup $G_{q,2}^j \subseteq \langle \mathrm{Aut}^+(T), G_{q,2}^i \rangle$.

There exists an element $\psi_* \in G_{q,2}^j$ with representative $\psi\colon T \setminus F_1 \to T \setminus F_2$ inducing the same bijection of trees as ϕ. Then $\psi^{-1}\phi\colon T \setminus F_1 \to T \setminus F_1$ preserves the trees of the forest $T \setminus F_1$, and therefore extends to an automorphism of T fixing pointwise the subtree F_1, and in particular the edge $e \subseteq F_1$. Thus, ϕ_* is a product of $\psi_* \in G_{q,2}^j$ and a type-preserving automorphism. $\square$

Since the embedded copies of $G_{q,2}$ are conjugate by elements of $\mathrm{Aut}^+(T)$, we may slightly abuse the notation and write $N_q = \langle \mathrm{Aut}^+(T), G_{q,2} \rangle$.

8.7 The simplicity of the Neretin group

In this section we present the proof of the simplicity of the Neretin group following Kapoudjian [5] using a method that was introduced by Epstein in [2]. We remark that the following lemmas apply to a large variety of examples and are thus worth recalling.

We begin by two general lemmas about actions on topological spaces. The setting we will consider will consist of a compact Hausdorff topological space X, and a faithful group action by homeomorphisms of a group G on X.

Lemma 8.10 *Let X and G be as above, let $\mathscr{U}$ be a basis of X on which G acts transitively, and let $1 \neq H \lhd G$ be a non-trivial normal subgroup of G. Then, for all $g \in G$ such that $\mathrm{supp}(g) \subseteq V \in \mathscr{U}$ there exists an element $\rho \in H$ such that $\rho|_V = g|_V$.*

Proof Let $1 \neq \alpha \in H$ be any non-trivial element of H. Let $x \in X$ be a point for which $\alpha^{-1}(x) \neq x$. One can find a basis set $V_0 \in \mathscr{U}$ such that $\alpha^{-1}(V_0) \cap V_0 = \varnothing$.

Assume first that $V = V_0$, and consider $\rho = [g, \alpha] = g\alpha g^{-1}\alpha^{-1}$. Since $\alpha \in H$ and H is normal in G we get that $\rho \in H$. Moreover, by our assumption $\mathrm{supp}(g) \subseteq V$, thus when restricted to V we see that $\rho|_V = g|_V$, as required.

More generally, if $V \neq V_0$, we may find $h \in G$ such that $hV = V_0$. Now the element $g' = hgh^{-1}$ satisfies $\mathrm{supp}(hgh^{-1}) \subseteq V_0$ and thus by the above we can find $\rho' \in H$ that satisfies $\rho'|_{V_0} = g'|_{V_0}$. Since H is normal $\rho = h^{-1}\rho h \in H$, and ρ satisfies $\rho|_V = g|_V$ $\qquad\square$

Lemma 8.11 *Let X and G be as above, let $\mathscr{U}$ be a basis of X on which G acts transitively, and let $1 \neq H \lhd G$ be a non-trivial normal subgroup of G in which there exists $\alpha_1, \alpha_2 \in H$ such that for some $x \in X$ the points $x, \alpha_1(x), \alpha_2(x)$ are distinct. Then, for all $g_1, g_2 \in G$ such that $\mathrm{supp}(g_1), \mathrm{supp}(g_2) \subseteq V \in \mathscr{U}$ there exist elements $\rho_1, \rho_2 \in H$ such that $[g_1, g_2] = [\rho_1, \rho_2]$.*

Proof Let $x \in X$ and $\alpha_1, \alpha_2 \in H$ be as assumed. One can find a basis set $V_0 \in \mathscr{U}$ such that $V_0, \alpha_1(V_0), \alpha_2(V_0)$ are pairwise disjoint.

As in the proof of the previous lemma we may assume up to conjugation that $V = V_0$, and consider $\rho_1 = [g, \alpha_1]$ and $\rho_2 = [g, \alpha_2]$. Again we have $\rho_1, \rho_2 \in H$ as required. Moreover, by our assumption $\mathrm{supp}(g_1), \mathrm{supp}(g_2) \subseteq V$, thus when restricted to V we see that $[\rho_1, \rho_2]|_V = [g_1, g_2]|_V$. Moreover, ρ_i ($i = 1, 2$) preserves the pairwise disjoint sets $V_0, \alpha_1(V_0), \alpha_2(V_0)$ and is supported on $V_0 \cup \alpha_i(V_0)$. It follows that $[\rho_1, \rho_2]|_{X\setminus V} = \mathrm{id} = [g_1, g_2]|_{X\setminus V}$. Thus overall we get $[g_1, g_2] = [\rho_1, \rho_2]$. $\qquad\square$

Remark 8.12 Note that in order to find such α_1, α_2 as in Lemma 8.11, by Lemma 8.10 it is enough to find two such elements in G that are supported on a basis set.

We now apply the previous lemmas to prove the simplicity of the Neretin group.

Theorem 8.13 *The Neretin group N_q is simple.*

Proof Let $1 \neq H \lhd N_q$ be a non-trivial subgroup of the Neretin group. From Lemma 8.9 the Neretin group is generated by $\mathrm{Aut}^+(T_q)$ and $G_{q,2}$. In fact, it is enough to take the commutator subgroup $G'_{q,2}$ of $G_{q,2}$, since $\mathrm{Aut}^+(T_q) \cap (G_{q,2} \setminus G'_{q,2}) \neq \varnothing$ whenever $G_{q,2} \neq G'_{q,2}$. From Theorems 8.8 and 8.7 the subgroups $\mathrm{Aut}^+(T_q)$ and $G'_{q,2}$ are simple. Thus, in order to prove the claim it is enough to show that $H \cap \mathrm{Aut}^+(T_q) \neq 1$ and $H \cap G'_{q,2} \neq 1$.

The group $\mathrm{Aut}(T_q)$ acts transitively on oriented edges of T_q, and thus acts transitively on **half-trees** (i.e. the connected components of complements of edges). Let U be the set of ends in ∂T_q of some fixed half-tree. The orbit $\mathrm{Aut}(T_q).U$, which consists of all sets of ends of half-trees, forms a basis for the topology of the boundary. Therefore, the Neretin group, which acts faithfully by homeomorphisms on the boundary of the tree T_q, also acts transitively on a basis, namely $N_q.U$.

The desired conclusion follows from Lemma 8.11 and Remark 8.12 since one can find two pairs of non-commuting elements in $\mathrm{Aut}^+(T_q)$ and $G_{q,2}$ that are supported on the ends of a half-tree. $\qquad\Box$

Acknowledgements. During the work on this paper the first author was supported by a scholarship of the Foundation for Polish Science and by the grant 2012/06/A/ST1/00259 of the National Science Center. The work was conducted during the first author's internship at the Warsaw Center of Mathematics and Computer Science.

References for this chapter

[1] Matthew Brin. Higher dimensional Thompson groups. *Geom. Dedicata*, 108:163–192, 2004.

[2] D. B. A. Epstein. The simplicity of certain groups of homeomorphisms. *Compos. Math.*, 22:165–173, 1970.

[3] G. Higman. *Finitely Presented Infinite Simple Groups*. Dept. of Pure Mathematics, Dept. of Mathematics, I.A.S., Australian National University, 1974.

[4] Bruce Hughes. Local similarities and the Haagerup property (with an appendix by Daniel S. Farley). *Groups, Geometry, and Dynamics*, 3(2):299–315, 2009.

[5] Christophe Kapoudjian. Simplicity of Neretin's group of spheromorphisms. *Annales de l'institut Fourier*, 49(4):1225–1240, 1999.

[6] Yu. A. Neretin. On Combinatorial Analogs of the Group of Diffeomorphisms of the Circle. *Russian Academy of Sciences. Izvestiya Mathematics (in Russian)*, 41(2):337–349, April 1993.

[7] Jacques Tits. Sur le groupe des automorphismes d'un arbre. In *Essays on Topology and Related Topics*, pages 188–211. Springer Berlin Heidelberg, 1970.

9

The scale function and tidy subgroups

Albrecht Brehm, Maxime Gheysens, Adrien Le Boudec and Rafaela Rollin

Abstract

This is an introduction to the structure theory of totally disconnected locally compact groups initiated by Willis in 1994. The two main tools in this theory are the scale function and tidy subgroups, for which we present several properties and examples.

As an illustration of this theory, we give a proof of the fact that the set of periodic elements in a totally disconnected locally compact group is always closed, and that such a group cannot have ergodic automorphisms as soon as it is non-compact.

9.1	Introduction	145
9.2	The structure of minimising subgroups	147
9.3	Examples	154
9.4	Ergodic automorphisms of locally compact groups	157
	References for this chapter	160

9.1 Introduction

Let G be a locally compact totally disconnected group. Recall that this means that G is a group endowed with a group topology that is locally compact, and such that connected components are singletons. Examples of such groups include for instance the automorphism group $\mathrm{Aut}(T_d)$ of a regular tree T_d of degree $d \geq 3$, or the general linear group $\mathrm{GL}_n(\mathbf{Q}_p)$ endowed with its p-adic topology. We will denote by $\mathrm{Aut}(G)$ the group of bicontinuous automorphisms of the group G.

Scale function

Recall that according to van Dantzig's theorem, compact open subgroups of G exist and form a basis of neighbourhoods of the identity in G. We will denote by $\mathscr{B}(G)$ the collection of compact open subgroups of G. For $U \in \mathscr{B}(G)$ and $\alpha \in \mathrm{Aut}(G)$, observe that $\alpha(U) \cap U$ is open in G, and since U is compact, this implies that $\alpha(U) \cap U$ has finite index in $\alpha(U)$. The **scale function** is defined as

$$s\colon \mathrm{Aut}(G) \to \mathbf{N}, \quad \alpha \mapsto \min_{U \in \mathscr{B}(G)} |\alpha(U) : \alpha(U) \cap U|.$$

By identifying an element $g \in G$ with the inner automorphism $x \mapsto gxg^{-1}$, we obtain a function on G

$$s\colon G \to \mathbf{N}, \quad g \mapsto \min_{U \in \mathscr{B}(G)} |gUg^{-1} : gUg^{-1} \cap U|.$$

We will call a compact open subgroup $U \in \mathscr{B}(G)$ **minimising** for $\alpha \in \mathrm{Aut}(G)$ if

$$s(\alpha) = |\alpha(U) : \alpha(U) \cap U|,$$

i.e. if the minimum is attained at U.

The scale function was introduced by Willis in [8], and was subsequently used to answer questions in different fields of mathematics. We refer the reader to the introduction of [11] and to references therein for more details. Here we aim to give an account of this theory, with an emphasis on examples (see Section 9.3).

Tidy subgroups

For $\alpha \in \mathrm{Aut}(G)$ and $U \in \mathscr{B}(G)$, we let

$$U_+ = \bigcap_{n \geq 0} \alpha^n(U) \text{ and } U_- = \bigcap_{n \geq 0} \alpha^{-n}(U).$$

Note that by definition $U \cap \alpha(U_+) = U_+$. We therefore have an injective map

$$\chi\colon \alpha(U_+)/U_+ \to \alpha(U)/\alpha(U) \cap U,$$

and in particular $|\alpha(U_+) : U_+| \leq |\alpha(U) : \alpha(U) \cap U|$. This motivates the following definition.

Definition 9.1 A compact open subgroup U is called **tidy above** for α if $|\alpha(U_+) : U_+| = |\alpha(U) : \alpha(U) \cap U|$.

We also let

$$U_{++} = \bigcup_{n \geq 0} \alpha^n(U_+) \text{ and } U_{--} = \bigcup_{n \geq 0} \alpha^{-n}(U_-).$$

Note that these are increasing unions, and it follows that U_{++} and U_{--} are subgroups of G.

Definition 9.2 A compact open subgroup U is called **tidy below** for α if U_{++} and U_{--} are closed in G.

Definition 9.3 A compact open subgroup is called **tidy** for α if it is both tidy above and tidy below for α.

One of the reasons why this notion of tidy subgroups is relevant in this context is that it precisely describes the properties that minimising subgroups must have in common (see Theorem 9.10).

Remark 9.4 As mentioned earlier, one can consider an element $g \in G$ as the induced inner automorphism of G. Conversely, an automorphism α can be interpreted as conjugation in the group $\mathbf{Z} \ltimes_\alpha G$, so that we will freely switch between the two terminologies.

9.2 The structure of minimising subgroups

9.2.1 First properties of the scale function

In this section we establish some elementary properties of the scale function, using hardly more than its definition. The proofs are mainly taken from [4].

Recall that if μ is a left-invariant Haar measure on G, then for every $g \in G$ there exists a unique positive real number $\Delta(g)$ such that $\mu(Ag^{-1}) = \Delta(g)\mu(A)$ for every Borel subset A of G. The function Δ is called the modular function on G, and is a continuous group homomorphism from G to the multiplicative group of positive real numbers. There is an easy way to express Δ using a compact open subgroup of G.

Lemma 9.5 *For all $g \in G$ and for all $U \in \mathscr{B}(G)$, the modular function is given by*

$$\Delta(g) = \frac{|U : U \cap g^{-1}Ug|}{|U : U \cap gUg^{-1}|}.$$

Proof   The proof is a simple computation:

$$|U : U \cap g^{-1}Ug| = |gUg^{-1} : gUg^{-1} \cap U|$$

$$= \frac{\mu(gUg^{-1})}{\mu(gUg^{-1} \cap U)}$$

$$= \frac{\mu(U)\Delta(g)}{\mu(gUg^{-1} \cap U)}$$

$$= |U : U \cap gUg^{-1}| \cdot \Delta(g).$$

□

Next we will use this representation of the modular function to connect it to the scale.

Proposition 9.6   *The relation* $\dfrac{s(g)}{s(g^{-1})} = \Delta(g)$ *holds for every* $g \in G$.

Proof   Let $g \in G$ and $U_1, U_2 \in \mathscr{B}(G)$ be such that $s(g) = |U_1 : U_1 \cap g^{-1}U_1g|$ and $s(g^{-1}) = |U_2 : U_2 \cap gU_2g^{-1}|$. Since U_1 and U_2 minimise these indices, we have

$$\frac{s(g)}{s(g^{-1})} \geq \frac{|U_1 : U_1 \cap g^{-1}U_1g|}{|U_1 : U_1 \cap gU_1g^{-1}|} = \Delta(g) = \frac{|U_2 : U_2 \cap g^{-1}U_2g|}{|U_2 : U_2 \cap gU_2g^{-1}|} \geq \frac{s(g)}{s(g^{-1})},$$

using Lemma 9.5 to put the modular function in the middle of the inequality.

□

Furthermore the proof of Proposition 9.6 shows (keeping notation) that U_1 and U_2 are minimising for *both* g and g^{-1}. So the next corollary follows directly.

Corollary 9.7   *For* $g \in G$ *and* $U \in \mathscr{B}(G)$ *we have* $s(g) = |gUg^{-1} : gUg^{-1} \cap U|$ *if and only if* $s(g^{-1}) = |g^{-1}Ug : g^{-1}Ug \cap U|$.

The property that g and g^{-1} have the same minimising subgroups can be used to prove the following property of the scale function.

Corollary 9.8   *An element* $g \in G$ *normalises some* $U \in \mathscr{B}(G)$ *if and only if* $s(g) = 1 = s(g^{-1})$.

Proof   The "only if" direction holds due to the definition of the scale function. For the "if" direction we use Corollary 9.7 to see that there exists $U \in \mathscr{B}(G)$ such that $|U : U \cap g^{-1}Ug| = 1 = |U : U \cap gUg^{-1}|$. So $g^{-1}Ug$ and gUg^{-1} both contain U and hence are equal to U.

□

Example 9.9 Let $G = \mathrm{Aut}(T_d)$ be the automorphism group of a regular tree T_d of degree $d \geq 3$. Recall that an element $\varphi \in \mathrm{Aut}(T_d)$ either stabilises a point or an edge, in which case it is called elliptic, or there exists a unique axis along which φ is a translation, and we say that φ is hyperbolic. One can easily see that φ normalises a compact open subgroup of $\mathrm{Aut}(T_d)$ if and only if φ is elliptic, in which case we have $s(\varphi) = s(\varphi^{-1}) = 1$. The scale function of a hyperbolic element will be computed in Section 9.3.

9.2.2 Tidiness witnesses the scale

Given $g \in G$, the properties shared by all the compact open subgroups at which the minimum is attained in the definition of the scale of g are exactly the tidiness criteria.

Theorem 9.10 (Theorem 3.1 in [10]) *Let U be a compact open subgroup of G and $g \in G$. Then U is minimising for g if and only if U is tidy for g.*

On the proof We first assume that the forward implication is proved, and explain how to deduce the converse implication. As explained in [7], which says that the quantity $|V : V \cap g^{-1}Vg|$ does not depend on the choice of V as soon as V is tidy for α. Now choose a minimising subgroup U, and a tidy subgroup V. According to the forward implication of the statement the subgroup U is also tidy, and by the previous lemma one has $|V : V \cap g^{-1}Vg| = |U : U \cap g^{-1}Ug| = s(g)$. Therefore V is minimising as well, which proves the claim.

The forward direction requires more work. This can be found in [10, Theorem 3.1] or [7]. $\square$

9.2.3 Subgroups being tidy above

In this section we give several characterisations of the notion of being tidy above, and explain how to construct subgroups tidy above for a given automorphism.

Proposition 9.11 *For every $\alpha \in \mathrm{Aut}(G)$ and every $U \in \mathcal{B}(G)$, the following statements are equivalent:*

(i) $U = U_+ U_-$,
(ii) U is tidy above for α,
(iii) $U = U_+ (U \cap \alpha^{-1}(U))$.

Proof As observed earlier, the map
$$\chi \colon \alpha(U_+)/U_+ \to \alpha(U)/\alpha(U) \cap U, \quad xU_+ \mapsto x(\alpha(U) \cap U)$$

is well defined and injective. To see that (i) implies (ii) we only have to show surjectivity of χ. This follows immediately from

$$\alpha(U)(\alpha(U) \cap U) = \alpha(U_+)\alpha(U_-)(\alpha(U) \cap U) \subseteq \alpha(U_+)(\alpha(U) \cap U)$$

and the disjointness of the cosets.

Now we show that (ii) forces (iii). Since χ is injective and by assumption the domain and the codomain have the same finite cardinality, χ has to be surjective. But this yields already

$$\alpha(U) \subseteq \alpha(U)(\alpha(U) \cap U) = \alpha(U_+)(\alpha(U) \cap U),$$

and by applying α^{-1} on both sides we get the claim.

In order to prove that (iii) implies (i) we calculate by using the hypothesis

$$\alpha^{-1}(U) \cap U = \alpha^{-1}\big(U_+(\alpha^{-1}(U) \cap U)\big) \cap U_+(\alpha^{-1}(U) \cap U).$$

The last term is a set of the form

$$L_1(M_1 \cap N_1) \cap L_2(M_2 \cap N_2), \quad L_i \subseteq N_i, \ M_2 = N_1, \ L_1 \subseteq L_2.$$

One can easily check that this set can be written as $L_1(M_1 \cap M_2 \cap N_2)$. Using this we obtain the following expression for U:

$$U = U_+(\alpha^{-1}(U) \cap U) = U_+\alpha^{-1}(U_+) \underbrace{\big(\alpha^{-2}(U) \cap \alpha^{-1}(U) \cap U\big)}_{\alpha^{-1}(\alpha^{-1}(U) \cap U) \cap (\alpha^{-1}(U) \cap U)}$$

and by induction $U = U_+ \bigcap_{n=0}^{k} \alpha^{-n}(U)$ for every $k \geq 1$.

The conclusion then follows from a compactness argument, expressing the idea that $\bigcap_{n=0}^{k} \alpha^{-n}(U)$ tends to U_- when k goes to infinity (see [8, Lemma 1] or [7]). □

The proof of the following proposition yields an explicit procedure to exhibit subgroups that are tidy above for a given automorphism.

Proposition 9.12 (Existence of subgroups being tidy above) *For every $\alpha \in$ Aut(G), there is a compact open subgroup that is tidy above for α.*

Proof The idea is to start from an arbitrary compact open subgroup of G and to consider $U_n = \bigcap_{i=0}^{n} \alpha^i(U)$ for $n \geq 0$. By construction there is an embedding

$$\chi_n \colon \alpha(U_n)/\alpha(U_n) \cap U_n \hookrightarrow \alpha(U_{n-1})/\alpha(U_{n-1}) \cap U_{n-1}.$$

Hence these quotients form a chain with decreasing cardinalities. The compact open property of $U_0 = U$ guarantees that U_0/U_1 is finite and therefore the chain has to become stationary. So there exists $N \in \mathbf{N}$ such that

$$|\alpha(U_j)/U_{j+1}| = |\alpha(U_N)/U_{N+1}|$$

for every $j \geq N$. We will show that $V = U_N$ is tidy above. By the choice of N the inclusions $\chi_{j,N} \colon \alpha(U_j)/\alpha(U_j) \cap U_j \hookrightarrow \alpha(U_N)/\alpha(U_N) \cap U_N$ are even bijections for all $j \geq N$. So, by surjectivity of the map $\chi_{j,N}$, we obtain

$$
\begin{aligned}
\alpha(V) &\subseteq \alpha(U_N)U_{N+1} \\
&= \alpha(U_j)U_{N+1} \\
&= \alpha(U_j)(U_N \cap \alpha^{N+1}(U)) \\
&= \alpha(U_j)(V \cap \alpha(V)).
\end{aligned}
$$

A compactness argument similar to the one of the end of the proof of Proposition 9.11 yields $\alpha(V) = \alpha(V_+)(V \cap \alpha(V))$. Applying α^{-1} on both sides, we obtain that V is tidy above for α, which was the claim. $\qquad\square$

An element $g \in G$ is called *periodic* if $\overline{\langle g \rangle}$ is compact. We denote the set of all periodic elements of G by $P(G)$. For example the reader can check as an exercise that $\varphi \in \mathrm{Aut}(T_d)$ is periodic if and only if φ is elliptic.

The following theorem was proved by Willis using tidy subgroups and properties of the scale function such as continuity or Corollary 9.8 (see [9, Theorem 2]). Following [7], we present here a proof using only tidiness above.

Theorem 9.13 *The set $P(G)$ is closed in G.*

We will need the following easy lemma.

Lemma 9.14 *If U is tidy above for $g \in G$, then we have $(UgU)^n = Ug^nU$ and $(Ug^{-1}U)^n = Ug^{-n}U$ for all $n \geq 1$.*

Proof It is sufficient to show the first equation, since g and g^{-1} have the same tidy above subgroups. We prove this equation by induction on n. For $n = 1$ there is nothing to prove. For $n+1$ we get $(UgU)^{n+1} = (UgU)^n UgU = (Ug^nU)gU$ by induction hypothesis. The last term is equal to $Ug^nU_-U_+gU$ since U is tidy above for g. Using the definition of U_+ and U_- we see that $Ug^nU_- = Ug^n$ and $U_+gU = gU$. Altogether we conclude that $(UgU)^{n+1} = Ug^nU_-U_+gU = Ug^ngU = Ug^{n+1}U$, which completes the induction step. $\qquad\square$

Proof of Theorem 9.13 We give the proof under the additional assumption that G is second countable, which allows us to use sequences to check that $P(G)$ is closed.

Let $(g_i)_{i \geq 0}$ be a sequence in $P(G)$ with limit $g \in G$. We need to prove that g is contained in $P(G)$. Let $U \in \mathcal{B}(G)$ be tidy above for g, and fix $i \geq 0$ such that $g_i \in UgU$ (which is an open neighbourhood of g). Then there are $u, v \in U$ with $g_i = ugv$.

Now we want to prove the existence of an integer $n \geq 0$ such that $g_i^n \in U$. For this consider $(g_i^j)_{j \geq 0}$ which is a sequence in the compact set $\overline{\langle g_i \rangle}$ and thus has a convergent subsequence. Let $(g_i^{j_k})_{k \geq 0}$ be this convergent subsequence and h its limit point. Since $h \in \overline{\langle g_i \rangle}$, there is an integer $m \in \mathbf{N}$ such that $h \in g_i^m U$, which is an open neighbourhood of g_i^m. So we can build a new sequence $(g_i^{j_k - m})_{k \geq 0}$ converging to $g_i^{-m} h \in U$ which gives us the existence of the desired n with $g_i^n = (ugv)^n \in U$.

Due to Lemma 9.14, there are $u', v' \in U$ such that $u' g^n v' = (ugv)^n \in U$, so already g^n has to be contained in U. Therefore $\overline{\langle g^n \rangle}$ is compact as a closed subset of the compact set U, and $\overline{\langle g \rangle}$ is compact, since it has $\overline{\langle g^n \rangle}$ as compact subgroup of finite index. Hence g is periodic. $\qquad \square$

In [9], Willis mentions the following application of this result. In 1981 Hofmann asked if, for a general group G such that $P(G)$ is dense in G, already $P(G) = G$ holds. In general the answer is no. A counterexample is the group E of motions of the Euclidean plane, for which one can check that $P(E)$ consists of all rotations and reflections, and is a proper dense subgroup of E. From Theorem 9.13 we can deduce a positive answer to Hofmann's question for totally disconnected locally compact groups.

9.2.4 Construction of tidy subgroups

In this section we explain how to construct tidy subgroups. Roughly the idea of the procedure is, given a subgroup that is tidy above for $g \in G$, to add a g-invariant compact subgroup.

For $g \in G$ and $U \in \mathscr{B}(G)$, we let

$$\mathscr{L}_U = \left\{ x \in G \ : \ x \in g^n U g^{-n} \text{ for all but finitely many } n \in \mathbf{Z} \right\}.$$

The following proof appears in [7].

Lemma 9.15 *The subset $\mathscr{L}_U$ is a subgroup normalised by g that is relatively compact.*

Proof The verification of the fact that $\mathscr{L}_U$ is a subgroup normalised by g is easy, and we leave it to the reader. We prove that $\mathscr{L}_U$ is contained in a compact subset of G.

Since $gU_+ g^{-1}$ is compact and U is open, the intersection $gU_+ g^{-1} \cap U$ has finite index in $gU_+ g^{-1}$. But this intersection is equal to U_+ by definition, and it follows that $U_+ \cap \mathscr{L}_U$ has finite index in $gU_+ g^{-1} \cap \mathscr{L}_U$. We let X be a finite set of left coset representatives, so that

$$gU_+ g^{-1} \cap \mathscr{L}_U = X (U_+ \cap \mathscr{L}_U).$$

We claim that for every $n \geq 1$, one has

$$g^n U_+ g^{-n} \cap \mathscr{L}_U = \left(\prod_{i=n-1}^{0} g^i X g^{-i} \right) U_+ \cap \mathscr{L}_U. \tag{9.1}$$

We argue by induction. The case $n = 1$ is nothing but the definition of the set X. Assume that (9.1) holds for some $n \geq 1$. Since $\mathscr{L}_U$ is normalised by g, we have

$$g^{n+1} U_+ g^{-(n+1)} \cap \mathscr{L}_U = g \left(g^n U_+ g^{-n} \cap \mathscr{L}_U \right) g^{-1}.$$

Using the induction hypothesis we obtain

$$g^{n+1} U_+ g^{-(n+1)} \cap \mathscr{L}_U = g \left(\prod_{i=n-1}^{0} g^i X g^{-i} \right) (U_+ \cap \mathscr{L}_U) g^{-1}$$

$$= \left(\prod_{i=n-1}^{0} g^{i+1} X g^{-(i+1)} \right) g (U_+ \cap \mathscr{L}_U) g^{-1}$$

$$= \left(\prod_{i=n}^{1} g^i X g^{-i} \right) X (U_+ \cap \mathscr{L}_U),$$

where the last equality follows from the definition of X. So we obtain that

$$g^{n+1} U_+ g^{-(n+1)} \cap \mathscr{L}_U = \left(\prod_{i=n}^{0} g^i X g^{-i} \right) U_+ \cap \mathscr{L}_U,$$

and the proof of the induction step is complete.

Since X is finite, there exists some integer $n_0 \geq 1$ such that X lies inside $g^{-n_0} U_- g^{n_0}$. It follows that for every $n \geq n_0$, we have

$$g^n X g^{-n} \subset g^{n-n_0} U_- g^{-(n-n_0)} \subset U_- \subset U.$$

Using (9.1) we obtain that for every $n \geq n_0$,

$$g^n U_+ g^{-n} \cap \mathscr{L}_U \subset U \left(\prod_{i=n_0-1}^{0} g^i X g^{-i} \right) U_+ \cap \mathscr{L}_U \subset U \left(\prod_{i=n_0-1}^{0} g^i X g^{-i} \right) U =: K.$$

We obtain that $\mathscr{L}_U$, which is the increasing union for $n \geq n_0$ of $g^n U_+ g^{-n} \cap \mathscr{L}_U$, is contained in K. The latter being compact since X is finite and U is compact, this finishes the proof. $\qquad\square$

For $g \in G$ and $U \in \mathscr{B}(G)$, we denote by L_U the closure of $\mathscr{L}_U$ in G. The reason why the definition of $\mathscr{L}_U$ is relevant in this context comes from the following result.

Proposition 9.16 *Let $g \in G$ and $U \in \mathscr{B}(G)$ being tidy above for g. Then U is tidy for g if and only if $L_U \leq U$.*

Proof See for instance [7]. □

The following gives a recipe to construct tidy subgroups.

Proposition 9.17 *For every $g \in G$, there exists $U \in \mathscr{B}(G)$ that is tidy for g.*

Proof The strategy consists essentially in two steps.

Step 1. Given a compact open subgroup U of G, there exists $n \geq 0$ such that $\bigcap_{i=0}^{n} g^{i} U g^{-i}$ is tidy above for g. The existence of such an integer has been shown in the proof of Proposition 9.12.

Step 2. Assume we have a compact open subgroup U tidy above for g. Let $U' = \bigcap_{x \in L_U} x U x^{-1}$. Since L_U is compact according to Lemma 9.15, it follows that $U' = \bigcap_{i=1}^{k} x_i U x_i^{-1}$ for some elements $x_1, \ldots, x_k \in L_U$, and therefore U' is open in G. Since by definition U' is normalised by L_U, it follows that $U' L_U$ is a compact open subgroup of G. Now going back to Step 1, we can find some integer $n \geq 0$ such that $U'' = \bigcap_{i=0}^{n} g^{i} U' L_U g^{-i}$ is tidy above for g. Since L_U is normalised by g, clearly $L_U \subset U''$. We refer the reader to [7] or [10, Lemmas 3.3–3.8] to see that this property implies that U'' is also tidy below for g. □

9.3 Examples

The goal of this section is to illustrate with several examples the notions and results from the previous section.

9.3.1 Infinite direct product of a finite group

Let F be a non-trivial finite group and $G = F^{\mathbf{Z}}$ endowed with the product topology. We define $\alpha \in \mathrm{Aut}(G)$ by $\alpha((f_n)) = (f_{n+1})$. Since G is compact, it is itself a tidy subgroup for α. For every subset $I \subseteq \mathbf{Z}$, define

$$E_I = \{(f_n) \in G : f_n = e \text{ for every } n \in I\}.$$

Let us consider the compact open subgroup $U = E_{[-5,3]}$. By definition we have $U_- = E_{[-5,\infty[}$ and $U_+ = E_{]-\infty,3]}$. We observe that

$$U \subseteq E_{]-\infty,3]} E_{[-5,\infty[} = U_+ U_- \subseteq U$$

and therefore U is tidy above for α. See Figure 9.1.

More generally, if I is a finite non-empty subset of $\mathbf{Z}$ whose minimum and

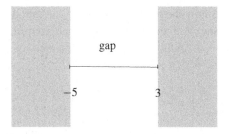

Figure 9.1 This figure illustrates the subgroup $U = E_{[-5,3]}$ of $F^{\mathbf{Z}}$. The fact that U is tidy above corresponds to the existence of only one "gap".

maximum are denoted respectively by a and b and $U = E_I$, then we observe that $U_- = E_{[a,\infty[}$ and $U_+ = E_{]-\infty,b]}$, hence $U_+U_- = E_{[a,b]}$. Therefore U is tidy above for α if and only if $I = [a,b]$. Following the strategy used in the proof of Proposition 9.12, we see that the intersection $\bigcap_{i=0}^{n} \alpha^i(E_I)$ will indeed be equal to some $E_{[a',b]}$ for n large enough (see Figure 9.2).

Figure 9.2 A compact open subgroup of $F^{\mathbf{Z}}$ which is not tidy above for α. The successive intersections $\bigcap_{i=0}^{n} \alpha^i(E_I)$ will gradually exhibit fewer and fewer "gaps".

Furthermore we observe that U_{++} (resp. U_{--}) is the subgroup of sequences that are trivial below (resp. above) some index, which is a proper dense subgroup, hence no E_I is tidy below for α. We can also see that $\mathscr{L}_U$ is the dense subgroup of sequences with finite support, which is another reason why E_I cannot be a tidy subgroup by Proposition 9.16. Moreover, we observe that the tidy subgroup given by the recipe of Proposition 9.17 is the whole group G.

9.3.2 A p-adic Lie group

We let p be a prime number, $G = \mathrm{GL}_2(\mathbf{Q}_p)$ be the group of invertible 2×2 matrices over the field of p-adic numbers and $g \in G$ be the diagonal matrix

$g = \text{diag}(p, 1)$. We follow the recipe to construct a compact open subgroup of G tidy for g. Let us start with the compact open subgroup $U = \text{SL}_2(\mathbf{Z}_p)$. Since conjugation by g^n, $n \in \mathbf{Z}$, multiplies the upper right entry of a matrix by p^n and the lower left entry by p^{-n}, it follows that

$$U_+ = \begin{pmatrix} * & 0 \\ * & * \end{pmatrix} \cap \text{SL}_2(\mathbf{Z}_p) \text{ and } U_- = \begin{pmatrix} * & * \\ 0 & * \end{pmatrix} \cap \text{SL}_2(\mathbf{Z}_p).$$

Now we claim that $U_+ U_- \subsetneq U$. Indeed, argue by contradiction and assume that $U_+ U_- = U$. Using the natural morphism $\text{SL}_2(\mathbf{Z}_p) \twoheadrightarrow \text{SL}_2(\mathbf{F}_p)$, we obtain that a similar equality holds in the finite group $\text{SL}_2(\mathbf{F}_p)$, where $\mathbf{F}_p$ is the field with p elements. But this is impossible for counting reasons, since the set of products xy in $\text{SL}_2(\mathbf{F}_p)$ with x lower triangular and y upper triangular has cardinality $(p-1)p^2$, whereas the group $\text{SL}_2(\mathbf{F}_p)$ has cardinality $(p-1)p(p+1)$.

So it follows that $U = \text{SL}_2(\mathbf{Z}_p)$ is not tidy above for g. Following the construction on the proof of Proposition 9.12, we let

$$U' = U \cap g U g^{-1} = \begin{pmatrix} * & p\mathbf{Z}_p \\ * & * \end{pmatrix} \cap \text{SL}_2(\mathbf{Z}_p).$$

Similarly U'_+ and U'_- are the sets of matrices of determinant one of the form

$$U'_+ = \begin{pmatrix} \mathbf{Z}_p^\times & 0 \\ \mathbf{Z}_p & \mathbf{Z}_p^\times \end{pmatrix} \text{ and } U'_- = \begin{pmatrix} \mathbf{Z}_p^\times & p\mathbf{Z}_p \\ 0 & \mathbf{Z}_p^\times \end{pmatrix}.$$

Now observe that the upper left entry of any element of U' must be a unit in $\mathbf{Z}_p$, and a trivial computation shows that

$$\begin{pmatrix} a & pb \\ c & d \end{pmatrix} = \begin{pmatrix} a & 0 \\ c & a^{-1} \end{pmatrix} \begin{pmatrix} 1 & pba^{-1} \\ 0 & 1 \end{pmatrix},$$

so one has $U' = U'_+ U'_-$ and U' is tidy above for g. Moreover

$$U'_{++} = \begin{pmatrix} \mathbf{Z}_p^\times & 0 \\ \mathbf{Q}_p & \mathbf{Z}_p^\times \end{pmatrix} \text{ and } U'_{--} = \begin{pmatrix} \mathbf{Z}_p^\times & \mathbf{Q}_p \\ 0 & \mathbf{Z}_p^\times \end{pmatrix}$$

are closed in G, so U' is also tidy below for g. Therefore U' is minimising for g, and it follows that the scale of g is the index of U'_+ in gU'_+g^{-1}, which is equal to p.

9.3.3 The automorphism group of a regular tree

We let $G = \text{Aut}(T_d)$ be the automorphism group of the regular tree T_d of degree $d \geq 3$, and we compute the scale function and describe tidy subgroups of any element of G (see also [8, Section 3]).

If $\varphi \in G$ is elliptic then φ normalises either an edge-stabiliser or a vertex-stabiliser. Such a compact open subgroup is necessarily tidy for φ, and φ has scale one.

Now assume that φ is a hyperbolic element whose axis is defined by the sequence of vertices (v_n), $n \in \mathbf{Z}$, indexed so that φ translates in the positive direction. We denote by $\xi^-, \xi^+ \in \partial T_d$ the repelling and attracting endpoints of φ. Let v be a vertex of T_d, and let $U \in \mathscr{B}(G)$ be the stabiliser of v. Then U_+ (resp. U_-) is the subgroup of elements fixing pointwise the set of vertices $\varphi^n(v)$ for $n \geq 0$ (resp. $n \leq 0$). In particular any element of U_+U_- fixes pointwise the unique geodesic between v and the axis of φ. It follows that if v does not lie on the axis of φ then U cannot be tidy above for φ. When v lies on the axis of φ, a similar argument shows that an element of U_+U_- cannot send the edge emanating from v and pointing toward ξ^- to the one pointing toward ξ^+, so again $U_+U_- \subsetneq U$.

Now let U be the stabiliser of v_0 for example, and let $U' = U \cap \varphi U \varphi^{-1}$ be the stabiliser of the geodesic between v_0 and v_ℓ, where $\ell \geq 1$ is the translation length of φ. Note that any element of U' fixes the two half-trees of T_d obtained by cutting the edge (v_0, v_1). Again U'_+ (resp. U'_-) is the pointwise stabiliser of the geodesic ray $(\varphi^n v_0)_{n \geq 0}$ (resp. $(\varphi^n v_\ell)_{n \leq 0}$), and an easy verification shows that $U' = U'_+U'_-$. Moreover U'_{++} (resp. U'_{--}) is the subgroup of the stabiliser in G of ξ^+ (resp. ξ^-) consisting of elliptic isometries, and therefore is closed. So U' is tidy for g.

A direct computation shows that the stabiliser of $v_{2\ell}$ has index $(d-1)^\ell$ in U', so it follows that the scale of φ is equal to $(d-1)^\ell$.

9.4 Ergodic automorphisms of locally compact groups

We conclude with an application of tidy subgroups to ergodic theory, due to Previts and Wu [5]. In 1955 Halmos asked the following question (see [2, p. 29]): *Can an automorphism of a locally compact but non-compact group be an ergodic measure-preserving transformation?* Recall that a transformation T is called **ergodic** relatively to a measure μ if for any measurable set E such that $\mu(T^{-1}E \triangle E) = 0$, one has $\mu(E) = 0$ or $\mu(E^c) = 0$. Here, ergodicity has to be understood with respect to a left-invariant Haar measure on the group.

The answer to Halmos's question is negative, as it was progressively proved by Juzvinskiĭ, Rajagopalan, Kaufman, Wu and Aoki in several works ranging from 1965 to 1985. See the introduction of [5] for more details on this topic.

We give here only a part of this solution, namely the result of Aoki for the totally disconnected case, following the proof given in [5].

Theorem 9.18 ([1, Theorem 1]) *A totally disconnected locally compact group admitting an ergodic automorphism must be compact.*

Conversely, a compact group may have ergodic automorphisms: consider the Bernoulli shift σ on the profinite group $(\mathbf{Z}/2\mathbf{Z})^{\mathbf{Z}}$, defined by $\sigma((g_n)) = (g_{n-1})$.

It is clear that a non-trivial discrete group cannot have an ergodic automorphism, so we may assume the group under consideration to be non-discrete.

Before going to the proof of Theorem 9.18, let us mention that, by a result of Rajagopalan ([6, Theorem 1]), a continuous ergodic automorphism of a locally compact group is automatically bicontinuous.

9.4.1 Reduction of the problem

We will first gradually reduce the problem to a more tractable case. More precisely, we will show that we may assume the group to be second countable and the automorphism to have a dense orbit.

Lemma 9.19 *If a locally compact group G admits an ergodic automorphism f, then G is σ-compact.*

Proof Indeed, let V be a symmetric compact neighbourhood of the identity. The subgroup $H = \langle V \rangle = \bigcup_n V^n$ is obviously open and σ-compact. Hence so is the subgroup H' generated by the f-translates of H. But H' is clearly f-invariant, hence $H' = G$ by ergodicity. $\square$

Lemma 9.20 *To prove Theorem 9.18, we may assume that the group is separable and metrisable (in particular, second countable).*

Proof By the previous lemma, the group G is σ-compact. Therefore, by the Kakutani–Kodaira theorem (see [3, Theorem 8.7]), there exists a normal subgroup K such that the quotient G/K is separable and metrisable. Let $K_n = \bigcap_{-n}^{n} f^j(K)$. The quotient G/K_n embeds into the product

$$\prod_{j=-n}^{n} G/f^j(K),$$

hence is also separable and metrisable. Therefore, so is $G' = G/K_\infty$ (where $K_\infty = \bigcap_{-\infty}^{\infty} f^j(K)$), which is the inverse limit of the G/K_n. Obviously, the automorphism f descends to G' and is still ergodic; moreover, G is compact if and only if G' is so (since K_∞ is compact). $\square$

Lemma 9.21   *If f is an ergodic automorphism of a locally compact second countable group, then f has a dense orbit.*

Proof   Let $\{O_n\}$ be a countable basis of open sets. Suppose by contradiction that no point of G has a dense orbit. Then for any $x \in G$, there is an integer n_x such that x does not belong to the orbit of O_{n_x}. The latter is an f-invariant set containing an open subset, hence its complement has measure zero, by ergodicity. Therefore G could be written as the countable union of the sets $G \setminus \bigcup_j f^j(O_n)$, hence would have measure zero, which is absurd.   □

9.4.2 Tidy subgroups come into play

We will need the following result about tidy subgroups.

Proposition 9.22   *Let G be a totally disconnected locally compact group. If U is a tidy subgroup for $\alpha \in \mathrm{Aut}(G)$, then the set $U^\star = \bigcup_{i \geqslant 0} \alpha^i(U)$ is (open and) closed.*

Proof   See [8, Proposition 1].   □

We finally need the following easy lemma.

Lemma 9.23   *Let X be a locally compact non-discrete space and f an automorphism with a dense orbit. If $A \subseteq X$ is a non-empty open and closed subset such that $f(A) \subseteq A$, then $A = X$.*

Proof   Let x be a point in A with dense orbit, which exists because A is open. By density of the orbit and non-discreteness of the space, there is a strictly monotone sequence of integers n_i such that x is the limit of $f^{n_i}(x)$. We may assume all the n_i to be of the same sign, say positive (the negative case being proved similarly). Let $k \in \mathbf{N}$. As $f(A) \subseteq A$, $f^k(x) \in A$. Moreover, $f^{-k}(x)$ is the limit of the points $f^{n_i - k}(x)$, which are in A for i big enough. Hence $f^{-k}(x)$ is also in A, as the latter is closed. Therefore A is a closed set containing the orbit of x, which is dense, hence $A = X$.   □

Proof of Theorem 9.18   Let U be a tidy subgroup for the automorphism f. By Proposition 9.22, $U^\star = \bigcup_{i \geqslant 0} f^i(U)$ is closed and open. By Lemma 9.23, $U^\star = G$. Therefore, by compactness of U, $f^{-1}(U)$ must lie inside $U_m = \bigcup_{i \geqslant 0}^m f^i(U)$ for some m. Thus we have

$$U_m \subseteq U_{m+1} = U \cup f(U_m) \subseteq f(U_m)$$

and Lemma 9.23 again implies that $G = U_m$, hence G is compact.   □

References for this chapter

[1] N. Aoki, *Dense orbits of automorphisms and compactness of groups*, Topology Appl. **20** (1985), no. 1, 1–15.

[2] P. Halmos, *Lectures on ergodic theory*, Chelsea Publishing Co., New York, 1960.

[3] E. Hewitt and K. Ross, *Abstract harmonic analysis. Vol. I*, second ed., Grundlehren der Mathematischen Wissenschaften, vol. 115, Springer-Verlag, Berlin-New York, 1979.

[4] R. Möller, *Structure theory of totally disconnected locally compact groups via graphs and permutations*, Canad. J. Math. **54** (2002), no. 4, 795–827.

[5] W. Previts and T. Wu, *Dense orbits and compactness of groups*, Bull. Austral. Math. Soc. **68** (2003), no. 1, 155–159.

[6] M. Rajagopalan, *Ergodic properties of automorphisms of a locally compact group*, Proc. Amer. Math. Soc. **17** (1966), 372–376.

[7] P. Wesolek, *A course on totally disconnected locally compact Polish groups*, unpublished notes (2014) available at
http://people.math.binghamton.edu/wesolek/.

[8] G. Willis, *The structure of totally disconnected, locally compact groups*, Math. Ann. **300** (1994), no. 2, 341–363.

[9] G. Willis, *Totally disconnected groups and proofs of conjectures of Hofmann and Mukherjea*, Bull. Austral. Math. Soc. **51** (1995), no. 3, 489–494.

[10] G. Willis, *Further properties of the scale function on a totally disconnected group*, J. Algebra **237** (2001), no. 1, 142–164.

[11] G. Willis, *The scale and tidy subgroups for endomorphisms of totally disconnected locally compact groups*, Math. Ann. **361** (2015), no. 1-2, 403–442.

10

Contraction groups and the scale

Phillip Wesolek

Abstract

We study contraction groups and their relation to the scale function and the nub group. The primary reference is [1]. We also discuss an application and questions around topologically simple groups. We presuppose a familiarity with the theory of tidy subgroups. For the background material on tidy subgroups, we direct the reader to [3] and to Chapter 9 in this volume.

10.1	Introduction	161
10.2	Properties of contraction groups	164
10.3	The nub and the closure of the contraction group	167
10.4	Application: The Tits core	169
	References for this chapter	169

10.1 Introduction

10.1.1 Notations

We use "td" and "lc" for "totally disconnected" and "locally compact", respectively. The results of these notes hold for all tdlc groups; however, for ease of discourse, we tacitly assume our groups are also second countable. An automorphism of a topological group is a group automorphism that is also a homeomorphism; we denote the group of automorphisms of a topological group G by $\mathrm{Aut}(G)$. The set of compact open subgroups of a group G is denoted $\mathscr{U}(G)$. When discussing lc groups, μ always denotes the left invariant Haar measure, and Δ denotes the modular function for μ. We also make use of the notation $\forall^{\infty}$ which reads "for all but finitely many".

10.1.2 Contraction groups

Definition 10.1   Let G be a tdlc group and $\alpha \in \mathrm{Aut}(G)$. The **contraction group** associated to α is defined to be

$$\mathrm{con}(\alpha) := \{x \in G \mid \alpha^n(x) \to 1 \text{ as } n \to \infty\}.$$

The set $\mathrm{con}(\alpha)$ is a subgroup; however, the group $\mathrm{con}(\alpha)$ is not in general closed. An example demonstrating this phenomenon is presented later.

The contraction group is related to another subgroup.

Definition 10.2   Let G be a tdlc group and $\alpha \in \mathrm{Aut}(G)$. The **parabolic subgroup** associated to α is defined to be

$$P_\alpha := \{x \in G \mid \{\alpha^n(x) : n \in \mathbf{N}\} \text{ is relatively compact}\}.$$

It is easy to verify P_α is again a subgroup. Applying some tidy subgroup theory, we argue that P_α is also closed.

For $\alpha \in \mathrm{Aut}(G)$ and $U \in \mathscr{U}(G)$, we put

$$U_+ := \bigcap_{i \geqslant 0} \alpha^i(U) \text{ and } U_- := \bigcap_{i \geqslant 0} \alpha^{-i}(U).$$

We further define

$$U_{++} := \bigcup_{i \geqslant 0} \alpha^i(U_+) \text{ and } U_{--} := \bigcup_{i \geqslant 0} \alpha^{-i}(U_-).$$

A subgroup $U \in \mathscr{U}(G)$ is said to be **tidy** for $\alpha \in \mathscr{U}(G)$ if $U = U_+ U_-$ and both U_{++} and U_{--} are closed. The first condition is called **tidy from above** for α, and the second is called **tidy from below** for α. A compact open subgroup satisfying the first condition is occasionally called **semi-tidy** for α.

Fact 10.3 (see [3])   *If $U \in \mathscr{U}(G)$ is tidy for $\alpha \in \mathrm{Aut}(G)$, then*

(i) $\overline{\mathrm{Tlim}}_{n \to \infty} \alpha^n(U) = U_{++}$. *That is to say, there is $(u_i)_{i \in \mathbf{N}} \subseteq U$ and $(n_i)_{i \in \mathbf{N}} \subseteq \mathbf{N}$ so that $\alpha^{n_i}(u_i) \to w$ if and only if $w \in U_{++}$.*

(ii) *If $n < m$, then $\alpha^n(U) \cap \alpha^m(U) = \bigcap_{i=n}^{m} \alpha^i(U)$.*

Proposition 10.4 (Willis, [4])   *If G is a tdlc group and $\alpha \in \mathrm{Aut}(G)$, then P_α is a closed subgroup.*

Proof Take $U \in \mathscr{U}(G)$ tidy for α and suppose $(x_i)_{i \in \mathbf{N}} \subseteq P_\alpha$ converges to some x. We may find $N \geqslant 0$ such that $x_i^{-1} x_j \in U$ for all $i, j \geqslant N$. Fix $i, j \geqslant N$. Since it is also the case that $x_i^{-1} x_j \in P_\alpha$, there is some sequence $(n_k)_{k \in \mathbf{N}}$ of natural numbers and $w \in G$ so that $\alpha^{n_k}(x_i^{-1} x_j) \to w$ as $k \to \infty$. Fact 10.3 implies $w \in U_{++}$; say $m \geqslant 0$ is so that $w \in \alpha^m(U_+)$. For all sufficiently large k, it now follows from a second application of Fact 10.3 that

$$x_i^{-1} x_j \in U \cap \alpha^{m-n_k}(U) = \bigcap_{i=m-n_k}^{0} \alpha^i(U).$$

Since the n_k tend to infinity, we conclude that $x_i^{-1} x_j \in U_-$.
Letting $j \to \infty$, we now see that $x \in x_i U_- \subseteq P_\alpha$, and therefore, P_α is closed. $\square$

We require one further definition.

Definition 10.5 The **Levi factor** of α is defined to be $M_\alpha := P_\alpha \cap P_{\alpha^{-1}}$.

10.1.3 Examples

We pause for a moment to give examples.

Example 10.6 Fix F a non-trivial finite group and form $G := F^{\mathbf{Z}}$. Let α be the automorphism defined by $\alpha(f)(n) := f(n-1)$; that is, α is the so-called right shift. Then $P_\alpha = G$, and

$$\mathrm{con}(\alpha) = \{ f \in G \mid \forall^\infty n \leqslant 0 \; f(n) = 1 \}.$$

Observe that $\mathrm{con}(\alpha)$ is *not* closed but is relatively compact. It is also worth noting $s(\alpha^{-1}) = 1$ where s is the scale function.

Our next example requires a basic technique from tdlc group theory.

Definition 10.7 Suppose A is a countable set, $(G_a)_{a \in A}$ is a sequence of tdlc groups and, for each $a \in A$, there is a distinguished $U_a \in \mathscr{U}(G_a)$. The **local direct product** of $(G_a)_{a \in A}$ over $(U_a)_{a \in A}$ is defined to be

$$\bigoplus\nolimits_{a \in A} (G_a, U_a) :=$$

$$\left\{ f : A \to \bigsqcup\nolimits_{a \in A} G_a \mid \forall a \in A, \, f(a) \in G_a \text{ and } \forall^\infty a \in A, \, f(a) \in U_a \right\}.$$

There is a canonical tdlc group topology on $\bigoplus_{a \in A}(G_a, U_a)$ that makes the Cartesian product $\prod_{a \in A} U_a$ a compact open subgroup; see for example, [3]. When we refer to the local direct product, we always assume the product is endowed with the aforementioned topology.

Example 10.8 Fix F a non-trivial finite group and let F_i for $i \in \mathbf{Z}$ list copies of F. For $i \leqslant 0$, let $U_i = F_i$ and for $i > 0$, let $U_i := \{1\}$. We may then form the local direct product

$$G := \bigoplus_{i \in \mathbf{Z}} (F_i, U_i).$$

The group G admits a left shift: define $\alpha : G \to G$ by $\alpha(f)(n) := f(n + 1)$. One now verifies $P_\alpha = G$, $P_{\alpha^{-1}} = \{1\}$, $\mathrm{con}(\alpha) = G$ and $\mathrm{con}(\alpha^{-1}) = \{1\}$. Observe here that $s(\alpha^{-1}) = |F|$ and that $\mathrm{con}(\alpha)$ is not relatively compact.

10.2 Properties of contraction groups

Definition 10.9 Suppose $H \leqslant G$ is a subgroup of a tdlc group G. We say a sequence $(x_i)_{i \in \mathbf{N}} \subseteq G$ **converges to 1 modulo** H if for all $U \in \mathscr{U}(G)$, there is $N \geqslant 0$ so that $x_i \in UH$ for all $i \geqslant N$.

Let G be a tdlc group and $\alpha \in \mathrm{Aut}(G)$. For $U \in \mathscr{U}(G)$, put

$$U_s := \bigcap_{i \in \mathbf{Z}} \alpha^i(U).$$

Lemma 10.10 *Suppose* $x \in G$, $K \leqslant G$ *is compact and* α-*invariant, and the sequence* $(\alpha^n(x))_{n \in \mathbf{N}}$ *converges to* 1 *modulo* K. *For all* $L \in \mathscr{U}(G)$, *there is* $h \in K$ *so that* $\forall^\infty n \, \alpha^n(xh) \in L$. *Further,* $(\alpha^n(xh))_{n \in \mathbf{N}}$ *converges to* 1 *modulo* L_s.

Proof By passing to an open subgroup of L if necessary, we may assume $L \cap K$ is tidy from above for α in K; that is to say, $(L \cap K) = (L \cap K)_-(L \cap K)_+$. See, for example, [3].

Find $U \leqslant_o L$ so that $\alpha(U) \leqslant L$. Since $\alpha^n(x)$ converges to 1 modulo K, there is $N > 0$ so that $\alpha^n(x) \in UK$ for all $n \geqslant N$; fix such an N. We now inductively build a sequence $(h_i)_{i \in \mathbf{N}}$ so that for each i,

(i) $h_i \in K$ and
(ii) $\alpha^{N+j}(xh_i) \in U(L \cap K)$ for all $0 \leqslant j \leqslant i$.

For the base case, there is $k \in K$ so that $\alpha^N(x)k \in U$. The α-invariance of K implies $h_0 := \alpha^{-N}(k) \in K$. The element h_0 therefore satisfies the base case.

Suppose we have built our sequence up to h_k. The induction hypothesis implies

$$\alpha^{N+k+1}(xh_k) \in \alpha(U)\alpha(L \cap K) \subseteq L\alpha((L \cap K)_-)\alpha((L \cap K)_+)$$
$$= L\alpha((L \cap K)_+).$$

There is then $s_{k+1} \in (L \cap K)_+$ so that $\alpha(\alpha^{N+k}(xh_k)s_{k+1}^{-1}) \in L$. Put $h_{k+1} :=$

$h_k \alpha^{-N-k}(s_{k+1}^{-1})$. We now check h_{k+1} has the desired properties. Since $h_k \in K$ and $s_{k+1} \in (L \cap K)_+$, it is immediate that $h_{k+1} \in K$. For each $0 \leqslant j < k+1$, we see that (ii) holds as

$$\alpha^{N+j}(xh_{k+1}) = \alpha^{N+j}(xh_k)\alpha^{-k+j}(s_{k+1}^{-1}) \in U(L \cap K)(L \cap K)_+ = U(L \cap K).$$

For the case $j = k+1$, $\alpha^{N+k+1}(xh_{k+1}) \in L \cap UK \subseteq U(L \cap K)$. The element h_{k+1} therefore behaves as required finishing the inductive construction.

The sequence $(h_i)_{i \in \mathbf{N}}$ is contained in K, so there is a convergent subsequence; say $h_{k_i} \to h$. For each $n > N$, we have $\alpha^{N+n}(xh) = \lim_i \alpha^{N+n}(xh_{k_i})$. By construction, $\forall^\infty i\ \alpha^{N+n}(xh_{k_i}) \in L$, and therefore $\alpha^{N+n}(xh) \in L$. We conclude that $\alpha^{N+n}(xh) \in L$ for all $n \geqslant N$ verifying the first claim of the lemma.

For the second claim, fix $V \leqslant_o L$. We argue that $\forall^\infty n\ \alpha^n(xh) \in VL_s$. Indeed, suppose not, so there is a sequence $\alpha^{n_i}(xh)$ such that each term avoids VL_s. Since L is compact, we may assume $\alpha^{n_i}(xh) \to w \in L$. However, the set of limit points of such sequences is an α-invariant set implying that $w \in L_s$ — an absurdity. □

Proposition 10.11 *If $V \in \mathscr{U}(G)$, then for all $x \in V_{--}$, there is $h \in V_s$ so that $xh \in \mathrm{con}(\alpha)$.*

Proof Let $(W_i)_{i \in \mathbf{N}}$ be a $\subseteq$-decreasing basis at 1 for V_{--} so that $W_0 \supseteq V_-$. Fixing $x \in V_{--}$, we now inductively build a sequence of elements $(y_k)_{k \in \mathbf{N}}$ of V_{--} and natural numbers $(N_k)_{k \in \mathbf{N}}$ so that the following hold:

(1) $y_0 = x$ and $y_{k+1} \in y_k(W_k)_s$;
(2) $\alpha^n(y_k) \in W_k$ for all $n \geqslant N_k$; and
(3) $\alpha^n(y_k)$ converges to 1 modulo $(W_k)_s$.

For the base case, fix $N_0 \geqslant 0$ so that $x \in \alpha^{-N_0}(V_-)$. It is immediate that $\alpha^n(x) \in W_0$ for all $n \geqslant N_0$. We also must check $\alpha^n(x)$ converges to 1 modulo V_s. This is also immediate since $\alpha^n(V_-) \to V_s$ in the Vietoris topology as $n \to \infty$.

Suppose we have built y_k. By condition (3), we infer there is $N_{k+1} > 0$ so that $\alpha^n(y_k) \in W_{k+1}(W_k)_s$ for all $n \geqslant N_{k+1}$. Lemma 10.10 implies there is $h \in (W_k)_s$ so that $\alpha^n(y_k h) \in W_{k+1}$ for all $n \geqslant N_{k+1}$. Putting $y_{k+1} = y_k h$, the induction is completed by Lemma 10.10.

The sequence $(y_k(W_k)_s)_{k \in \mathbf{N}}$ is $\subset$-decreasing with singleton intersection; say $\{y\} = \bigcap y_k(W_k)_s$. For all $n \geqslant N_k$, we see that

$$\alpha^n(y) \in \alpha^n(y_k(W_k)_s) \subseteq W_k,$$

where the last equality follows since $(W_k)_s$ is α-invariant. Therefore, $\alpha^n(y) \to 1$. An easy induction argument further shows that $h := x^{-1}y \in V_s$, and we have proved the proposition. □

Corollary 10.12 *If $V \in \mathcal{U}(G)$, then $V_{--} = \mathrm{con}(\alpha)V_s$.*

We note a further property of V_{--}.

Lemma 10.13 *If $\alpha \in \mathrm{Aut}(G)$ and $V \in \mathcal{U}(G)$, then $\mathrm{con}(\alpha) \trianglelefteq V_{--}$.*

Proof For $x \in \mathrm{con}(\alpha)$, we may find N so that $\alpha^n(x) \in V$ for all $n \geqslant N$. It follows that $\alpha^N(x) \in V_-$, so $\mathrm{con}(\alpha) \leqslant V_{--}$.

We now check that $\mathrm{con}(\alpha)$ is also normal in V_{--}. Take $x \in \mathrm{con}(\alpha)$ and $v \in V_{--}$; say $v \in \alpha^{-N}(V_-)$. The set $K := \overline{\{\alpha^n(v) \mid n \geqslant 0\}}$ is contained in $\alpha^{-N}(V)$ and is therefore compact. Letting $(W_i)_{i \in \mathbf{N}}$ be an $\subseteq$-decreasing basis at 1 for G of compact open subgroups and fixing j, we claim that for all but finitely many indices i,

$$W_i^K := \{kwk^{-1} \mid w \in W_i \text{ and } k \in K\} \subseteq W_j.$$

Indeed, suppose this fails. There is thus a sequence $(k_i w_i k_i^{-1})_{i \in \mathbf{N}}$ avoiding W_j. Since K and W are compact, we may pass to a subsequence so that $k_i \to k$ and $w_i \to w$. However, the W_i form a $\subseteq$-decreasing basis implying that $w = 1$. This is absurd since on the other hand $kwk^{-1} \notin W_j$.

We now consider $\alpha^n(vxv^{-1}) = \alpha^n(v)\alpha^n(x)\alpha(v^{-1})$. Fix W_j from our basis above. Our previous discussion gives that $\forall^\infty i\, W_i^K \subseteq W_j$, so we may find i so that $W_i^K \subseteq W_j$. For all n large enough, it is the case that $\alpha^n(x) \in W_i$, and by definition, $\alpha^n(v) \in K$. Therefore,

$$\alpha^n(vxv^{-1}) \in W_i^K \subseteq W_j$$

for all sufficiently large n. We conclude that $\alpha^n(vxv^{-1}) \to 1$ verifying the proposition. $\quad\square$

We now consider the scale function and its relation with the contraction group. This requires some additional notation. Suppose $\alpha \in \mathrm{Aut}(G)$ and $H \leqslant G$ is a closed α-invariant subgroup of G. We may thus compute the scale function of α either as an automorphism of G or as an automorphism of H. To distinguish these, we write s_G and s_H to denote the scale of α as an automorphism of G and as an automorphism of H, respectively.

Lemma 10.14 *Let G be a tdlc group and $V \in \mathcal{U}(G)$ be tidy for $\alpha \in \mathrm{Aut}(G)$. Then, $s_G(\alpha^{-1}) = s_{V_{--}}(\alpha^{-1})$.*

Proof Using observations from [3], we have $V \cap V_{--} = V_-$, so V_- is a compact open subgroup of V_{--}. The group V_- is tidy from above for α in V_{--} since $(V_-)_- = V_-$. To see V_- is tidy, one verifies that $(V_-)_{--} = V_{--}$ and $(V_-)_{++} = V_-$. From the properties of tidy subgroups, we infer the desired lemma:

$$s_G(\alpha^{-1}) = |\alpha^{-1}(V_-) : V_-| = s_{V_{--}}(\alpha^{-1}). \qquad \Box$$

Theorem 10.15 (Baumgartner, Willis [1]) *Let G be a tdlc group and $\alpha \in$ Aut(G). Then con(α) is relatively compact if and only if $s_G(\alpha^{-1}) = 1$.*

Proof Suppose con(α) is relatively compact and let $V \in \mathscr{U}(G)$ be tidy for α. Since con(α) is relatively compact, Corollary 10.12 implies V_{--} is a product of compact groups and therefore compact, hence $\Delta_{V_{--}}(\alpha) = 1$. The properties of the scale function give that

$$\Delta_{V_{--}}(\alpha) = \frac{s_{V_{--}}(\alpha^{-1})}{s_{V_{--}}(\alpha)},$$

so $s_{V_{--}}(\alpha) = s_{V_{--}}(\alpha^{-1})$. The group V_- is a compact open subgroup of V_{--}, and it follows that $s_{V_{--}}(\alpha) = 1$. Therefore, $s_{V_{--}}(\alpha^{-1}) = 1$. Applying Lemma 10.14, we conclude that $s_G(\alpha^{-1}) = 1$.

Conversely, fix $U \in \mathscr{U}(G)$ tidy for α, so $|\alpha^{-1}(U) : U \cap \alpha^{-1}(U)| = 1$. Therefore, $\alpha^{-1}(U) \leqslant U$, and it follows that $U_{--} \leqslant U$ and, therefore, that U_{--} is compact. Since con$(\alpha) \leqslant U_{--}$, we have proved the converse. $\qquad \Box$

Corollary 10.16 *If $\Delta(\alpha) > 1$ for $\alpha \in$ Aut(G), then con(α) is not relatively compact and, in particular, non-trivial.*

10.3 The nub and the closure of the contraction group

Of course, to study the contraction group as a locally compact topological group, we must pass to the closure. Understanding the structure of the closure requires introducing a new subgroup.

Definition 10.17 Let G be a tdlc group and $\alpha \in$ Aut(G). The **nub** group is defined to be

$$\text{nub}(\alpha) := \bigcap \{V \mid V \text{ is tidy for } \alpha\}.$$

Similar to the proof of Corollary 10.12, one can show

Proposition 10.18 *For G a tdlc group and $\alpha \in$ Aut(G), $M_\alpha \, \text{con}(\alpha) = P_\alpha$.*

Proposition 10.18 allows us to prove a factorization of the closure of the contraction group. Let G be a tdlc group and $\alpha \in \mathscr{U}(G)$.

Lemma 10.19 $\text{nub}(\alpha) = \overline{\text{con}(\alpha)} \cap M_\alpha$.

Proof Take $g \in \mathrm{nub}(\alpha)$. Recalling that V tidy for α implies $\alpha(V)$ is tidy for α, we infer that $\alpha^n(g) \in \mathrm{nub}(\alpha)$ for all $n \in \mathbf{Z}$. The set $\{\alpha^n(g) \mid n \in \mathbf{Z}\}$ is then relatively compact, hence $g \in M_\alpha$. Suppose for contradiction $g \notin \mathrm{con}(\alpha)$. For some $W \in \mathscr{U}(G)$, it is then the case that $gW \cap \overline{\mathrm{con}(\alpha)} = \varnothing$. Hence,

$$g \notin \overline{\left(\mathrm{con}(\alpha)W_s\right)} = \overline{W_{--}};$$

the latter equality follows from Proposition 10.11. Via [1, Corollary 3.25], $\overline{W_{--}} = V_{--}$ where V is the tidy subgroup built from W. This results in an absurdity since $g \in V_{--}$.

Conversely, take $x \in \overline{\mathrm{con}(\alpha)} \cap M_\alpha$ and let W be tidy for α. Since $x \in \overline{\mathrm{con}(\alpha)}$, Corollary 10.12 implies $x \in \overline{W_{--}}$. On the other hand, $x \in M_\alpha$, so $\{\alpha^n(x) \mid n \in \mathbf{Z}\}$ is a compact α-invariant subset of $\overline{W_{--}}$. It follows $\overline{\{\alpha^n(x) \mid n \in \mathbf{Z}\}}$ is contained in W_s; cf. [3]. We conclude that $x \in W$ and therefore that $x \in \mathrm{nub}(\alpha)$. $\square$

Theorem 10.20 (Baumgartner, Willis [1]) $\overline{\mathrm{con}(\alpha)} = \mathrm{con}(\alpha)\,\mathrm{nub}(\alpha)$.

Proof Fix V tidy for α, then

$$
\begin{aligned}
\overline{\mathrm{con}(\alpha)} &= V_{--} \cap \overline{\mathrm{con}(\alpha)} \\
&= (\mathrm{con}(\alpha)V_s) \cap \overline{\mathrm{con}(\alpha)} \\
&\subseteq \mathrm{con}(\alpha)\left(V_s \cap \overline{\mathrm{con}(\alpha)}\right) \\
&\subseteq \mathrm{con}(\alpha)\left(M_\alpha \cap \overline{\mathrm{con}(\alpha)}\right) \\
&= \mathrm{con}(\alpha)\,\mathrm{nub}(\alpha) \\
&\subseteq \overline{\mathrm{con}(\alpha)}.
\end{aligned}
$$

$\square$

As a corollary, we may characterize when the contraction group is closed.

Corollary 10.21 *The contraction group* $\mathrm{con}(\alpha)$ *is closed if and only if the nub* $\mathrm{nub}(\alpha)$ *is the trivial subgroup.*

Proof The reverse direction is immediate from Theorem 10.20. For the forward implication, let W be any compact open subgroup. There is an open subgroup $W' \leq W$ so that W' is tidy from above for α. Furthermore, $\overline{\left(\mathrm{con}(\alpha)W_s'\right)} = \overline{W'_{--}}$, hence W'_{--} is closed. This implies that W'_{++} is also closed; see [4, Lemma 3] or [3]. Therefore, W' is tidy for α. We conclude that α admits arbitrarily small tidy subgroups, whereby $\mathrm{nub}(\alpha) = \{1\}$. $\square$

10.4 Application: The Tits core

Definition 10.22 Let G be a tdlc group. The **Tits core** of G is defined to be

$$G^\dagger := \left\langle \overline{\mathrm{con}(g)} \mid g \in G \right\rangle.$$

One can verify that $Aut^+(\mathscr{T}_n)$ for $\mathscr{T}_n$ the n-regular tree and $G^+(k)$ for G a simple k-algebraic group with k a non-archimedean field are exactly the respective Tits cores. Moreover, some of the theorems from these special cases generalize.

Theorem 10.23 (Caprace, Reid, Willis, [2]) *Let G be a tdlc group and D a dense subgroup. If $G^\dagger$ normalises D, then $G^\dagger \leqslant D$.*

Corollary 10.24 *Let G be a tdlc group. If G is topologically simple, then $G^\dagger$ is abstractly simple.*

The results of P-E. Caprace, C. Reid and G. Willis show the presence of non-trivial contraction groups has strong structural consequences for topologically simple groups. One naturally asks if we always have non-trivial contraction groups.

Question 10.25 Does every compactly generated tdlc group that is topologically simple and non-discrete have at least one non-trivial contraction group?

Recall $g \in G$ is **periodic** if $\langle g \rangle$ is relatively compact; this is the topological analogue of torsion. A positive answer to the following question would imply a negative answer to the previous.

Question 10.26 Is there a compactly generated tdlc group that is topologically simple, non-discrete and periodic? That is to ask, do simple topological Burnside groups exist?

In another direction, the work [2] constructs a new invariant. Recall for a tdlc group G the collection of closed subgroups, denoted $S(G)$, comes with a canonical compact topology. This topological space is often called the **Chabauty space** of G.

Theorem 10.27 (Caprace, Reid, Willis, [2]) *Let G be a tdlc group. The map $G \to S(G)$ defined by $g \mapsto \overline{\mathrm{con}(g)}$ is continuous.*

References for this chapter

[1] U. Baumgartner and G. Willis, *Contraction groups and scales of automorphisms of totally disconnected locally compact groups*, Israel J. Math. **142** (2004), 221–248.

[2] Pierre-Emmanuel Caprace, Colin D. Reid and George A. Willis, *Limits of contraction groups and the Tits core*, J. Lie Theory **24** (2014), no. 4, 957–967.

[3] Phillip Wesolek, *A course on totally disconnected locally compact Polish groups*, unpublished notes (2014) available at
`http://people.math.binghamton.edu/wesolek/`.

[4] George Willis, *The structure of totally disconnected, locally compact groups*, Math. Ann. **300** (1994), no. 2, 341–363.

11

The Bader–Shalom normal subgroup theorem

Światosław Gal

Abstract

We explain the Bader–Shalom normal subgroup theorem in the case of lattices in products of simple groups. This special case is sufficient to describe the various ingredients of the proof.

11.1	Introduction	171
11.2	Amenability and property (T) of Kazhdan	172
11.3	Proof of Theorem 11.1	174
11.4	Properties of Furstenberg–Poisson boundaries	174
11.5	Proof of Theorem 11.3	176
	References for this chapter	177

11.1 Introduction

Recall that a group Γ is **just infinite** if every proper quotient of Γ is finite.

A **lattice** Γ in G is a discrete subgroup such that G/Γ carries G-invariant probability measure. This is automatically the case when the discrete subgroup is cocompact; we then speak of a **uniform** (or cocompact) lattice. A lattice Γ in $G_1 \times G_2$ is **irreducible**. if the projections of Γ to both G_1 and G_2 have dense images. We will define amenability and property (T) later on.

The aim of this article is to sketch the proof of the following result of Uri Bader and Yehuda Shalom [4] generalizing Margulis' Normal Subgroup Theorem (see [6]).

Theorem 11.1 (Normal Subgroup Theorem)   *Let G_1 and G_2 be topologically simple non-discrete compactly generated locally compact groups.*
Let $\Gamma < G_1 \times G_2$ be an irreducible uniform lattice.
Then Γ is just infinite.

As in the pattern established by Margulis, the proof relies basically on the following two results. (We note that both Theorems 11.1 and 11.2 can be stated more generally by suitably relaxing the simplicity assumption on G_i.)

Theorem 11.2 (Property (T) Half [9, Theorem 0.1])   *Let G_1, G_2 and Γ be as in Theorem 11.1.*
Then, for any non-trivial normal subgroup N of Γ, the quotient Γ/N has property (T) of Kazhdan.

Theorem 11.3 (Amenability Half)   *Let G_1 and G_2 be locally compact groups and let $\Gamma < G = G_1 \times G_2$ be an irreducible lattice. Let N be a normal subgroup of Γ.*
Then Γ/N is amenable if (and only if) for both i the quotient $G_i/\overline{\mathrm{pr}_i(N)}$ is amenable.

We shall not discuss the proof of Theorem 11.2. Instead we will concentrate on Theorem 11.3.

11.2 Amenability and property (T) of Kazhdan

In the proof of Theorem 11.1 we will assume that N is a non-trivial normal subgroup of Γ and prove that Γ/N is both amenable and has property (T) of Kazhdan.

Let us first discuss **amenability** for discrete groups. The following theorem characterizes those.

Theorem 11.4   *Let Γ be a countable discrete group. The following properties are equivalent.*

  (i)   *For any finite subset S of Γ and a number $\varepsilon > 0$ there exists $f \in \ell^1\Gamma$ such that for each $s \in S$ one has $\|sf - f\|_1 < \varepsilon\|f\|_1$.*
 (ii)   *For any finite subset S of Γ and a number $\varepsilon > 0$ there exists $f \in \ell^2\Gamma$ such that for each $s \in S$ one has $\|sf - f\|_2 < \varepsilon\|f\|_2$.*
(iii)   *There exists an invariant state on $\mathscr{B}\ell^2\Gamma$.*
 (iv)   *There exists an invariant state on $\ell^\infty\Gamma$.*
  (v)   *The Čech–Stone compactification $\beta\Gamma$ of Γ supports a Γ-invariant probability measure.*

(vi) *Every compact space K on which Γ acts supports a Γ-invariant probability measure.*

(vii) *Every compact metrizable space K on which Γ acts supports a Γ-invariant probability measure.*

Proof ($i \Rightarrow ii$) The Mazur map $Mf(g) = \text{sign}(f(g))|f(g)|^{1/2}$ between unit spheres in $\ell^1\Gamma$ and $\ell^2\Gamma$ is known to be continuous (actually Hölder with exponent two) [7].

($ii \Rightarrow iii$) Let $\{f_i\}_{i=1}^{\infty}$ be a sequence on almost invariant vectors as S exhausts Γ and ε tends to zero. Then any weak-$*$ limit of $T \mapsto \langle f_i|T f_i \rangle$ is an invariant state.

($iii \Rightarrow iv$) Obvious, as $\ell^\infty\Gamma$ is a subalgebra of diagonal operators in $\mathscr{B}\ell^2\Gamma$.

($iv \Rightarrow i$) It is a result of Day [1]. Let m be an invariant mean. Since $\ell^1\Gamma$ is dense in $\ell^\infty\Gamma$ with weak-$*$ topology there exists a sequence of probability measures μ_i converging to m. This shows that if S is a finite set then the weak-$*$ closure

$$\overline{\{(s_*\mu - \mu)_{s \in S} \mid \mu \in \text{Proba}(\Gamma)\}} \in \text{Proba}(\Gamma)^S$$

contains zero. Note that the weak-$*$ topology, when restricted to Proba^S, is the weak topology. Therefore the above set, being convex, is, by Hahn–Banach theorem, also norm-closed, and (i) follows.

($iv \Leftrightarrow v$) Every bounded function f on Γ extends uniquely to a continuous (thus integrable) function F on $\beta\Gamma$ and we can define a measure μ on $\beta\Gamma$ by a state on $\ell^\infty\Gamma$ and vice versa by a formula $m(f) = \int_{\beta\Gamma} F \mu$.

($v \Leftrightarrow vi$) This follows from the fact that any compact Γ-set admits a Γ-equivariant map from $\beta\Gamma$.

($vi \Leftrightarrow vii$) (cf. [6, Chapter IV, Lemma 4.2]) Let K be a compact Γ set without a Γ-invariant probability measure. We need to construct another compact Γ-set K' which is also metrizable without such a measure. For each continuous function f on K define a closed (thus compact) subset Z_f of $\text{Proba}(K)$ consisting of measures μ such that for all $\gamma \in \Gamma$ one has $\int f \mu = \int f \gamma^* \mu$. This family of compact sets has empty intersection. Therefore there exists a finite subfamily $Q \subset C(K)$ such that $\bigcap_{f \in Q} Z_f = \varnothing$.

We have a canonical map $\pi \colon K \to [0,1]^{\Gamma Q}$ which is moreover Γ-equivariant. Thus $K' := \pi(K)$ is a metrizable (as a subset of a Hilbert cube) Γ-set. Notice that the functions from Q and their translates by Γ are pull-backs of coordinate functions. If μ was an invariant measure on K' then any of its lifts would belong to the intersection of Z_f as f runs through Q which is absurd. $\square$

In the proof of Theorem 11.1 we would conclude with (vii) for Γ/N. Then we want to use the implication ($vii \Rightarrow ii$) to show that $\ell^2(\Gamma/N)$ contains a sequence of more and more invariant vectors.

A group Δ has **Kazhdan's property (T)** if for every unitary representation $\mathcal{H}$ of Δ, such that for any finite subset $S \subset \Delta$ and $\varepsilon > 0$ there exists $\xi \in \mathcal{H}$ such that for all $s \in S$ one has $\|s\xi - \xi\| < \varepsilon\|\xi\|$, $\mathcal{H}$ contains a Δ-invariant vector.

We conclude that $\ell^2(\Gamma/N)$ contains an invariant vector, which must be a constant function, therefore the counting measure on Γ/N is finite, so is Γ/N.

11.3 Proof of Theorem 11.1

Let $N \lhd \Gamma$ be a non-trivial normal subgroup.

First, we argue that $\mathrm{pr}_i(N)$ is non-trivial for both i. Suppose indeed that it is trivial, say for $i = 1$. In particular, N projects injectively to a discrete subgroup of G_2, which we still denote by N. Since $\mathrm{pr}_2(\Gamma)$ is dense in G_2 and normalises N, it follows that N is normal in G_2. Since G_2 is simple, this implies $N = G_2$, contradicting that G_2 is non-discrete.

To show that Γ/N is finite we show that it is both amenable and Kazhdan. The first is immediate by Theorem 11.3 and simplicity of G_1 and G_2. The second follows from Theorem 11.2.

11.4 Properties of Furstenberg–Poisson boundaries

In what follows we would consider measures on groups. By an **admissible measure** μ on a topological locally continuous group G we understand a probability measure which is continuous with respect to the Haar measure and is not supported on a proper closed subsemigroup.

In such a context we say that a probability space (X, η) is a G-space if G acts in a *measure class preserving manner*, that is, for any $g \in G$ the measure ν and its image $g_*\nu$ define the same ideal of measure null sets.

We say that (X, η) is a (G, μ)-space if $\mu * \eta = \eta$ where $\mu * \eta$ is the push-forward of $\mu \times \eta$ by the action map $G \times X \to X$.

By a **factor** $\psi \colon (X, \eta) \to (Y, \upsilon)$ we understand a measurable map such that $\psi_*\eta = \upsilon$. We also say that in this situation (Y, υ) is a factor of (X, η).

Given an admissible measure μ on G we define $\mathcal{H}(G, \mu)$ the Banach space (with supremum norm) of bounded harmonic functions on G, that is, functions invariant under the right convolution with μ, i.e.

$$\mathscr{H}(G,\mu) := \left\{ \varphi \mid \varphi * \mu = \varphi \right\},$$

where we recall $(\varphi * \mu)(h) = \int_G \varphi(hg^{-1})d\mu(g)$.

Given a (G,μ)-space (X,ν), one defines the Poisson transform $L^\infty(X) \to \mathscr{H}(G,\mu)$, $f \mapsto \hat{f}$ by the formula $\hat{f}(g) = \int_X f\, dg_*\nu$. It is straightforward to see that the Poisson transform takes only constant values if and only if the measure ν is invariant under the G-action.

Theorem 11.5 ([2])   *For any (G,μ) there exists a G-space (B,ν) such that the Poisson transform is an isometry.*

Such a space is called the **Furstenberg–Poisson boundary**. Obviously, it is unique up to measurable G-isomorphism. We say that (X,η) is a **boundary** of (G,μ) if (X,η) is a factor of the Furstenberg–Poisson boundary.

A modern treatment of theory of boundaries can be found in [3].

Observe that for any (G,μ)-space (X,η) there exists a map $\varphi\colon B \to \mathrm{Proba}(X)$ such that $\int_B f\nu = \eta$. Indeed, we have a commutative diagram

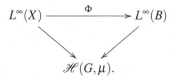

Let $\varphi\colon \mathrm{Proba}(B) \to \mathrm{Proba}(X)$ be the dual of Φ. Then, for any $f \in L^\infty(X)$ we have

$$\int_X f\, d\eta = \hat{f}(e) = \widehat{\Phi(f)}(e) = \int_B \Phi(f)\, d\nu = \int_X f\, d\varphi(\nu).$$

We may view B as Dirac measures inside $\mathrm{Proba}(B)$, and $\varphi(\nu) = \int_B \varphi\, d\nu$ as φ is affine.

We say that the G-action on Y is **amenable** if for any compact metrizable space K with continuous G-action there exists a G-equivariant map $Y \to K$.

Lemma 11.6   *If K is a compact metrizable space, than the space $\mathrm{Proba}(K)$ of Borel probability measures on K is such.*

Sketch of proof   Consider a linearly dense sequence of continuous uniformly bounded functions f_i on K. Define a metric on Borel measures by $d(\eta_1,\eta_2) = \sum_{i\in\mathbb{N}} |\eta_1(f_i) - \eta_2(f_i)|/2^i$. □

Lemma 11.7   *Every compact metrizable space K with continuous G-action supports a μ-stationary measure for any measure μ on G.*

Proof Let δ be any probability measure on Z (e.g. Dirac measure supported by any point) then any accumulation point η of the sequence $\frac{1}{n}\sum_{k=0}^{n-1}(\mu^k * \delta)$ is μ-stationary. □

Thus we observed the following.

Observation 1 The action of any group G on any of its Furstenberg–Poisson boundaries is amenable.

The first ingredient in the proof of Theorem 11.3 is the following result of Zimmer.

Theorem 11.8 ([10, Theorem 4.3.5]) *Assume H is a closed subgroup of G. Assume that the action of G on Y is amenable. Then the action of H on Y is amenable.*

The second ingredient is the solution to the Furstenberg conjecture.

Theorem 11.9 ([5, 8]) *G is amenable if and only if there exists a measure μ such that the Furstenberg–Poisson boundary of (G,μ) is one point (equivalently, every bounded μ-harmonic function on G is constant).*

We will use the non-trivial "only if" part of the above result. However, let us notice that the "if" part is obvious. Indeed, let K be a compact G-space. If B is a point then B carries a G-invariant measure η (actually, the converse is also true). If $\varphi\colon \mathrm{Proba}(B) \to \mathrm{Proba}(K)$ is the map discussed just before Lemma 11.6, then $\varphi^*(\eta)$ is also a G-invariant measure on K.

The third ingredient is the following result.

Theorem 11.10 (Intermediate Factor Theorem) *Let μ_i be measures on G_i for $i \in \{1,2\}$. Set $G = G_1 \times G_2$ and $\mu = \mu_1 \times \mu_2$. Let (B,v) be the Furstenberg–Poisson boundary of (G,μ). Let (X,ξ) be a probability space with measure preserving G-action. Assume that both G_1 and G_2 act on X ergodically. Assume that (Y,η) is a G-space and the projection $B \times X \to X$ factors through Y. Then there exists a factor $(B,v) \to (C,v_C)$ such that (Y,η) is isomorphic to $(C \times X, v_C \times \xi)$. Furthermore C decomposes as a product of factors for G_1 and G_2.*

11.5 Proof of Theorem 11.3

Let K be a compact metrizable Γ/N-space, that is Γ-space on which N acts trivially. We need to show that K carries a Γ-invariant probability measure.

Let $X = G/\Gamma$. It is a probability space with measure preserving G-action (since Γ is a lattice). Moreover left-G_1-invariant functions on $(G_1 \times G_2)/\Gamma$ can be identified with right-Γ-invariant functions on $G_1 \backslash (G_1 \times G_2) = G_2$. Since Γ projects densely to G_2 we see that all such functions are essentially constant, thus G_1 acts on X ergodically. Likewise G_2.

Let $i \in 1, 2$. Assume that $G_i' = G_i/\overline{\mathrm{pr}_i(N)}$ are amenable. By Theorem 11.9 we may choose admissible measures μ_i' on G_i' such that Furstenberg–Poisson boundaries of (G_i', μ_i') are trivial. Let us take μ_i on G such that they project to μ_i' on factors. Put, as usual, $G = G_1 \times G_2$ and $\mu = \mu_1 \times \mu_2$.

By Theorem 11.8 we may find a Γ-map $\varphi \colon B \to \mathrm{Proba}(K) =: Z$. Let $\zeta = \varphi^* \nu$. Let us construct an *induced space* $Y = (G \times Z)/\Gamma$ where Γ acts on $G \times Z$ by $(g,z)\gamma = (g\gamma, \gamma^{-1}d)$. We would write $g\Gamma z$ for the class of (g,z). The formula $g(g'\Gamma z) = (gg')\Gamma z$ defines G-action on Y. Let $\tilde{\xi}$ be a measure on G such that $\pi_*^X \tilde{\xi} = \xi$ where $\pi^X \colon G \to X$ is the natural projection. Equip Y with the measure $\eta := \pi_*^Y (\tilde{\xi} \times \zeta)$ where $\pi^Y \colon G \times Z \to Y$ is the natural projection.

Obviously the maps $X \times B \ni (g\Gamma, b) \mapsto (g\Gamma\varphi(b)) \in Y$ and $Y \ni (g\Gamma z) \mapsto g\Gamma \in X$ are G-equivariant. Applying Intermediate Factor Theorem (Theorem 11.10) we get a map $\Psi \colon Y \to G \times C$ such that the diagram

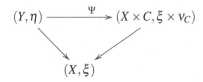

commutes. That is, Ψ is a map of the form $\Psi(g\Gamma z) = (g\Gamma, \psi(g,z))$. Moreover since Ψ is equivariant $\psi(g,z) = g\psi(e,z) =: g\psi_0(z)$, and $\psi_0 \colon (Z, \zeta) \to (C, \nu_C)$ is Γ-equivariant.

Since $\overline{\mathrm{pr}_i(N)}$ acts trivially on Z it acts trivially on C_i. Therefore C_i is G_i'-space and (C_i, ν_i) is a (G_i', μ_i')-boundary, therefore trivial. In particular (Z, ζ) is measurably isomorphic to trivial space, i.e. ζ is supported by a single point, namely an invariant probability measure on K.

References for this chapter

[1] M.M. Day, *Amenable semigroups*, Illinois J. Math. **1** (1957), 509–544.

[2] H. Furstenberg, *A Poisson formula for semi-simple Lie groups*, Ann. Math. **77** (1963), no. 2, 335–386

[3] A. Furman, *Random walks on groups and random transformations*, Handbook of dynamical systems, vol. 1A, pp. 931–1014. Amsterdam: North-Holland, 2002.

[4] U. Bader and Y. Shalom, *Factor and normal subgroup theorems for lattices in products of groups*, Invent. Math. **163** (2006), no. 2, 415–454.

178 *Gal*

[5] V. Kaimanovich and A. Vershik, *Random walks on discrete groups: boundary and entropy*, Ann. Probab. **11** (1983), 457–490.

[6] G. Margulis, *Discrete subgroups of semisimple Lie groups*, Berlin, Springer, 1991.

[7] S. Mazur, *Une remarque sur lhomomorphie des champs fonctionnels*, Studia Math. **1** (1929), 83–85.

[8] J. Rosenblatt, *Ergodic and mixing random walks on locally compact groups*, Math. Ann. **257** (1981), 31–42.

[9] Y. Shalom, *Rigidity of commensurators and irreducible lattices*, Invent. Math. **141** (2000), no. 1, 1–54.

[10] R.J. Zimmer, *Ergodic theory and semisimple groups*, Monogr. Math. vol. 81, Basel: Birkhäuser, 1984.

12
Burger–Mozes' simple lattices
Laurent Bartholdi

Abstract

We explain the methods that allowed Burger and Mozes to construct the first examples of (non-trivial) finitely presented torsion-free simple groups.

12.1	Introduction	179
12.2	Lattices in products of trees	180
12.3	A non-separable lattice	182
12.4	A non-residually finite lattice	183
12.5	A simple lattice	184
	References for this chapter	185

12.1 Introduction

The purpose of this note is to present a complete proof (modulo some computer calculations) of the following:

Theorem (Burger–Mozes, [3]) *There exist groups Γ that are*

(i) *finitely presented;*
(ii) *simple;*
(iii) *torsion-free;*
(iv) *biautomatic;*
(v) *fundamental groups of non-positively curved 2-dimensional complexes; hence of cohomological dimension 2;*
(vi) *presentable as amalgams $F *_E F$ for finitely generated free groups E, F and finite-index inclusions of E in F.*

These groups Γ will appear as lattices in a product of "universal groups" $U(F_v) \times U(F_h)$, for finite permutation groups F_v, F_h; see Chapter 6 in this volume.

The main ingredient is a variant of Margulis' "normal subgroup theorem", proven by Burger–Mozes. This result has been further extended by Bader and Shalom, as explained in Chapter 11 in this volume, to which we refer for a proof:

Theorem 12.1 (Burger–Mozes, [3, Corollary 5.1]) *Let $d_v, d_h \geq 3$ be integers, let F_v, F_h respectively be 2-transitive subgroups of* $\mathrm{Sym}(d_v), \mathrm{Sym}(d_h)$, *and let* $\Gamma \leq U(F_v) \times U(F_h)$ *be a cocompact lattice with dense projections:* $\overline{\mathrm{pr}_v(\Gamma)} \supseteq U(F_v)^+$ *and* $\overline{\mathrm{pr}_h(\Gamma)} \subseteq U(F_h)^+$.

Then Γ is just infinite; i.e. all non-trivial normal subgroups of Γ have finite index.

It follows that, if Γ is not residually finite, then $\bigcap_{1 \neq N \lhd \Gamma} N$ is simple: it is again a lattice to which Theorem 12.1 applies, and has no non-trivial normal subgroup.

The second ingredient is a step-by-step construction of a non-residually finite lattice. This can be done by pure thought, as in the original article of Burger and Mozes. In this note we have rather invoked techniques from combinatorial group theory when they were simple, and relied on Rattaggi's method of computer calculation, when they were not.

12.2 Lattices in products of trees

Consider two regular trees T_v, T_h of degree d_v, d_h respectively. The product $T_v \times T_h$ is naturally a 2-complex: its vertices are of the form (x, y) for vertices $x \in T_v, y \in T_h$; it has "horizontal" edges of the form $\{x\} \times f$ for a vertex $x \in T_v$ and an edge $f \subset T_h$, "vertical" edges of the form $e \times \{y\}$, and square 2-cells of the form $e \times f$ for edges $e \subset T_v, f \subset T_h$.

Consider next a group Γ acting freely on $T_v \times T_h$, with a single orbit of vertices. Thus, by choosing a base vertex $(o_v, o_h) \in T_v \times T_h$, the set of vertices of $T_v \times T_h$ is naturally in bijection with Γ.

Γ may then be described as the fundamental group of the geometric object $K_\Gamma := \Gamma \backslash (T_v \times T_h)$. This is a 2-complex consisting of a single vertex; a collection of "vertical" and "horizontal" oriented loops at that vertex, respectively written S_v, S_h, and in bijection with the Γ-orbits of geometric edges in T_v, T_h; and a collection of "squares" whose perimeter reads a vertical, horizontal, vertical, horizontal edge in sequence. Furthermore, the universal covering of K_Γ

being a product of trees, the link of the vertex must be the complete bipartite graph with vertex set $S_v^{\pm 1} \sqcup S_h^{\pm 1}$; we denote by $S_v^{\pm 1}$ the set of oriented vertical edges, of cardinality d_v, and similarly for $S_h^{\pm 1}$. Thus the squares' labels are such that, for every $s_v \in S_v^{\pm 1}$ and every $s_h \in S_h^{\pm 1}$, there exists a single corner of a square two-cell at which the two incident edges carry the labels s_v, s_h with correct orientation. In particular, K_Γ is made of a single zero-cell, $(d_v + d_h)/2$ geometric one-cells and $d_v d_h/4$ two-cells. Algebraically, this amounts to a presentation

$$\Gamma = \langle S_v \sqcup S_h \mid \text{relations } s_v s_h s_v' s_h' \rangle. \tag{12.1}$$

Fix a basepoint $(o_v, o_h) \in T_v \times T_h$, and consider

$$\Gamma_v = \left\{ \gamma \in \Gamma : \gamma(T_v \times \{o_h\}) = T_v \times \{o_h\} \right\},$$

the stabilizer of o_h in the action of Γ on T_h. On the one hand, Γ_v is a group acting freely and without inversions on T_v, and therefore is a free group. In fact, it is the subgroup of Γ generated by S_v, of rank $d_v/2$, and T_v may be identified with the Cayley graph of Γ_v. On the other hand, Γ_v acts on T_h fixing the basepoint o_h. This action may be described quite concretely, in terms of the presentation of Γ, as follows. Given $\gamma \in \Gamma_v$ and $x \in T_h$, consider the paths from o_v to γo_v in T_v and from o_h to x in T_h; these are the left and bottom edges of a unique rectangle in $T_v \times T_h$:

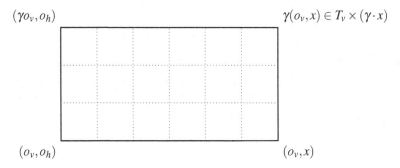

$(\gamma o_v, o_h)$ $\gamma(o_v, x) \in T_v \times (\gamma \cdot x)$

(o_v, o_h) (o_v, x)

By the "link" condition on the complex $\Gamma \backslash (T_v \times T_h)$, there is a single way of filling in this rectangle with labeled squares, and the top label gives the image of x under γ. Naturally the same considerations lead to a subgroup $\Gamma_h = \langle S_h \rangle$ acting on the rooted tree (T_v, o_v).

It is in general difficult to check whether a lattice, given e.g. by its presentation (12.1), has dense projections. However, the actions on rooted trees of Γ_v, Γ_h are computable, and this leads to a computable, sufficient condition:

Lemma 12.2 ([3, Proposition 5.2])   *Assume in the notation of Theorem 12.1*

*that the groups F_v, F_h are 2-transitive, with simple, non-abelian point stabi-
lizers $\cong L_v, L_h$ respectively. Consider the action of Γ_v on the ball of radius 2
around o_h; choose an edge e at o_h and let K_h denote the fixator of the ball
of radius 1 around e. Define similarly K_v. Then either $(K_v, K_h) = (1,1)$, or
$(K_v, K_h) \cong (L_v^{d_v-1}, L_h^{d_h-1})$ and the projections of Γ to $U(F_v)^+, U(F_h)^+$ are both
dense.*

The proof of this lemma is difficult (for the compiler), and relies on the
following results:

Theorem 12.3 (Thompson–Wielandt; see [1])   *Let T be a tree and let G be
a discrete, vertex-transitive subgroup of $\mathrm{Aut}(T)$. Fix a vertex $o \in T$ and let S
denote the sphere of radius 1 around o.*

*If the permutation group of S induced by G_o is primitive, then the fixator of
S is a p-group.*

Theorem 12.4 (Essentially [2, Proposition 3.3.1])   *Let T be a regular tree
and let G be a closed non-discrete vertex-transitive subgroup of $\mathrm{Aut}(T)$. Let F
be the permutation group induced on a sphere of radius 1 in T.*

*If F is 2-transitive and the stabilizer in F of a point is simple non-abelian,
then $G = U(F)$.*

12.3 A non-separable lattice

We now construct explicitly some lattices. The first step is to produce a lattice
Γ' that is not "subgroup separable", namely there exist a subgroup Δ and $g \in
\Gamma' \setminus \Delta$ such that, in every finite quotient of Γ', the image of g belongs to the
image of Δ. The lattice is given by $S_v' = \{a,b,c\}$, $S_h' = \{x,y\}$, and squares

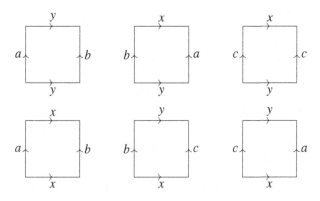

We first claim that the element $c \in \Gamma'_v$ acts transitively on $\{x,y\}^n \subset T_h$ for all $n \in \mathbf{N}$. This is proven by induction, using the following criterion: the action of c on $\{x,y\}^n$ must have odd signature for all $n \geq 1$. Indeed then, since $\{x,y\}^{n-1}$ is a single $\langle c \rangle$-orbit, there exists $w \in \{x,y\}^{n-1}$ such that $c^{2^{n-1}} \cdot wx = wy$ so that $\{x,y\}^n$ is a single orbit. For $\gamma \in \Gamma'_v$, let $\sigma_n(\gamma)$ denote the signature of the action of γ on $\{x,y\}^n$. Now the squares above give the relations

$$\sigma_n(a) = \sigma_{n-1}(b)^2 = 1, \quad \sigma_n(b) = \sigma_n(c) = \sigma_{n-1}(c)\sigma_{n-1}(a), \quad \sigma_1(c) = -1$$

from which $\sigma_n(c) = -1$ for all $n \geq 1$.

We then claim that, for all $m \geq 1$, the element $x^{-1}y$ belongs to $\Gamma'_v \langle c^m \rangle^{\Gamma'}$, and more precisely can be written as the product of an element of Γ'_v with a conjugate of a power of c. Indeed, since c acts transitively on $\{x,y\}^m$ for all m, it has infinite order; so there exists $w \in \{x,y\}^*$ such that $c^m \cdot wx = wy$. Consider the corresponding rectangle in $T_v \times T_h$:

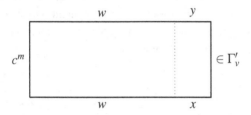

It gives in Γ' the relation $(c^m)^{wx} x^{-1} y \in \Gamma'_v$, as desired.

Lemma 12.5 (Wise, [6, Corollary 6.4]) *For every homomorphism* $\pi \colon \Gamma' \to Q$ *to a finite group,* $\pi(x^{-1}y) \in \pi(\Gamma'_v)$.

Proof Since Q is finite, there exists $m \geq 1$ such that $\pi(c^m) = 1$; then $\pi(x^{-1}y) \in \pi(\Gamma'_v \langle c^m \rangle^{\Gamma'}) = \pi(\Gamma'_v)$. $\qquad\square$

12.4 A non-residually finite lattice

We next construct a lattice that is not residually finite. It has presentation

$$\Gamma'' = \langle S'_v \sqcup S'_h \sqcup \overline{S'_h} \mid \text{two copies of the squares from } \Gamma' \rangle.$$

Note the following automorphism θ of Γ'': it fixes S'_v, and exchanges the two copies $S'_h, \overline{S'_h}$ by $x \leftrightarrow \bar{x}, y \leftrightarrow \bar{y}$. Its fixed point set is precisely Γ''_v. The claim follows from

Lemma 12.6 (Long–Niblo [4]) *Let G be a residually finite group, and let θ be an automorphism of G. Then* Fix(θ) *is separable in G.*

Proof Choose $g \in G \setminus \mathrm{Fix}(\theta)$; so $g^{-1}\theta(g) \neq 1$. Thus there exists $\pi \colon G \to Q$ with Q finite and $\pi(g^{-1}\theta(g)) \neq 1$. Define $\phi \colon G \to Q \times Q$, $\phi(g) = (\pi(g), \pi(\theta g))$. Note then that $\mathrm{Fix}(\theta)$ maps to the diagonal of $Q \times Q$, while g does not. $\square$

The lemma applies as follows: if Γ'' were residually finite, then $\Gamma''_v = \mathrm{Fix}(\theta)$ would be separable in Γ''; so Γ'_v would be separable in Γ', contradicting Lemma 12.5.

In fact, we obtain a bit more than non-residual finiteness: there is a specific element $y^{-1}x\overline{x}^{-1}\overline{y} \in \Gamma''_h$ that belongs to every finite-index subgroup of Γ''.

12.5 A simple lattice

We finally imbed the group Γ'' into a larger group $\Gamma \leq U(A_{10}) \times U(A_{10})$ by imbedding the complex associated with Γ'' into a complex of degrees $(d_v, d_h) = (10, 10)$. Many examples are possible, but we content ourselves with a single one, constructed by Rattaggi [5]. Set $S_v = S'_v \sqcup \{d, e\}$ and $S_h = S'_h \sqcup \overline{S'_h} \sqcup \{z\}$; the first two rows of squares are those of Γ'':

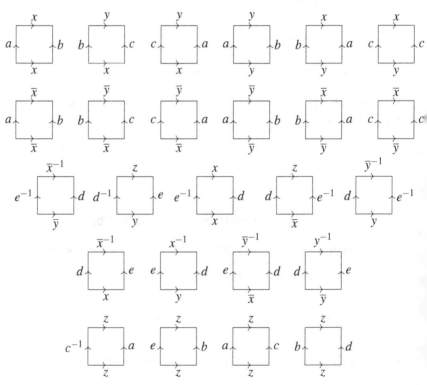

It is easy for a computer to check that Γ_v, Γ_h act by A_{10} on 1-balls, and that the finite groups K_v, K_h defined in Lemma 12.2 are non-trivial; so that both projections of Γ in $U(A_{10})$ have dense image in $U(A_{10})^+$, so that Γ is just infinite by Theorem 12.1.

The group Γ cannot be simple: there is always a homomorphism

$$\sigma \colon \Gamma \to \{\pm 1\} \times \{\pm 1\}, \quad S_v \mapsto (-1, 1), \quad S_h \mapsto (1, -1).$$

Set $\Gamma_0 = \ker(\sigma)$.

Proof of the main theorem Set $\Gamma_1 = \bigcap_{1 \neq N \triangleleft \Gamma} N$. Then Γ_1 is non-trivial, because it contains $w = y^{-1} x \bar{x}^{-1} y$, so it is simple and has finite index in Γ. A computer algebra program such as GAP can compute the normal closure of w in Γ, and check that it has index 4 in Γ, whence it coincides with Γ_0. Therefore $\Gamma_0 = \Gamma_1 = \langle w \rangle^\Gamma$ so Γ_0 is simple.

To see that Γ_0 is torsion-free, it suffices to note that it is the fundamental group of a complex with contractible universal cover $T_v \times T_h$.

The decomposition of Γ_0 as amalgam comes from its action on T_h with fundamental domain an edge. The group F is in fact Γ_v, and E is the stabilizer of an edge touching o_h in T_h. □

Note that GAP is used as a discovery tool, in letting us find short expressions of the generators of Γ_0 as products of conjugates of w. With these expressions in hand, a human being can check their validity, bypassing the requirement of a computer.

On the other hand, the original examples by Burger–Mozes were found, and proven to be simple, using a clever and purely silicon-free method; see the last chapter of [3].

References for this chapter

[1] van Bon, John, *Thompson-Wielandt-like theorems revisited*, Bull. Lond. Math. Soc. **35**, no. 1, 30–36 (2003).

[2] Burger, Marc and Mozes, Shahar, *Groups acting on trees: From local to global structure*, Inst. Hautes Études Sci. Publ. Math. **92**, 113–150 (2001).

[3] Burger, Marc and Mozes, Shahar, *Lattices in product of trees*, Inst. Hautes Études Sci. Publ. Math. **92**, 151–194 (2001).

[4] Long, Darren D. and Niblo, Graham A., *Subgroup separability and 3-manifold groups*, Math. Z. **207**, no. 2, 209–215 (1991).

[5] Rattaggi, Diego, *Three amalgams with remarkable normal subgroup structures*, J. Pure Appl. Algebra **210**, no. 2, 537–541 (2007).

[6] Wise, Daniel T., *Complete square complexes*, Comment. Math. Helv. **82**, no. 4, 683–724 (2007).

13

A lecture on invariant random subgroups

Tsachik Gelander

Abstract

Invariant random subgroups (IRS) are conjugacy invariant probability measures on the space of subgroups in a given group G. They can be regarded both as a generalization of normal subgroups as well as a generalization of lattices. As such, it is intriguing to extend results from the theories of normal subgroups and of lattices to the context of IRS. Another approach is to analyse and then use the space $\mathrm{IRS}(G)$ as a compact G-space in order to establish new results about lattices. The second approach has been taken in the work [1], sometimes refered to as the seven samurai paper.[1] In these lecture notes we shall try to give a taste of both approaches.

13.1	The Chabauty space of closed subgroups	187
13.2	Invariant measures on Sub_G	189
13.3	IRS and lattices	192
13.4	The Gromov–Hausdorff topology	197
13.5	The Benjamini–Schramm topology	197
13.6	Higher rank and rigidity	200
	References for this chapter	203

---

[1] Note added in proof: the preprint [1] has later been split into two papers [3] and [4]; however, in this chapter, we shall keep refereeing to the original manuscript [1].

13.1 The Chabauty space of closed subgroups

13.1.1 Definition

Let G be a locally compact group. We denote by Sub_G the space of closed subgroups of G equipped with the Chabauty topology. This topology is generated by sets of the following two types:

(i) $O_1(U) := \{H \in \mathrm{Sub}_G : H \cap U \neq \varnothing\}$ with $U \subset G$ an open subset, and
(ii) $O_2(K) := \{H \in \mathrm{Sub}_G : H \cap K = \varnothing\}$ with $K \subset G$ a compact subset.

Exercise 13.1 Show that a sequence $H_n \in \mathrm{Sub}_G$ converges to a limit H iff

(i) for any $h \in H$ there is a sequence $h_n \in H_n$ such that $h = \lim h_n$, and
(ii) for any sequence $h_{n_k} \in H_{n_k}$, with $n_{k+1} > n_k$, which converges to a limit, we have $\lim h_{n_k} \in H$.

Exercise 13.2 (Suggested by Ian Biringer, see [11]) The space Sub_G is metrisable when G is so. Indeed, let d be a proper metric on G and d_H the corresponding Hausdorff distance between compact subsets of G. Show that

$$\rho(H_1, H_2) := \int_0^\infty d_{\mathrm{H}}\big(H_1 \cap B_r(id_G), H_2 \cap B_r(id_G)\big) e^{-r} dr$$

is a compatible metric on Sub_G.

Example 13.3

(i) $\mathrm{Sub}_{\mathbf{R}} \cong [0, \infty]$. Indeed, every proper non-trivial closed subgroup of $\mathbf{R}$ is of the form $\alpha \mathbf{Z}$ for some $\alpha > 0$. When $\alpha \to 0$ the corresponding group tends to $\mathbf{R}$ and when $\alpha \to \infty$ it tends to $\{0\}$.
(ii) $\mathrm{Sub}_{\mathbf{R}^2}$ is homeomorphic to the sphere S^4 (this was proved by Hubbard and Pourezza, see [20]).
(iii) *Question:* What can you say about $\mathrm{Sub}_{\mathbf{R}^n}$ (cf. [24])?

One direction which seems interesting to study is the case of semisimple Lie groups. Indeed, much is known about discrete and general closed subgroups of (semisimple) Lie groups, and it is possible to deduce information about the structure of Sub_G.

Problem 13.1.1 *What can you say about* Sub_G *for* $G = \mathrm{SL}_2(\mathbf{R})$?

13.1.2 Compactness

While the structure of Sub_G in general is highly complicated, we at least know that it is always compact:

Proposition 13.4 (Exercise)   *For every locally compact group G, the space* Sub_G *is compact.*

We can use Sub_G in order to compactify certain sets of closed subgroups. For instance one can study the Chabauty compactification of the space of lattices in G. In particular, it is interesting to determine the points of that compactification:

Problem 13.1.2   *Determine which subgroups of* $\text{SL}_3(\mathbf{R})$ *are limits of lattices.*

This problem might be more accessible if we replace $\text{SL}_3(\mathbf{R})$ with a group for which the congruence subgroup property is known for all lattices.

13.1.3 When is G isolated?

It is useful to know under which conditions G is an isolated point in Sub_G.

Exercise 13.5   A discrete group Γ is isolated as a point in Sub_Γ iff it is finitely generated.

Let us examine some non-discrete cases:

Proposition 13.6   *Let G be a connected simple Lie group. Then G is an isolated point in* Sub_G.

The idea is that if H is sufficiently close to G then it has points close to 1 whose logarithms generate the Lie algebra of G. This implies that the connected component of identity H° is normal in G, and as G is simple, it is either trivial or everything. Thus, it is enough to show that H° is non-trivial, i.e. that H is not discrete. This is proved in [21, 33] — more precisely it is shown there that a non-nilpotent connected Lie group is never a limit of discrete subgroups.

Exercise 13.7   Show that G is not a limit of discrete subgroups, relying on the classical:

Theorem 13.8 (Zassenhaus, see [30])   *A Lie group G admits an identity neighbourhood U such that for every discrete group $\Gamma \leq G$, the Lie algebra $\langle \log(\Gamma \cap U) \rangle$ is nilpotent.*

For more details about Proposition 13.6 as well as other results in this spirit, see [1, Section 2].

Example 13.9   The circle group S^1 is not isolated in its Chabauty space Sub_{S^1}. Indeed, one can approximate S^1 by finite cyclic subgroups.

More generally,

Exercise 13.10 Show that if G surjects on S^1, then G is not isolated in Sub_G.

In fact, using Theorem 13.8 one can show:

Proposition 13.11 *A connected Lie group is isolated in Sub_G iff it does not surject on S^1 (i.e. has no non-trivial characters).*

Note that a connected Lie group G does not surject on the circle iff its commutator G' is dense in G. Such groups are called *topologically perfect*.

Exercise 13.12 Let G be a Lie group. Deduce from Theorem 13.8 that if $H \in \mathrm{Sub}_G$ is a limit of discrete groups, then the identity connected component $H°$ of H is nilpotent.

A similar result can be proved for semisimple groups over non-archimedean local fields:

Exercise 13.13 Consider $G = \mathrm{SL}_n(\mathbf{Q}_p)$ and show that G is an isolated point in Sub_G.

Hint: Use the following facts:

• $\mathrm{SL}_n(\mathbf{Z}_p)$ is a maximal subgroup of $\mathrm{SL}_n(\mathbf{Q}_p)$.
• The Frattini subgroup of $\mathrm{SL}_n(\mathbf{Z}_p)$ is open, i.e. of finite index.

The following result is proved in [17].

Theorem 13.14 *Let G be a semisimple analytic group over a local field, then G is isolated in Sub_G.*

13.2 Invariant measures on Sub_G

The group G acts on Sub_G by conjugation and it is natural to consider the invariant measures on this compact G-space.

Definition 13.15 An Invariant Random Subgroup (hereafter IRS) is a Borel regular probability measure on Sub_G which is invariant under conjugations.

13.2.1 First examples and remarks

(1) The Dirac measures correspond to normal subgroups. In view of this, one can regard IRSs as a generalization of normal subgroups.

(2) Let $\Gamma \le G$ be a lattice (or more generally a closed subgroup of finite co-volume). The map $G \to \Gamma^G \subset \mathrm{Sub}_G$, $g \mapsto g\Gamma g^{-1}$, factors through G/Γ. Hence

we may push the invariant probability measure on G/Γ to a conjugation invariant probability measure on Sub_G supported on the closure of the conjugacy class of Γ. In view of this, IRSs also generalize 'lattices' or more precisely finite volume homogeneous spaces G/Γ — as conjugated lattices give rise to the same IRS. We shall denote the IRS associated with (the conjugacy class of) Γ by μ_Γ.

For instance let Σ be a closed hyperbolic surface and normalise its Riemannian measure. Every unit tangent vector yields an embedding of $\pi_1(\Sigma)$ in $\mathrm{PSL}_2(\mathbf{R})$. Thus the probability measure on the unit tangent bundle corresponds to an IRS of type (2) above.

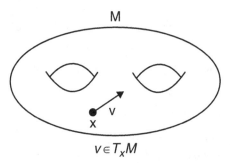

(3) Let again $\Gamma \leq G$ be a lattice in G, and let $N \lhd \Gamma$ be a normal subgroup of Γ. As in (2) the G-invariant probability measure on G/Γ can be used to choose a random conjugate of N in G. This is an IRS supported on the (closure of the) conjugacy class of N. More generally, every IRS on Γ can be induced to an IRS on G. Intuitively, the random subgroup is obtained by conjugating Γ by a random element from G/Γ and then picking a random subgroup in the corresponding conjugate of Γ. One can express the induction of $\mu \in \mathrm{IRS}(\Gamma)$ to $\mathrm{IRS}(G)$ by:

$$\mathrm{Ind}_\Gamma^G(\mu) = \frac{1}{m(\Omega)} \int_\Omega (i_g)_* \mu \, dm(g),$$

where m is Haar measure on G, Ω is a fundamental domain for G/Γ and $i_g :$ $\mathrm{Sub}_G \to \mathrm{Sub}_G$ corresponds to conjugation by g.

13.2.2 Connection with pmp actions

Let $G \curvearrowright (X,m)$ be a probability measure preserving action. By a result of Varadarajan, the stabilizer of almost every point in X is closed in G. Moreover, the stabilizer map $X \to \mathrm{Sub}_G$, $x \mapsto G_x$ is measurable, and hence one can push the measure m to an IRS on G. In other words the random subgroup is the stabilizer of a random point in X.

This reflects the connection between invariant random subgroups and pmp actions. Moreover, the study of pmp G-spaces can be divided into

- the study of stabilizers (i.e. IRS),
- the study of orbit spaces

and the interplay between the two.

The connection between IRS and pmp actions goes also in the other direction:

Theorem 13.16 *Let G be a locally compact group and μ an IRS in G. Then there is a probability space (X,m) and a measure preserving action $G \curvearrowright X$ such that μ is the push-forward of the stabilizer map $X \to \mathrm{Sub}_G$.*

This was proved in [5] for discrete groups and in [1, Theorem 2.4] for general G. The first thing that comes to mind is to take the given G action on (Sub_G, μ), but then the stabilizer of a point $H \in \mathrm{Sub}_G$ is $N_G(H)$ rather than H. To correct this one can consider the larger space Cos_G of all cosets of all closed subgroups, as a measurable G-bundle over Sub_G. Defining an appropriate invariant measure on $\mathrm{Cos}_G \times \mathbf{R}$ and replacing each fibre by a Poisson process on it, gives the desired probability space.

13.2.3 Topology

We shall denote by $\mathrm{IRS}(G)$ the space of G-invariant probability measures on $\mathrm{Sub}(G)$

$$\mathrm{IRS}(G) := \mathrm{Prob}(\mathrm{Sub}_G)^G$$

equipped with the w^*-topology. By Alaoglu's theorem $\mathrm{IRS}(G)$ is compact.

13.2.4 Existence

An interesting yet open question is whether this space is always non-trivial.

Question 13.2.1 Does every non-discrete locally compact group admit a non-trivial IRS?

A counterexample, if one exists, should in particular be a simple group without lattices. Currently the only known such example is the Neretin group and some close relatives (see [9], [25] and [26]). The question whether the Neretin group admits non-trivial IRS has two natural sub-questions:

Question 13.2.2

(i) Does the Neretin group admit a (non-discrete) closed subgroup of finite co-volume?

(ii) Does the Neretin group admit a non-trivial discrete IRS, i.e. an IRS with respect to which a random subgroup is a.s. discrete?

Remark 13.17   There are many discrete groups without non-trivial IRS, for instance $\mathrm{PSL}_n(\mathbf{Q})$, and also the Tarski Monsters.

13.2.5 Soficity of IRS

Definition 13.18   Let us say that an IRS μ is *co-sofic* if it is a weak-$*$ limit in $\mathrm{IRS}(G)$ of ones supported on lattices.

The following question can be asked for any locally compact group G, however we find the three special cases of $G = \mathrm{SL}_2(\mathbf{R}), \mathrm{SL}_2(\mathbf{Q}_p)$ and $\mathrm{Aut}(T)$ particularly intriguing:

Question 13.2.3   Is every IRS in G co-sofic?

Exercise 13.19   1. Show that the case $G = F_n$, the discrete rank n free group, is equivalent to the Aldous–Lyons conjecture that every unimodular network (supported on rank n Schreier graphs) is a limit of ones corresponding to finite Schreier graphs [6].

2. A Dirac mass δ_N, $N \lhd F_n$ is co-sofic iff the corresponding group $G = F_n/N$ is sofic.

13.3 IRS and lattices

Viewing IRS as a generalization of lattices there are two directions toward which one is urged to go:

(1) Extend classical theorems about lattices to general IRS.

(2) Use the compact space $\mathrm{IRS}(G)$ in order to study its special 'lattice' points.

Remarkably, the approach (2) turns out to be quite fruitful in the theory of asymptotic properties of lattices. We shall see later on (see Section 13.6) an example of how rigidity properties of G-actions yield interesting data of the geometric structure of locally symmetric spaces $\Gamma \backslash G / K$ when the volume tends to infinity. This approach is also useful for proving uniform statements regarding the set of all lattices (see Section 13.3.3).

The following sections will demonstrate in various forms both approaches (1) and (2) and some interactions between the two (for instance we will use the extension of Borel's density theorem for IRS in order to prove a strong version of the Kazhdan–Margulis theorem about lattices).

13.3.1 The IRS compactification of moduli spaces

One direction in the spirit of (2) which hasn't been applied yet (to the author's knowledge) is simply to obtain, using $\mathrm{IRS}(G)$, new compactifications of certain natural spaces.

Example 13.20 Let Σ be a closed surface of genus ≥ 2. As we have seen in 13.2.1(2), every hyperbolic structure on Σ corresponds to an IRS in $\mathrm{PSL}_2(\mathbf{R})$. Taking the closure in $\mathrm{IRS}(G)$ of the set of hyperbolic structures on Γ, one obtains an interesting compactification of the moduli space of Σ.

Problem 13.3.1 *Analyse the IRS compactification of Mod(Σ).*

Note that the resulting compactification is similar to (but is not exactly) the Deligne–Munford compactification.

13.3.2 Borel density theorem

Theorem 13.21 (Borel density theorem for IRS, [1]) *Let G be a connected non-compact simple (centre-free) Lie group. Let μ be an IRS on G without atoms. Then a random subgroup is μ-a.s. discrete and Zariski dense.*

Note that since G is simple, the only possible atoms are at the trivial group $\{1\}$ and at G. Since G is an isolated point in Sub_G, it follows that

$$\mathrm{IRS}_d(G) := \{\mu \in \mathrm{IRS}(G) : \text{a } \mu\text{-random subgroup is a.s. discrete}\}$$

is a compact space. We shall refer to the points of IRS_d as discrete IRS.

In order to prove Theorem 13.21 one first observes that there are only countably many conjugacy classes of non-trivial finite subgroups in G, hence the measure of their union is zero with respect to any non-atomic IRS. Then one can apply the same idea as in Furstenberg's proof of the classical Borel density theorem [14]. Indeed, taking the Lie algebra of $H \in \mathrm{sub}_G$ as well as of its Zariski closure induce measurable maps (see [17])

$$H \mapsto \mathrm{Lie}(H), \ H \mapsto \mathrm{Lie}(\overline{H}^Z)$$

As G is non-compact, Furstenberg's argument implies that the Grassman variety of non-trivial subspaces of $\mathrm{Lie}(G)$ does not carry an $\mathrm{Ad}(G)$-invariant measure. It follows that $\mathrm{Lie}(H) = 0$ and $\mathrm{Lie}(\overline{H}^Z) = \mathrm{Lie}(G)$ almost surely, and the two statements of the theorem follow.

Remark 13.22   The analogue of Theorem 13.21 holds, more generally, in the context of semisimple analytic groups over local fields, see [17].

13.3.3 Kazhdan–Margulis theorem

Definition 13.23   A family $\mathscr{F} \subset \mathrm{IRS}(G)$ of invariant random subgroups is said to be **weakly uniformly discrete** if for every $\varepsilon > 0$ there is an identity neighbourhood $U \subset G$ such that

$$\mu(\{\Gamma \in \mathrm{Sub}_G : \Gamma \cap U \neq \{1\}\}) < \varepsilon$$

for every $\mu \in \mathscr{F}$.

Theorem 13.24   *Let G be a connected non-compact simple Lie group. Then* $\mathrm{IRS}_d(G)$ *is weakly uniformly discrete.*

Assume, on the contrary, that for some $\varepsilon > 0$ we can find for every identity neighbourhood $U \subset G$ a discrete IRS μ_U such that

$$\mu_U\big(\{\Gamma \in \mathrm{Sub}_G : \Gamma \cap U \text{ non-trivially}\}\big) \geq \varepsilon.$$

Then letting U run over a suitable base of identity neighbourhoods and taking a weak limit μ, it would follow that μ is not discrete in contrast to the compactness of $\mathrm{IRS}_d(G)$.

As a straightforward consequence, taking $\varepsilon < 1$, we deduce the classical Kazhdan–Margulis theorem, and in particular the positivity of the lower bound on the volume of locally G/K-manifolds:

Corollary 13.25 (Kazhdan–Margulis theorem [23])   *There is an identity neighbourhood* $\Omega \subset G$ *such that for every lattice* $\Gamma \leq G$ *there is* $g \in G$ *such that* $g\Gamma g^{-1} \cap \Omega = \{1\}$.

Viewing the stabilizer of a random point in a probability measure preserving G-space as an IRS, Theorem 13.24 can be reformulated as follows:

Theorem 13.26 (pmp actions are uniformly weakly locally free)   *For every* $\varepsilon > 0$ *there is an identity neighbourhood* $U \subset G$ *such that for every non-trivial ergodic pmp G-space* (X, m) *there is a subset* $Y \subset X$ *with* $m(Y) > 1 - \varepsilon$ *such that* $u \cdot y \neq y$ *for all* $y \in Y$ *and* $u \in U$.

For complete proofs of the results of this subsection, in a more general setup, see [15].

13.3.4 Stuck–Zimmer rigidity theorem

Perhaps the first result about IRS, and certainly one of the most remarkable, is the Stuck–Zimmer rigidity theorem, which can be regarded as a (far reaching) generalization of Margulis' normal subgroup theorem.

Theorem 13.27 (Stuck–Zimmer [32]) *Let G be a connected simple Lie group of real rank ≥ 2. Then every ergodic pmp action of G is either (essentially) free or transitive.*

In view of Theorem 13.16, one can read Theorem 13.27 as: *every non-atomic ergodic IRS in G is of the form μ_Γ for some lattice $\Gamma \leq G$.*

In order to see that Margulis' normal subgroup theorem is a consequence of Theorem 13.27, suppose that $\Gamma \leq G$ is a lattice and $N \lhd \Gamma$ is a normal subgroup in Γ. As described in Example 13.2.1(3) there is an IRS μ supported on (the closure of) the conjugacy class of N. Since $G \curvearrowright G/\Gamma$ is ergodic and (Sub_G, μ) is a factor, it is also ergodic. In view of Theorem 13.16 there is an ergodic pmp G-space (X, m) such that the stabilizer of an m-random point is a μ-random conjugate of N. By Theorem 13.27, X is transitive, i.e. $X = G/N$. It follows that G/N carries a G-invariant probability measure. Thus N is a lattice, hence of finite index in Γ.

Remark 13.28

(i) Stuck and Zimmer proved the theorem for the wider class of higher rank semisimple groups with property (T), where in this case the assumption is that every factor is non-compact and acts ergodically. The situation for certain groups, such as $SL_2(\mathbf{R}) \times SL_2(\mathbf{R})$ is still unknown.

(ii) Recently A. Levit [27] proved the analogous result for analytic groups over non-archimedean local fields.

13.3.5 An exotic IRS in rank one

In the lack of Margulis' normal subgroup theorem there are IRSs supported on non-lattices. Indeed, if G has a lattice with an infinite index normal subgroup $N \lhd \Gamma$, arguing as in the previous section, one obtains an ergodic pmp space for which almost any stabilizer is a conjugate of N.

We shall now give a more interesting example, in the lack of rigidity:

Example 13.29 (An exotic IRS in $\mathrm{PSL}_2(\mathbf{R})$, [1])   Let A, B be two copies of a torus with 2 open discs removed. We choose hyperbolic metrics on A and B so that all 4 boundary components are geodesic circles of the same length, and such that A admits a closed geodesic of length much smaller than the injectivity radius at any point of B. We may agree that one boundary component of A (resp. of B) is 'on the left side' and the other is 'on the right side', and fix a special point on each boundary component, in order to specify a gluing pattern of a 'left' copy and a 'right' copy, each of either A or B.[2]

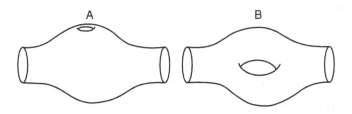

Figure 13.1 The hyperbolic building blocks

Now consider the space $\{A, B\}^{\mathbf{Z}}$ with the $(\frac{1}{2}, \frac{1}{2})$-Bernoulli measure, i.e. the measure giving probability $1/2$ to each copy of A or B independently. Any element α in this space is a two-sided infinite sequence of As and Bs and we can glue copies of A, B 'along a bi-infinite line' following this sequence. This produces a random surface M^{α}. Choosing a probability measure on the unit tangent bundle of A (resp. of B) we define an IRS in $\mathrm{PSL}_2(\mathbf{R})$ as follows. First choose M^{α} randomly, next choose a point and a unit tangent vector in the copy of A or B which lies at the place M_0^{α} (above 0 in the sequence α), then take the fundamental group of M^{α} according to the chosen point and direction.

Figure 13.2 A random surface

As the $\mathbf{Z}$-action on the Bernoulli space of sequences is ergodic one sees that the corresponding IRS is also ergodic. It can be shown that almost surely the corresponding group is not contained in a lattice in $\mathrm{PSL}_2(\mathbf{R})$. However, this IRS is co-sofic. Analog constructions can be made in $\mathrm{SO}(n, 1)$ for all ns (see [1, Section 13]).

---

[2] All the nice figures in these lecture notes were made by Gil Goffer.

13.4 The Gromov–Hausdorff topology

Given a compact metric space X, the Hausdorff distance $\mathrm{Hd}_X(A,B)$ between two closed subsets is defined as

$$\mathrm{Hd}_X(A,B) := \inf\{\varepsilon : N_\varepsilon(A) \supset B \text{ and } N_\varepsilon(B) \supset A\},$$

where $N_\varepsilon(A)$ is the ε-neighbourhood of A. The space 2^X of closed subsets of X equipped with the Hausdorff metric is compact.

Given two compact metric spaces X, Y, the Gromov distance $\mathrm{Gd}(X,Y)$ is defined as

$$\mathrm{Gd}(X,Y) := \inf_Z\{\mathrm{Hd}_Z(i(X), j(Y))\},$$

over all compact metric spaces Z admitting isometric copies $i(X), j(Y)$ of X, Y respectively. If $(X,p), (Y,q)$ are pointed compact metric spaces, i.e. ones with a chosen point, we define the Gromov distance

$$\mathrm{Gd}((X,p),(Y,q)) := \inf_Z\{\mathrm{Hd}_Z(i(X), j(Y)) + d_Z(i(p), j(q))\}.$$

The Gromov–Hausdorff distance between two pointed proper (not necessarily bounded) metric spaces $(X,p), (Y,q)$ can be defined as

$$\mathrm{GHd}((X,p),(Y,q)) := \sum_{n\in\mathbf{N}} 2^{-n}\mathrm{Gd}((B_X(n),p),(B_Y(n),q)),$$

where $B_X(n)$ is the ball of radius n around p.

13.5 The Benjamini–Schramm topology

Let $\mathscr{M}$ be the space of all (isometry classes of) pointed proper metric spaces equipped with the Gromov–Hausdorff topology. This is a huge space and for many applications it is enough to consider compact subspaces of it obtained by bounding the geometry. That is, let $f(\varepsilon, r)$ be an integer valued function defined on $(0,1) \times \mathbf{R}^{>0}$, and let $\mathscr{M}_f$ consist of those spaces for which $\forall \varepsilon, r$, the ε-entropy of the r-ball $B_X(r,p)$ around the special point is bounded by $f(\varepsilon, r)$, i.e. no $f(\varepsilon, r) + 1$ points in $B_X(r,p)$ form an ε-discrete set. Then $\mathscr{M}_f$ is a compact subspace of $\mathscr{M}$.

In many situations one prefers to consider some variants of $\mathscr{M}$ which carry more information about the spaces. For instance when considering graphs, it may be useful to add colours and orientations to the edges. The Gromov–Hausdorff distance defined on these objects should take into account the colouring and orientation. Another example is smooth Riemannian manifolds, in which case it is better to consider framed manifolds, i.e. manifold with a cho-sen point and a chosen frame at the tangent space at that point. In that case, one replaces the Gromov–Hausdorff topology by the ones determined by (ε, r) relations (see [1, Section 3] for details), which remembers also the directions from the special point.

We define the Benjamini–Schramm space $\mathscr{BS} = \mathrm{Prob}(\mathscr{M})$ to be the space of all Borel probability measures on $\mathscr{M}$ equipped with the weak-$*$ topology. Given f as above, we set $\mathscr{BS}_f := \mathrm{Prob}(\mathscr{M}_f)$. Note that $\mathscr{BS}_f$ is compact.

The name of the space is chosen to hint that this is the same topology in-duced by 'local convergence', introduced by Benjamini and Schramm in [10], when restricting to measures on rooted graphs. Recall that a sequence of ran-dom rooted bounded degree graphs converges to a limiting distribution iff for every n the statistics of the n ball around the root (i.e. the probability vector corresponding to the finitely many possibilities for n-balls) converges to the limit.

The case of general proper metric spaces can be described similarly. A se-quence $\mu_n \in \mathscr{BS}_f$ converges to a limit μ iff for any compact pointed 'test-space' $M \in \mathscr{M}$, any r and arbitrarily small[3] $\varepsilon > 0$, the μ_n probability that the r ball around the special point is 'ε-close' to M tends to the μ-probability of the same event.

Example 13.30 An example of a point in $\mathscr{BS}$ is a measured metric space, i.e. a metric space with a Borel probability measure. A particular case is a finite volume Riemannian manifold — in which case we scale the Riemannian measure to be one, and then randomly choose a point (and a frame).

Thus a finite volume locally symmetric space $M = \Gamma \backslash G / K$ produces both a point in the Benjamini–Schramm space and an IRS in G. This is a special case of a general analogy that I'll now describe. Given a symmetric space X, let us denote by $\mathscr{M}(X)$ the space of all pointed (or framed) complete Rieman-nian orbifolds whose universal cover is X, and by $\mathscr{BS}(X) = \mathrm{Prob}(\mathscr{M}(X))$ the corresponding subspace of the Benjamini–Schramm space.

Let G be a non-compact simple Lie group with maximal compact subgroup $K \leq G$ and an associated Riemannian symmetric space $X = G/K$. There is a natural map

$$\{\text{discrete subgroups of } G\} \to \mathscr{M}(X), \ \Gamma \mapsto \Gamma \backslash X.$$

It can be shown that this map is continuous, hence inducing a continuous map

$$\mathrm{IRS}_d(G) \to \mathscr{BS}(X).$$

---

[3] This doesn't mean that it happens for all ε.

It can be shown that the later map is one to one, and since $\text{IRS}_d(G)$ is compact, it is a homeomorphism to its image (see [1, Corollary 3.4]).

Exercise 13.31 (Invariance under the geodesic flow)   Given a tangent vector $\bar{v}$ at the origin (the point corresponding to K) of $X = G/K$, define a map $\mathscr{F}_{\bar{v}}$ from $\mathscr{M}(X)$ to itself by moving the special point using the exponent of $\bar{v}$ and applying parallel transport to the frame. This induces a homeomorphism of $\mathscr{BS}(X)$. Show that the image of $\text{IRS}_d(G)$ under the map above is exactly the set of $\mu \in \mathscr{BS}(X)$ which are invariant under $\mathscr{F}_{\bar{v}}$ for all $\bar{v} \in T_K(G/K)$.

Thus we can view geodesic-flow invariant probability measures on framed locally X-manifolds as IRS on G and vice versa, and the Benjamini–Schramm topology on the first coincides with the IRS-topology on the second.

Exercise 13.32   Show that the analogy above can be generalized, to some extent, to the context of general locally compact groups. Given a locally compact group G, fixing a right invariant metric on G, we obtain a map $\text{Sub}_G \to \mathscr{M}$, $H \mapsto G/H$, where the metric on G/H is the induced one. Show that this map is continuous and deduce that it defines a continuous map $\text{IRS}(G) \to \mathscr{BS}$.

For the sake of simplicity we shall now forget 'the frame' and consider pointed X-manifolds, and $\mathscr{BS}(X)$ as probability measures on such. We note that while for general Riemannian manifolds there is a benefit for working with framed manifolds, in the world of locally symmetric spaces of non-compact type, pointed manifolds, and measures on such, behave nicely enough.

In order to examine convergence in $\mathscr{BS}(X)$ it is enough to use as 'test-space' balls in locally X-manifolds. Moreover, since X is non-positively curved, a ball in an X-manifold is isometric to a ball in X iff it is contractible.

For an X-manifold M and $r > 0$, we denote by $M_{\geq r}$ the r-thick part in M:

$$M_{\geq r} := \{x \in M : \text{InjRad}_M(x) \geq r\},$$

where $\text{InjRad}_M(x) = \sup\{\varepsilon : B_M(x, \varepsilon) \text{ is contractible}\}$.

Note that since X is a homogeneous space, all choices of a probability measure on X correspond to the same point in $\mathscr{BS}(X)$, and we shall denote this point by X, with a slight abuse of notations. We have the following simple characterisation of convergence to X.

Proposition 13.33   *A sequence M_n of finite volume X-manifolds BS-converges to X iff for every $r > 0$:*

$$\frac{\text{vol}((M_n)_{\geq r})}{\text{vol}(M_n)} \to 1.$$

13.6 Higher rank and rigidity

Suppose now that G is a non-compact simple Lie group of real rank at least 2. The following result from [1] can be interpreted as 'large manifolds are almost everywhere fat':

Theorem 13.34 ([1]) *Let $M_n = \Gamma_n \backslash X$ be a sequence of finite volume X-manifolds with $\mathrm{vol}(M_n) \to \infty$. Then $M_n \to X$ in the Benjamini–Schramm topology.*

This means that for any r and ε there is $V(r, \varepsilon)$ such that if M is an X-manifold of volume $v \geq V(r, \varepsilon)$ then $\frac{\mathrm{vol}(M_{\geq r})}{v} \geq 1 - \varepsilon$ (see Figure 13.3).

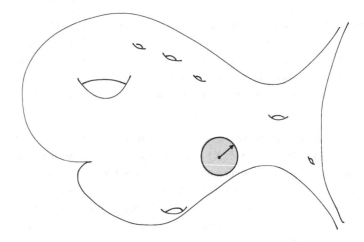

Figure 13.3 A large volume manifold with a random r-ball

Using the dictionary from the previous section we may reformulate Theorem 13.34 in the language of IRS:

Theorem 13.35 ([1]) *Let $\Gamma_n \leq G$ be a sequence of lattices with $\mathrm{vol}(G/\Gamma_n) \to \infty$ and denote by μ_n the corresponding IRS. Then $\mu_n \to \delta_{\{1\}}$.*

The proof makes use of the equivalence between the two formulations. The main ingredients in the proof are the Stuck–Zimmer rigidity theorem 13.27 and Kazhdan's property (T), which will be used at two places (in addition property (T) is used in the proof of 13.27).

Recall that by Kazhdan's theorem, G has property (T). This implies that a limit of ergodic measures is ergodic:

Theorem 13.36 ([19]) *Let G be a group with property (T) acting by homeomorphisms on a compact Hausdorff space X. Then the set of ergodic G-invariant probability Borel measures on X is w^*-closed.*

The idea is that if μ_n are probability measures converging to a limit μ and μ is not ergodic, then there is a continuous function f on X which, as a function in $L_2(\mu)$ is G-invariant, orthogonal to the constants and with norm 1. Thus for large n we have that f is almost invariant in $L_2(\mu_n)$, almost orthogonal to the constants and with norm almost 1. Since G has property (T) it follows that there is an invariant $L_2(\mu_n)$ function close to f, so μ_n cannot be ergodic.

Let now μ_n be a sequence as in 13.35, and let μ be a weak-$*$ limit of μ_n. Our aim is to show that $\mu = \delta_{\{1\}}$. Up to replacing μ_n by a subnet, we may suppose that $\mu_n \to \mu$. Let $M_n = \Gamma_n \backslash X$ be the corresponding manifolds, as in 13.34. By Theorem 13.36 we know that μ is ergodic. The following result is a consequence of Theorem 13.27:

Proposition 13.37 *The only ergodic IRSs on G are $\delta_G, \delta_{\{1\}}$ and μ_Γ for $\Gamma \leq G$ a lattice.*

Proof Let μ be an ergodic IRS on G. By Theorem 13.16 μ is the stabilizer of some pmp action $G \curvearrowright (X, m)$. By Theorem 13.27 the latter action is either essentially free, in which case $\mu = \delta_{\{1\}}$, or transitive, in which case the (random) stabilizer is a subgroup of co-finite volume. The Borel density theorem implies that in the latter case, the stabilizer is either G or a lattice $\Gamma \leq G$. $\square$

Thus, in order to prove Theorem 13.35 we have to exclude the cases $\mu = \delta_G$ and $\mu = \mu_\Gamma$. The case $\mu = \delta_G$ is impossible since G is an isolated point in Sub_G (see 13.6). Let us now suppose that $\mu = \mu_\Gamma$ for some lattice $\Gamma \leq G$ and aim towards a contradiction. For this, we will adopt the formulation of 13.34. Thus we suppose that $M_n \to M = \Gamma \backslash X$.

Recall that Property (T) of G implies that there is a lower bound $C > 0$ for the Cheeger constant of all finite volume X-manifolds. For our purposes, the Cheeger constant of a manifold M can be defined as the infimum of

$$\frac{\mathrm{vol}(N_1(S))}{\min\{\mathrm{vol}(M_i)\}},$$

where S is a subset which disconnects the manifold to two (open disjoint and nonempty) pieces M_1 and M_2, and $N_1(S)$ is the 1-neighbourhood of S.

Since $M = \Gamma \backslash X$ has finite volume we may pick a point $p \in M$ and r large enough so that the volume of $B_M(p, r-1)$, the $r-1$ ball around p in M, is greater than $\mathrm{vol}(M)(1-C)$ (note that if M is compact we may even take a ball that covers M). In particular, when taking $S = \{x \in M : d(x, p) = r\}$ we have that $\frac{\mathrm{vol}(N_1(S))}{\mathrm{vol}(B_M(p,r-1))} < C$. This on itself does not contradict property (T) since the complement of $B_M(p,r+1)$ is very small.

Now since M_n converges to M in the BS-topology, it follows that for large n, a random point q in M_n with positive probability satisfies that

$$\frac{\mathrm{vol}(B_{M_n}(q,r+1) \setminus B_{M_n}(q,r-1))}{\mathrm{vol}(B_{M_n}(q,r-1))} < C.$$

Bearing in mind that $\mathrm{vol}(M_n) \to \infty$, we get that for large n, the complement $M_n \setminus B_{M_n}(q,r+1)$ has arbitrarily large volume, and in particular greater than $\mathrm{vol}(B_{M_n}(q,r-1))$, see Figure 13.4. Now, this contradicts the assumption that C is the Cheeger constant of X. $\qquad\square$

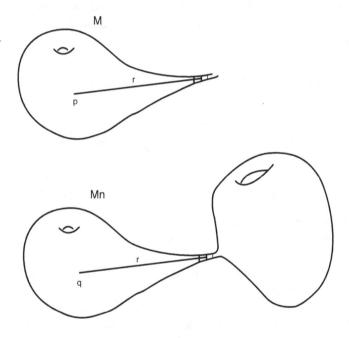

Figure 13.4 The Cheeger constant of M_n is too small.

Note that Theorems 13.34 and 13.35 can also be formulated as:

- The set of extreme points in IRS(G) (the ergodic IRS) is closed and equals $\{\delta_G, \delta_{\{1\}}, \mu_\Gamma, \Gamma \le G$ a lattice$\}$ and its unique accumulation point is $\delta_{\{1\}}$,

or as:

- The space of geodesic flow invariant probability measures on $\mathscr{M}(X)$ is compact and convex, its extreme points are the finite volume X-manifolds and the space X, and the later being the only accumulation point.

Finally let us note that Theorem 13.34 has many applications in the theory of asymptotic invariants, and in particular L_2-invariant, of arithmetic groups and locally symmetric manifolds. Most of the work [1] is dedicated to such asymptotic results and our main new ingredient is Theorem 13.34. For instance, one quite immediate application is that if M_n is a sequence of uniformly discrete (i.e. with a uniform lower bound on the injectivity radius) X-manifolds with volume $\mathrm{vol}(M_n) \to \infty$ then the normalised Betti numbers converge to a limit

$$\frac{b_i(M_n)}{\mathrm{vol}(M_n)} \to b_i^{(2)}(X),$$

and the limit $b_i^{(2)}(X)$ is computable, and vanishes for $i \neq \dim(X)/2$ (cf. [2]).

References for this chapter

[1] M. Abert, N. Bergeron, I. Biringer, T. Gelander, N. Nikolov, J. Raimbault, I. Samet, On the growth of L^2-invariants for sequences of lattices in Lie groups. ArXiv-preprint arXiv:1210.2961, 2012.

[2] M. Abert, N. Bergeron, I. Biringer, T. Gelander, N. Nikolov, J. Raimbault, I. Samet, On the growth of Betti numbers of locally symmetric spaces. C. R. Math. Acad. Sci. Paris, 349(15–16):831–835, 2011.

[3] M. Abert, N. Bergeron, I. Biringer, T. Gelander, N. Nikolov, J. Raimbault, I. Samet, On the growth of L^2-invariants for sequences of lattices in Lie groups. Ann. of Math. (2) **185**(3) (2017), 711–790.

[4] M. Abert, N. Bergeron, I. Biringer, T. Gelander, N. Nikolov, J. Raimbault, I. Samet, On the growth of L^2-invariants of locally symmetric spaces, II: exotic invariant random subgroups in rank one. ArXiv-preprint arXiv:1612.09510, 2016.

[5] M. Abert, Y. Glasner, B. Virag, Kesten's theorem for invariant random subgroups. Duke Math. J., 163(3):465–488, 2014.

[6] D. Aldous, R. Lyons, Processes on Unimodular Random Networks, Electron. J. Probab., 12, Paper 54 (2007), 1454–1508.

[7] M. Abert, T. Gelander, N. Nikolov, Rank, combinatorial cost and homology torsion growth in higher rank lattices. Duke Math. J. **166**(15): 2925–2964, 2017.

[8] U. Bader, B. Duchesne, J. Lecureux, Amenable invariant random sub-groups, Israel J. Math. **213**(1): 399–422, 2016.

[9] U. Bader, P.E. Caprace, T. Gelander, S. Mozes, Simple groups without lattices. Bull. Lond. Math. Soc., 44 (2012), no. 1, 5517.

[10] I. Benjamini and O. Schramm, Recurrence of distributional limits of finite planar graphs. Electron. J. Probab., 6:no. 23, 13 pp. (electronic), 2001.

[11] I. Biringer, Metrizing the Chabauty topology. ArXiv-preprint arXiv:1610.07396, 2016.

[12] L. Bowen, Invariant random subgroups of the free group. Groups Geom. Dyn. **9**(3): 891–916, 2015.

[13] L. Bowen, Random walks on random coset spaces with applications to Furstenberg entropy. Invent. Math., 196(2):485–510, 2014.

[14] H. Furstenberg, A note on Borel's density theorem. Proc. Amer. Math. Soc., 55(1):209–212, 1976.

[15] T. Gelander, The Kazhdan–Margulis theorem for IRS, Adv. Math., to appear.

[16] T. Gelander, Lectures on lattices and locally symmetric spaces. Geometric group theory, 249–282, IAS/Park City Math. Ser., 21, Amer. Math. Soc., Providence, RI, 2014.

[17] T. Gelander, A. Levit, Invariant random subgroups over non-archimedean local fields, preprint.

[18] Y. Glasner, Invariant random subgroups of linear groups. Israel J. Math. **219**(1): 215–270, 2017.

[19] E. Glasner, B. Weiss, Kazhdan's property T and the geometry of the collection of invariant measures. (English summary) Geom. Funct. Anal., 7 (1997), no. 5, 917–935.

[20] I. Pourezza, J. Hubbard, The space of closed subgroups of R^2. Topology, 18 (1979), no. 2, 143–146.

[21] M. Kuranishi, On everywhere dense imbedding of free groups in Lie groups. Nagoya Math. J., 2 (1951). 63–71.

[22] D.A. Kazhdan, Connection of the dual space of a group with the structure of its closed subgroups. Functional Analysis and Application, 1 (1967), 63–65.

[23] D.Kazhdan, G.Margulis, A proof of Selberg's hypothesis, Mat. Sb. (N.S.) 75(117) (1968) 163–168.

[24] B. Kloeckner, The space of closed subgroups of R^n is stratified and simply connected. J. Topol., 2 (2009), no. 3, 570–588.

[25] A. Le Boudec, Groups acting on trees with almost prescribed local action. Comment. Math. Helv., 91 (2016), no. 2, 253–293.

[26] W. Lederle, Coloured Neretin groups. Preprint (2017), arXiv:1701.03027.

[27] A. Levit, The Nevo–Zimmer intermediate factor theorem over local fields, Geom. Dedicata **186**: 149–171, 2017.

[28] J. Peterson, A. Thom, Character rigidity for special linear groups, J. Reine Angew. Math. **716**:207–228, 2016.

[29] V. Platonov, A. Rapinchuk, Algebraic Groups and Number Theory, Academic Press, 1994.

[30] M. S. Raghunathan, Discrete subgroups of Lie groups, Springer, New York (1972).

[31] J. Raimbault, On the convergence of arithmetic orbifolds, arXiv:1311.5375.

[32] G. Stuck and R.J. Zimmer, Stabilizers for ergodic actions of higher rank semisimple groups. Annals of Mathematics, 139 (1994) 723–747.

[33] H. Toyama, On discrete subgroups of a Lie group. Kodai Math. Sem. Rep., 1 (1949). no. 2, 36–37.

[34] R.J. Zimmer, Ergodic theory and semisimple groups. Springer, 1984.

14

L^2-Betti number of discrete and non-discrete groups

Roman Sauer

Abstract

In this survey we explain the definition of L^2-Betti numbers of locally compact groups — both discrete and non-discrete. Specific topics include the proportionality principle, Lück's dimension function, Petersen's definition for locally compact groups and some concrete examples.

14.1	Introduction	205
14.2	Continuous cohomology	207
14.3	von Neumann algebras, trace and dimension	213
14.4	L^2-Betti numbers of groups	219
References for this chapter		224

14.1 Introduction

This short survey addresses readers who, motivated by geometric group theory, seek a brief overview of the (algebraic) definitions of L^2-Betti numbers of discrete and non-discrete locally compact groups. We limit ourselves to the group case, mostly ignoring the theory of L^2-Betti numbers of equivariant spaces.

L^2-Betti numbers share many formal properties with ordinary Betti numbers, like Künneth and Euler–Poincaré formulas. But the powerful *proportionality principle* is a distinctive feature of L^2-Betti numbers. It also was a motivation to generalize the theory of L^2-Betti numbers to locally compact groups.

In its easiest form, the proportionality principle states that the L^2-Betti number in degree p, which is henceforth denoted as $\beta_p(\_)$, of a discrete group Γ

and a subgroup $\Lambda < \Gamma$ of finite index satisfies the relation

$$\beta_p(\Lambda) = [\Gamma : \Lambda] \beta_p(\Gamma). \tag{14.1}$$

The next instance of the proportionality principle involves lattices Γ and Λ in a semisimple Lie group G endowed with Haar measure μ. It says that their L^2-Betti numbers scale according to their covolume:

$$\beta_p(\Gamma)\mu(\Lambda\backslash G) = \beta_p(\Lambda)\mu(\Gamma\backslash G). \tag{14.2}$$

The proof of (14.2) becomes easy when we use the original, analytic definition of L^2-Betti numbers by Atiyah [1]. According to this definition, the p-th L^2-Betti number of Γ is given in terms of the heat kernel of the Laplace operator on p-forms on the symmetric space $X = G/K$:

$$\beta_p(\Gamma) = \lim_{t\to\infty} \int_{\mathscr{F}} \mathrm{tr}_{\mathbb{C}}(e^{-t\Delta^p}(x,x))\, d\mathrm{vol}.$$

Here $\mathscr{F} \subset X$ is a measurable fundamental domain for the Γ-action on X and $e^{-t\Delta^p}(x,x) : \mathrm{Alt}^p(T_x X) \to \mathrm{Alt}^p(T_x X)$ is the integral kernel — called **heat kernel** — of the bounded operator $e^{-t\Delta^p}$ obtained from the unbounded Laplace operator Δ^p on $L^2\Omega^p(X)$ by spectral calculus. Since G acts transitively on X by isometries, it is clear that the integrand in the above formula is constant in x. So there is a constant $c > 0$ only depending on G such that $\beta_p(\Gamma) = c\,\mathrm{vol}(\Gamma\backslash X)$, from which one deduces (14.2).

Gaboriau's theory of L^2-Betti numbers of measured equivalence relations elaborated in [12] (see also [28] for an extension to measured groupoids) greatly generalized the equation (14.2) to the setting of measure equivalence.

Definition 14.1 Let Γ and Λ be countable discrete groups. Assume that there is a Lebesgue measure space (Ω, ν) on which Γ and Λ act in a commuting, measure-preserving way such that both the Γ- and the Λ-action admit measurable fundamental domains of finite ν-measure. Then we say that Γ and Λ are **measure equivalent**. We say that (Ω, ν) is a **measure coupling** of Γ and Λ.

Theorem 14.2 (Gaboriau) *If (Ω, ν) is a measure coupling of Γ and Λ, then*

$$\beta_p(\Gamma)\nu(\Lambda\backslash\Omega) = \beta_p(\Lambda)\nu(\Gamma\backslash\Omega).$$

If Γ and Λ are lattices in the same locally compact G, then G endowed with its Haar measure is a measure coupling for the actions given by $\gamma \cdot g := \gamma g$ and $\lambda \cdot g := g\lambda^{-1}$ for $g \in G$ and $\gamma \in \Gamma$, $\lambda \in \Lambda$. So Gaboriau's theorem generalizes (14.2) to lattices in arbitrary locally compact groups.

The geometric analogue of measure equivalence is quasi-isometry. One may wonder whether L^2-Betti numbers of quasi-isometric groups are also proportional. This is not true! The groups $\Gamma = F_3 * (F_3 \times F_3)$ and $\Lambda = F_4 * (F_3 \times F_3)$ are quasi-isometric. Their Euler characteristics are $\chi(\Gamma) = 1$ and $\chi(\Lambda) = 0$. By the Euler–Poincaré formula (see Theorem 14.45) the L^2-Betti numbers of Γ and Λ are not proportional. We refer to [8, p. 106] and [13, 2.3] for a discussion of this example. However, the vanishing of L^2-Betti numbers is a quasi-isometry invariant. This is discussed in Section 14.4.2.

Motivated by (14.1) and (14.2), Henrik Petersen introduced L^2-Betti numbers $\beta_*(G,\mu)$ of an arbitrary second countable, locally compact unimodular group G endowed with a Haar measure μ. The first instances of L^2-invariants for non-discrete locally compact groups are Gaboriau's first L^2-Betti number of a unimodular graph [11], which is essentially one of its automorphism group, and the L^2-Betti numbers of buildings in the work of Dymara [9] and Davis–Dymara–Januszkiewicz–Okun [7], which are essentially ones of the automorphism groups of the buildings. We refer to [26] for more information on the relation of the latter to Petersen's definition.

If G is discrete, we always take μ to be the counting measure. If a locally compact group possesses a lattice, it is unimodular. Petersen [25] (if $\Gamma < G$ is cocompact or G is totally disconnected) and then Kyed–Petersen–Vaes [15] (in general) showed the following generalization of (14.1). A proof which is much easier but only works for totally disconnected groups can be found in [26].

Theorem 14.3 (Petersen, Kyed–Petersen–Vaes) *Let Γ be a lattice in a second countable, locally compact group G with Haar measure μ. Then*

$$\beta_p(\Gamma) = \mu(\Gamma \backslash G)\beta_p(G,\mu).$$

Structure of the paper

In Section 14.2 we explain the definition of continuous cohomology and we explain how to compute it for discrete and totally disconnected groups by geometric models. In Section 14.3 we start with the definition of the von Neumann algebra of a unimodular locally compact group and its semifinite trace. The goal is to understand Lück's dimension function for modules over von Neumann algebras. In Section 14.4 we define L^2-Betti numbers, comment on their quasi-isometry invariance and present some computations.

14.2 Continuous cohomology

Throughout, let G be a second countable, locally compact group, and let E be topological vector space with a continuous G-action, i.e. the action map $G \times E \to E$, $(g,e) \mapsto ge$, is continuous. A topological vector space with such a

G-action is called a G-**module**. A G-**morphism** of G-modules is a continuous, linear, G-equivariant map.

14.2.1 Definition via the bar resolution

Let $C(G^{n+1}, E)$ be the vector space of continuous maps from G^{n+1} to E. The group G acts from the left on $C(G^{n+1}, E)$ via

$$(g \cdot f)(g_0, \ldots, g_n) = g f(g^{-1} g_0, \ldots, g^{-1} g_n).$$

If G is discrete, the continuity requirement is void, and $C(G^{n+1}, E)$ is the vector space of all maps $G^{n+1} \to E$. The fixed set $C(G^{n+1}, E)^G$ is just the set of continuous equivariant maps. The **homogeneous bar resolution** is the chain complex

$$C(G, E) \xrightarrow{d^0} C(G^2, E) \xrightarrow{d^1} C(G^3, E) \to \ldots$$

with differential

$$d^n(f)(g_0, \ldots, g_{n+1}) = \sum_{i=0}^{n+1} (-1)^i f(g_0, \ldots, \hat{g}_i, \ldots, g_{n+1}).$$

The chain groups $C(G^{n+1}, E)$ are endowed with the compact-open topology turning them into G-modules. This is a non-trivial topology even for discrete G, where it coincides with pointwise convergence.

Since the differentials are G-equivariant we can restrict to the equivariant maps and obtain a chain complex $C(G^{*+1}, E)^G$.

Definition 14.4   The cohomology of $C(G^{*+1}, E)^G$ is called the **continuous cohomology** of G in the G-module E and denoted by $H^*(G, E)$.

The differentials are continuous but usually have non-closed image, which leads to a non-Hausdorff quotient topology on the continuous cohomology. Hence it is natural to consider the following.

Definition 14.5   The **reduced continuous cohomology** $\overline{H}^*(G, E)$ of G in E is defined as the quotient $\ker(d^n) / \mathrm{clos}(\mathrm{im}(d^{n-1}))$ of $H^n(G, E)$, where we take the quotient by the closure of the image of the differential.

In homological algebra it is common to compute derived functors, such as group cohomology, by arbitrary injective resolutions. The specific definition of continuous cohomology by the homogeneous bar resolution, which is nothing else than usual group cohomology if G is discrete, is the quickest definition in the topological setting. But there is also an approach in the sense of homological algebra, commonly referred to as **relative homological algebra**.

We call an injective G-morphism of G-modules **admissible** if it admits a linear, continuous (not necessarily G-equivariant) inverse.

Definition 14.6 A G-module E is **relatively injective** if for any admissible injective G-morphism $j : U \to V$ and a G-morphism $f : U \to E$ there is a G-morphism $\bar{f} : V \to E$ such that $\bar{f} \circ j = f$.

Example 14.7 Let E be a G-module. Then $C(G^{n+1}, E)$ is relatively injective. Let $j : U \to V$ be an admissible G-morphism. Let $s : V \to U$ be a linear continuous map with $s \circ j = \mathrm{id}_U$. Given a G-morphism $f : U \to C(G^{n+1}, E)$, the G-morphism

$$(\bar{f})(v)(g_0, \dots g_n) = f\big(g_0 s(g_0^{-1} v)\big)(g_0, \dots, g_n)$$

satisfies $\bar{f} \circ j = f$. Similarly, if K is a compact subgroup, then $C((G/K)^{n+1}, E)$ is relatively injective. Here the extension is given by

$$(\bar{f})(v)([g_0], \dots [g_n]) = \int_K f\big(g_0 k s(k^{-1} g_0^{-1} v)\big)([g_0], \dots, [g_n]) d\mu(k),$$

where μ is the Haar measure normalised with $\mu(K) = 1$.

As the analogue of the fundamental lemma of homological algebra we have:

Theorem 14.8 *Let E be a G-module. Let $0 \to E \to E^0 \to E^1 \to \dots$ be a resolution of E by relatively injective G-modules E^i. Then the cohomology of $(E^*)^G$ is (topologically) isomorphic to the continuous cohomology of G in E.*

14.2.2 Injective resolutions to compute continuous cohomology

The homogeneous bar resolution is useful for proving general properties of continuous cohomology. But other injective resolutions coming from geometry are better suited for computations.

Definition 14.9 Let G be totally disconnected (e.g. discrete). Let X be a cellular complex on which G acts cellularly and continuously. We require that for each open cell $e \subset X$ and each $g \in G$ with $ge \cap e \neq \varnothing$, the multiplication by g is the identity on e. We say that X is a **geometric model** of G if X is contractible, its G-stabilisers are open and compact, and the G-action on the n-skeleton $X^{(n)}$ is cocompact for every $n \in \mathbf{N}$.

For simplicial G-actions, the above requirement on open cells can always be achieved by passing to the barycentric subdivision. A cellular complex with cellular G-action that satisfies the above requirement on open cells and whose stabilisers are open is a G-**CW-complex** in the sense of [32, II.1]. This means that the n-skeleton $X^{(n)}$ is built from $X^{(n-1)}$ by attaching G-orbits of n-cells according to pushouts of G-spaces of the form:

$$
\begin{array}{ccc}
\bigsqcup_{U \in \mathscr{F}_n} G/U \times S^{n-1} & \longrightarrow & X^{(n-1)} \\
\big\downarrow & & \big\downarrow \\
\bigsqcup_{U \in \mathscr{F}_n} G/U \times D^n & \longrightarrow & X^{(n)}
\end{array}
$$

Here $\mathscr{F}_n$ is a set of representatives of conjugacy classes of stabilizers of n-cells. We require that each $X^{(n)}$ is cocompact. So $\mathscr{F}_n$ is finite. Each coset space G/U is discrete. Let us fix a choice of pushouts, which is not part of the data of a cellular complex and corresponds to an equivariant choice of orientations for the cells. The horizontal maps induces an isomorphism in relative homology by excision. Let $C_*^{\mathrm{cw}}(X)$ be the cellular chain complex with $\mathbf{C}$-coefficients. We obtain isomorphisms of discrete G-modules:

$$
\bigoplus_{U \in \mathscr{F}_n} \mathbf{C}[G/U] \cong H_n \Big(\bigsqcup_{U \in \mathscr{F}_n} G/U \times (D^n, S^{n-1}) \Big) \xrightarrow{\cong} H_n\left(X^{(n)}, X^{(n-1)} \right)
$$

$$
\overset{\mathrm{def}}{=} C_n^{\mathrm{cw}}(X). \qquad (14.1)
$$

The G-action $(gf)(x) = gf(g^{-1}x)$ turns

$$
\hom_{\mathbf{C}}(C_n^{\mathrm{cw}}(X), E) \cong \prod_{U \in \mathscr{F}_n} C(G/U, E) = \bigoplus_{U \in \mathscr{F}_n} C(G/U, E)
$$

into a G-module. By Example 14.7 it is relatively injective. Further, the contractibility of X implies that $\hom_{\mathbf{C}}(C_*^{\mathrm{cw}}(X), E)$ is a resolution of E. The next statement follows from Theorem 14.8.

Theorem 14.10 *Let X be a geometric model for a totally disconnected group G. Then*

$$
H^n(G, E) \cong H^n(\hom_{\mathbf{C}}(C_*^{\mathrm{cw}}(X), E)^G).
$$

Not every discrete or totally disconnected group has a geometric model. For discrete groups having a geometric model means that the group satisfies the finiteness condition F_∞. But by attaching enough equivariant cells to increase connectivity one can show [18, 1.2]:

Theorem 14.11 *For every totally disconnected G there is a contractible G-CW-complex whose stabilisers are open and compact.*

More is true but not needed here: Every totally disconnected group has a classifying space for the family of compact-open subgroups.

What about the opposite case of a connected Lie group G? There we lack geometric models to compute continuous cohomology, but we can use the infinitesimal structure. The van Est isomorphism relates continuous cohomology and Lie algebra cohomology. We won't discuss it here and instead refer to [14].

14.2.3 Definition and geometric interpretation of L^2-cohomology

Let G be a second countable, locally compact group. For convenience, we require that the Haar measure μ is unimodular, i.e. invariant under left and right translations. Unimodularity becomes a necessary assumption in Section 14.3.

We consider continuous cohomology in the G-module $L^2(G)$ which consists of measurable square-integrable functions on G modulo null sets. A left G-action on $L^2(G)$ is given by $(gf)(h) = f(g^{-1}h)$.

The G-module $L^2(G)$ also carries the right G-action $(fg)(h) = f(hg^{-1})$, which commutes with the left action. In Section 14.2 we ignore this right action. It becomes important in Sections 14.3 and 14.4 when we define L^2-Betti numbers.

Definition 14.12 We call $H^*(G, L^2(G))$ and $\overline{H}^*(G, L^2(G))$ the L^2-**cohomology** and the **reduced** L^2-**cohomology** of G, respectively.

Assume that G possesses a geometric model X. By Theorem 14.10 the L^2-cohomology of G can be expressed as the cohomology of the G-invariants of the chain complex $\hom_{\mathbb{C}}(C_*^{\mathrm{cw}}(X), L^2(G))$. In view of Equation (14.1), this is a chain complex of Hilbert spaces

$$\hom_{\mathbb{C}}(C_n^{\mathrm{cw}}(X), L^2(G))^G \cong \bigoplus_{U \in \mathscr{F}_n} C(G/U, L^2(G))^G \cong \bigoplus_{U \in \mathscr{F}_n} L^2(G)^U$$

with bounded differentials.

Let us rewrite this chain complex in a way so that the group G does not occur anymore: As a (non-equivariant) cellular complex the n-th cellular chain group $C_n^{\mathrm{cw}}(X)$ comes with a preferred basis B_n, given by n-cells, which is unique up to permutation and signs. We define the vector space

$$L^2 C_{\mathrm{cw}}^n(X) := \Big\{ f : C_n^{\mathrm{cw}}(X) \to \mathbb{C} \mid \sum_{e \in B_n} |f(e)|^2 < \infty \Big\}$$
$$\subset \hom_{\mathbb{C}}(C_n^{\mathrm{cw}}(X), \mathbb{C})$$

of L^2-cochains in the cellular cochains; it has the structure of a Hilbert space with Hilbert basis $\{f_e \mid e \in B_n\}$ where $f_e(e') = 1$ for $e' = e$ and $f_e(e') = 0$ for $e' \in B_n \backslash \{e\}$. For general cellular complexes the differentials in the cellular cochain complex are not bounded as operators, but in the presence of a cocompact group action on skeleta they are.

Definition 14.13   The **(reduced) L^2-cohomology** of X is defined as the (reduced) cohomology of $L^2C^*_{cw}(X)$ and denoted by $L^2H^*(X)$ or $L^2\overline{H}^*(X)$, respectively.

Proposition 14.14   *The L^2-cohomology of G and the L^2-cohomology of a geometric model of G are isomorphic. Similarly for the reduced cohomology.*

Proof   The maps

$$\hom_{\mathbb{C}}(C_n^{cw}(X), L^2(G))^G \to L^2C_{cw}^n(X)$$

that take f to the L^2-cochain that assigns to $e \in B_n$ the essential value of the essentially constant function $f(e)|_U$, where $U < G$ is the open stabiliser of e, form a chain isomorphism. The claim follows now from Theorem 14.10.   □

14.2.4   Reduced L^2-cohomology and harmonic cochains

Definition 14.15   Let $W^0 \xrightarrow{d^0} W^1 \xrightarrow{d^1} W^2 \ldots$ be a cochain complex of Hilbert spaces such that the differentials are bounded operators. The **Laplace-operator** in degree n is the bounded operator $\Delta^n = (d^n)^* \circ d^n + d^{n-1} \circ (d^{n-1})^* : W^n \to W^n$. Here $*$ means the adjoint operator. A cochain $c \in W^n$ is **harmonic** if $\Delta^n(c) = 0$.

Proposition 14.16   *Every harmonic cochain is a cocycle, and inclusion induces a topological isomorphism*

$$\ker(\Delta^n) \xrightarrow{\cong} \overline{H}^n(W^*) = \ker(d^n)/\operatorname{clos}(\operatorname{im}(d^{n-1})).$$

Proof   Since Δ^n is a positive operator, we have $c \in \ker(\Delta^n)$ if and only if

$$\langle \Delta^n(c), c \rangle = \|d^n c\|^2 + \|(d^{n-1})^*(c)\|^2 = 0.$$

Hence $\ker(\Delta^n) = \ker(d^n) \cap \ker((d^{n-1})^*)$. The second statement follows from

$$
\begin{aligned}
W^p &= (\ker(d^n) \cap \operatorname{im}(d^{n-1})^\perp) \oplus \operatorname{clos}(\operatorname{im}(d^{n-1})) \oplus \ker(d^n)^\perp \\
&= (\ker(d^n) \cap \ker((d^{n-1})^*)) \oplus \operatorname{clos}(\operatorname{im}(d^{n-1})) \oplus \ker(d^n)^\perp \\
&= \ker(\Delta^n) \oplus \operatorname{clos}(\operatorname{im}(d^{n-1})) \oplus \ker(d^n)^\perp.
\end{aligned}
$$
   □

As a consequence of Propositions 14.14 and 14.16, we obtain:

Corollary 14.17   *Let X be a geometric model of G. The space of harmonic n-cochains of $L^2C^*_{cw}(X)$ is isomorphic to $\overline{H}^n(G, L^2(G))$.*

Example 14.18   The 4-regular tree is a geometric model of the free group F of rank two. Let $d : L^2 C^0_{cw}(X) \to L^2 C^1_{cw}(X)$ be the differential. Since there are no 2-cells, the first Laplace operator Δ^1 is just dd^*. A 1-cochain is harmonic if and only if it is in the kernel of d^*. We choose a basis B of $C^{cw}_1(X)$ by orienting 1-cells in the following way:

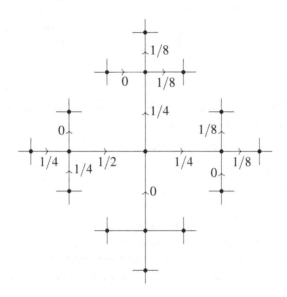

For $e \in B$ let $e(-)$ be the starting point and $e(+)$ the end point of e. Then:

$$d(f)(e) = f(e(+)) - f(e(-))$$
$$d^*(g)(v) = \sum_{e(+)=v} g(e) - \sum_{e(-)=v} g(e).$$

In the second equation the sums run over all edges whose starting or end point is v. The picture indicates a non-vanishing $c \in L^2 C^1_{cw}(X)$ with $d^*(c) = 0$, thus $\Delta^1(c) = 0$. Therefore $\overline{H}^1(F, \ell^2(F)) \neq 0$.

14.3 von Neumann algebras, trace and dimension

Throughout, we fix a second countable, locally compact group G with left invariant Haar measure μ. We require that G is unimodular, that is, μ is also right invariant.

14.3.1 The von Neumann algebra of a locally compact group

The continuous functions with compact support $C_c(G)$ form a C-algebra with involution through the convolution product

$$(f * g)(s) = \int_G f(t)g(t^{-1}s)d\mu(t)$$

and the involution $f^*(s) = \overline{f(s^{-1})}$. Taking convolution with $f \in C_c(G)$ is still defined for a function $\phi \in L^2(G)$. It follows from the integral Minkowski inequality that

$$\|f * \phi\|_2 \leq \|f\|_1 \|\phi\|_2,$$

where $\|_-\|_p$ denotes the L^p-Norm. We obtain a $*$-homomorphism into the algebra of bounded operators on $L^2(G)$:

$$\lambda : C_c(G) \to \mathscr{B}(L^2(G)), \ \lambda(f)(\phi) = f * \phi.$$

Similarly, we obtain a $*$-anti-homomorphism

$$\rho : C_c(G) \to \mathscr{B}(L^2(G)), \ \rho(f)(\phi) = \phi * f.$$

Definition 14.19   A **von Neumann algebra** is a subalgebra of the bounded operators of a Hilbert space that is closed in the weak operator topology and closed under taking adjoints. The **von Neumann algebra $L(G)$ of the group** G is the von Neumann algebra defined as the weak closure of $\text{im}(\lambda)$.

The weak closure $R(G)$ of $\text{im}(\rho)$ in $\mathscr{B}(L^2(G))$ is the commutant of $L(G)$ inside $\mathscr{B}(L^2(G))$.

Remark 14.20   Consider the operator $u_g \in \mathscr{B}(L^2(G))$ such that

$$u_g(\phi)(s) = \phi(g^{-1}s).$$

Similarly, we define $r_g(\phi)(s) = \phi(sg)$. We claim that $u_g \in L(G)$. If G is discrete, then the Kronecker function δ_g on G is continuous, so $\delta_g \in C_c(G)$, and we have $u_g := \lambda(\delta_g) \in L(G)$. If G is not discrete, we can choose a sequence $f_n \in C_c(G)$ of positive functions with $\|f_n\|_1 = 1$ whose supports tend to g. Then $(\lambda(f_n))_{n\in\mathbf{N}}$ converges strongly, thus weakly, to u_g, implying $u_g \in L(G)$. One can also show that $L(G)$ is the weak closure of the span of $\{u_g \mid g \in G\}$. Similarly it follows that $r_g \in R(G)$.

Remark 14.21   Let $j : L^2(G) \to L^2(G)$ be the conjugate linear isometry with

$$j(\phi)(s) = \phi^*(s) = \overline{\phi(s^{-1})}.$$

Then

$$J : L(G) \to R(G), \ J(T) = j \circ T^* \circ j$$

is a ∗-anti-isomorphism, that is, $J(T^*) = J(T)^*$ and $J(T \circ S) = J(S) \circ J(T)$. Furthermore, we have $J(\lambda_g) = r_g$ and $J(\lambda(f)) = J(\rho(f))$ for $f \in C_c(G)$.

Definition 14.22 ($L(G)$-module structures) The Hilbert space $L^2(G)$ is naturally a left $L(G)$-module via $T \cdot \phi = T(\phi)$ for $T \in L(G)$ and $\phi \in L^2(G)$. The left $L(G)$-module structure restricts to the left G-module structure via $g \cdot \phi := u_g(\phi)$. The Hilbert space $L^2(G)$ becomes a right $L(G)$-module by the anti-isomorphism J, explicitly $\phi \cdot T := J(T)(\phi)$. In the sequel, $L^2(G)$ will be regarded as a bimodule endowed with the left G-module structure and the right $L(G)$-module structure.

14.3.2 Trace

We explain the semifinite trace on the von Neumann algebra of a locally compact group. This a bit technical, but indispensable for the dimension theory. We start by discussing traces on an arbitrary von Neumann algebra $\mathscr{A}$.

Let $\mathscr{A}_+$ be the subset of positive operators in $\mathscr{A}$. For $S, T \in \mathscr{A}_+$ one defines $S \leq T$ by $T - S \in \mathscr{A}_+$; it is a partial order on $\mathscr{A}_+$. Every bounded totally ordered subset of $\mathscr{A}_+$ has a supremum in $\mathscr{A}_+$.

Definition 14.23 A **trace** on a von Neumann algebra $\mathscr{A}$ is a function $\tau : \mathscr{A}_+ \to [0, \infty]$ such that

(i) $\tau(S) + \tau(T) = \tau(S+T)$ for $S, T \in \mathscr{A}_+$;
(ii) $\tau(\lambda S) = \lambda \tau(S)$ for $S \in \mathscr{A}_+$ and $\lambda \geq 0$;
(iii) $\tau(SS^*) = \tau(S^*S)$ for $S \in \mathscr{A}$.

Let $\mathscr{A}_+^\tau = \{S \in \mathscr{A}_+ \mid \tau(S) < \infty\}$. A trace τ is **faithful** if $\tau(T) > 0$ for every $T \in \mathscr{A}_+ \setminus \{0\}$. It is **finite** if $\mathscr{A}_+^\tau = \mathscr{A}_+$. It is **semifinite** if $\mathscr{A}_+^\tau$ is weakly dense in $\mathscr{A}_+$. It is **normal** if the supremum of traces of a bounded totally ordered subset in $\mathscr{A}_+$ is the trace of the supremum.

If $\mathscr{A} \subset \mathscr{B}(H)$ is a von Neumann algebra with trace τ, then the $n \times n$-matrices $M_n(\mathscr{A}) \subset \mathscr{B}(H^n)$ are a von Neumann algebra with trace

$$(\tau \otimes \mathrm{id}_n)(S) := \tau(S_{11}) \ldots + \tau(S_{nn}).$$

Remark 14.24 Let $\mathscr{A}^\tau$ be the linear span of $\mathscr{A}_+^\tau$. It is clear that τ extends linearly to $\mathscr{A}^\tau$. If τ is finite, we have $\mathscr{A}^\tau = \mathscr{A}$, so τ is defined on all of $\mathscr{A}$. Further, $\mathscr{A}^\tau$ is always an ideal in $\mathscr{A}$ [31, p. 318]. The **trace property**

$$\tau(ST) = \tau(TS) \text{ for all } S \in \mathscr{A}^\tau \text{ and } T \in \mathscr{A}$$

holds true. Its deduction from the third property in Definition 14.23 takes a few lines and uses polarisation identities. See [31, Lemma 2.16 on p. 318].

After this general discussion we turn again to the von Neumann algebra of G. We call $\phi \in L^2(G)$ **left bounded** if there is a bounded operator, denoted by λ_ϕ on $L^2(G)$ such that $\lambda_\phi(f) = \phi * f$ for every $f \in C_c(G)$. Of course, every element $f \in C_c(G)$ is left bounded and $\lambda_f = \lambda(f)$. Define for an element $S^*S \in L(G)_+$ (every positive operator can be written like this):

$$\tau_{(G,\mu)}(S^*S) = \begin{cases} \|\phi\|_2^2 & \text{if there is a left bounded } \phi \in L^2(G) \text{ with } \lambda_\phi = S; \\ \infty & \text{otherwise.} \end{cases}$$

$$(14.1)$$

Notation 14.25 The Haar measure is only unique up to scaling. Hence we keep μ in the notation $\tau_{(G,\mu)}$. If $G = \Gamma$ is discrete, we always take the counting measure as Haar measure and simply write τ_Γ.

See [23, 7.2.7 Theorem] for a proof of the following fact.

Theorem 14.26 $\tau_{(G,\mu)}$ *is a faithful normal semifinite trace on* $L(G)$.

Let us try to obtain a better understanding of Definition (14.1). To this end, we first consider the case that $G = \Gamma$ is a discrete group. Then $\delta_e \in C_c(\Gamma) = \mathbb{C}[\Gamma] \subset L^2(\Gamma)$. For every $S \in L(G)$ it is $S(\delta_e) \in L^2(\Gamma)$ and $S = \lambda_{S(\delta_e)}$. This implies that every element in $L(\Gamma)_+$ has finite τ_Γ-trace. Hence τ_Γ is finite. Further, from

$$\tau_\Gamma(S^*S) = \|S(\delta_e)\|_2^2 = \langle S(\delta_e), S(\delta_e) \rangle_{L^2(\Gamma)} = \langle S^*S(\delta_e), \delta_e \rangle_{L^2(\Gamma)}$$

we conclude and record:

Remark 14.27 (Trace for discrete groups) If $G = \Gamma$ is discrete, then τ_Γ is finite and thus everywhere defined. For every $T \in L(\Gamma)$ we have $\tau_\Gamma(T) = \langle T(\delta_e), \delta_e \rangle_{L^2(\Gamma)}$.

Remark 14.28 (Trace for totally disconnected groups) Let G be totally disconnected. Then we have a decreasing neighbourhood basis (K_n) by open-compact subgroups. Then the element $\frac{1}{\mu(K_n)}\lambda(\chi_{K_n}) \in L(G)$ is a projection (see Example 14.39). Let $S \in L(G)$. Then $\phi_n := S(\chi_{K_n}) \in L^2(G)$ is a left bounded element such that $\lambda_{\phi_n} = S \circ \lambda(\chi_{K_n})$. From that and (14.1) it is easy to see that

$$\tau_{(G,\mu)}(S^*S) = \lim_{n \to \infty} \frac{1}{\mu(K_n)^2} \cdot \|S(\chi_{K_n})\|_2^2 \in [0, \infty].$$

Now back to general G:

Remark 14.29 (Trace for arbitrary groups) Because of $\lambda_f = \lambda(f)$ for every $f \in C_c(G)$ we obtain that $\tau_{(G,\mu)}(\lambda(f)^*\lambda(f)) = \|f\|_2^2$. One quickly verifies that the latter is just evaluation at the unit element: $\|f\|_2^2 = (f^*f)(e)$. It turns out that $C_c(G) \subset L(G)^{\tau(G,\mu)}$, and the trace $\tau_{(G,\mu)}$ is evaluation at $e \in G$ on $C_c(G)$.

14.3.3 Dimension

We explain first Lück's dimension function over a von Neumann algebra endowed with a finite trace. We refer to [16, 17] for proofs. Lück's work was a major advance in creating a general and algebraic theory of L^2-Betti numbers. An alternative algebraic approach was developed by Farber [10]. After that we discuss Petersen's generalization to von Neumann algebras with semifinite traces [24].

Finite traces

Let τ be a finite normal faithful trace on a von Neumann algebra $\mathscr{A}$. According to Remark 14.24, τ is a functional on $\mathscr{A}$ which satisfies the trace property $\tau(ST) = \tau(TS)$ for $S, T \in \mathscr{A}$. We start by explaining the dimension for finitely generated projective right $\mathscr{A}$-modules. Let P be such a module. This means that P is isomorphic to the image of left multiplication $l_M : \mathscr{A}^n \to \mathscr{A}^n$ with an idempotent matrix $M \in M_n(\mathscr{A})$. In general the sum $\sum_{i=1}^n M_{ii}$ of the diagonal entries of M depends not only on P but on the specific choice of M. However, Hattori and Stallings observed that its image in the quotient $\mathscr{A}/[\mathscr{A},\mathscr{A}]$ by the additive subgroup generated by all commutators is independent of the choices [4, Chapter IX]. This is true for an arbitrary ring $\mathscr{A}$. The trace τ is still defined on the quotient (see Remark 14.24). Therefore one defines:

Definition 14.30 The **Hattori–Stallings rank** $\mathrm{hs}(P) \in \mathscr{A}/[\mathscr{A},\mathscr{A}]$ of a finitely generated projective right $\mathscr{A}$-module P is defined as the image of $\sum_{i=1}^n M_{ii}$ in $\mathscr{A}/[\mathscr{A},\mathscr{A}]$ for any idempotent matrix $M \in M_n(\mathscr{A})$ with $P \cong \mathrm{im}(l_M)$. The **dimension** of P is defined as

$$\dim_\tau(P) = \tau(\mathrm{hs}(P)) = (\tau \otimes \mathrm{id}_n)(M) \in [0,\infty).$$

Henceforth $\mathscr{A}$-modules are understood to be right $\mathscr{A}$-modules.

Remark 14.31 Let $L^2(\mathscr{A},\tau)$ be the GNS-construction of $\mathscr{A}$ with respect to τ, that is the completion of the pre-Hilbert space $\mathscr{A}$ with inner product $\langle S,T \rangle_\tau = \tau(ST^*)$. By an observation of Kaplansky [17, Lemma 6.23 on p. 248], every finitely generated projective $\mathscr{A}$-module can be described by a projection matrix M, that is a matrix M with $M^2 = M$ *and* $M^* = M$. Then M yields a right Hilbert $\mathscr{A}$-submodule of $L^2(\mathscr{A},\tau)^n$, namely the image of left multiplication

$L^2(\mathscr{A}, \tau)^n \to L^2(\mathscr{A}, \tau)^n$ with A. And $\dim_\tau(M)$ coincides with the von Neumann dimension of this Hilbert $\mathscr{A}$-submodule. See [17, Chapter 1] for more information on Hilbert $\mathscr{A}$-modules. The classic reference is [21].

The idea of how to extend $\dim_\tau$ to arbitrary modules is almost naive; the difficulty lies in showing its properties.

Definition 14.32   Let M be an arbitrary $\mathscr{A}$-module. Its **dimension** is defined as

$$\dim_\tau(M) = \sup\{\dim_\tau(P) \mid P \subset M \text{ fin. gen. projective submodule}\} \in [0, \infty].$$

First of all, using the same notation $\dim_\tau$ as before requires a justification. But indeed, the new definition coincides with the old one on finitely generated projective modules. And so $\dim_\tau(\mathscr{A}) = \tau(1_\mathscr{A})$ which we usually normalise to be 1. The following two properties are important and, in the end, implied by the additivity and normality of τ.

Theorem 14.33 (Additivity)   *If $0 \to M_1 \to M_2 \to M_3 \to 0$ is a short exact sequence of $\mathscr{A}$-modules, then $\dim_\tau(M_2) = \dim_\tau(M_1) + \dim_\tau(M_3)$. Here $\infty + a = a + \infty = \infty$ for $a \in [0, \infty]$ is understood.*

Theorem 14.34 (Normality)   *Let M be an $\mathscr{A}$-module, and let M be the union of an increasing sequence of $\mathscr{A}$-submodules M_i. Then we have $\dim_\tau(M) = \sup_{i \in \mathbf{N}} \dim_\tau(M_i)$.*

Example 14.35   The only drawback of $\dim_\tau$ — in comparison to the dimension of vector spaces — is that $\dim_\tau(N) = 0$ does not, in general, imply $N = 0$. Here is an example. Take a standard probability space (X, ν). Then the ν-integral is a finite trace τ on $\mathscr{A} = L^\infty(X)$. Let $X = \bigcup_{i=1}^n X_i$ be an increasing union of measurable sets such that $\nu(X_i) < 1$ for every $i \in \mathbf{N}$. Each characteristic function χ_{X_i} is a projection in $\mathscr{A}$ with trace $\nu(X_i)$. And each $\chi_{X_i}\mathscr{A} = L^\infty(X_i)$ is a finitely generated projective module of dimension $\tau(\chi_{X_i}) = \nu(X_i)$. Let $M \subset \mathscr{A}$ be the increasing union of submodules $\chi_{X_i}\mathscr{A}$. Then $\mathscr{A}/M \neq 0$ but $\dim_\tau(\mathscr{A}/M) = \dim_\tau(\mathscr{A}) - \dim_\tau(M) = 0$ by additivity and normality.

Semifinite traces

Let $\mathscr{A}$ be a von Neumann algebra with a semifinite trace τ. The definition of the dimension function only needs small modifications, and the proof of being well defined and other properties run basically like the one in the finite case [25, Appendix B].

We say that a projective $\mathscr{A}$-module P is τ-**finite** if P is finitely generated and if one, then any, representing projection matrix $M \in M_n(\mathscr{A})$, $P \cong \operatorname{im}(l_M :$

$\mathscr{A}^n \to \mathscr{A}^n)$, satisfies $(\tau \otimes \mathrm{id}_n)(M) < \infty$. The **dimension** $\dim_\tau(P) \in [0, \infty)$ of a τ-finite projective $\mathscr{A}$-module $P \cong \mathrm{im}(l_M)$ is defined as $(\tau \otimes \mathrm{id}_n)(M)$.

Definition 14.36 Let M be an arbitrary $\mathscr{A}$-module. Its **dimension** is defined as

$$\dim_\tau(M) = \sup\{\dim_\tau(P) \mid P \subset M \ \tau\text{-finite projective submodule}\} \in [0, \infty].$$

Similarly as before, the notation is consistent with the one for τ-finite projective modules, and additivity and normality hold true.

Remark 14.37 While the situation in the semifinite case is quite similar to the finite case, there is one important difference: If τ is not finite, then $\dim_\tau(\mathscr{A}) = \tau(1_\mathscr{A}) = \infty$.

Notation 14.38 We write $\dim_{(G,\mu)}$ instead of $\dim_{\tau_{(G,\mu)}}$ and $\dim_\Gamma$ instead of $\dim_{\tau_\Gamma}$.

Example 14.39 Let $K < G$ be an open-compact subgroup. Then the characteristic function χ_K is continuous and $P = \lambda(\frac{1}{\mu(K)}\chi_K)$ is a projection in $L(G)$. According to the remark at the end of Subsection 14.3.2, we have $\tau_{(G,\mu)}(P) = 1/\mu(K)$, thus the dimension of the projective $L(G)$-module $P \cdot L(G)$ is $1/\mu(K)$. The $L(G)$-module $P \cdot L^2(G)$ is not projective, but, by an argument involving rank completion, one can show [25, B.25 Proposition] that

$$\dim_{(G,\mu)}(P \cdot L^2(G)) = \dim_{(G,\mu)}(P \cdot L(G)) = \frac{1}{\mu(K)}.$$

One easily verifies that $P \cdot L^2(G)$ consists of all left K-invariant functions, so

$$P \cdot L^2(G) = L^2(G)^K.$$

14.4 L^2-Betti numbers of groups

Throughout, G denotes a second countable, locally compact, unimodular group.

14.4.1 Definition

In the definition of L^2-cohomology we only used the left G-module structure. The right $L(G)$-module structure survives the process of taking G-invariants of the bar resolution, and so $H^*(G, L^2(G))$ inherits a right $L(G)$-module structure from the one of $L^2(G)$. The following definition is due to Petersen for second countable, unimodular, locally compact G. It is modelled after and coincides with Lück's definition for discrete G.

Definition 14.40   The L^2-Betti numbers of G with Haar measure μ are defined as

$$\beta_p(G,\mu) := \dim_{(G,\mu)}\left(H^p(G,L^2(G))\right) \in [0,\infty].$$

Remark 14.41   In Lück's book [17], where only the case of discrete G is discussed, the L^2-Betti numbers are defined as $\dim_G\left(H_p(G,L(G))\right)$. By [27, Theorem 2.2] Lück's definition coincides with the one above.

The homological algebra in Section 14.2 can be carried through such that the additional right $L(G)$-module structure is respected. In particular, we obtain (see Theorem 14.10):

Theorem 14.42   *Let G be totally disconnected and X a geometric model of G. Then*

$$\beta_p(G,\mu) = \dim_{(G,\mu)}\left(H^p\left(\mathrm{hom}_{\mathbb{C}}(C_*^{\mathrm{cw}}(X),L^2(G))^G\right)\right).$$

For totally disconnected groups we can compute the L^2-Betti numbers also through the reduced cohomology [25]:

Theorem 14.43   *Let G be totally disconnected. Then*

$$\beta_p(G,\mu) = \dim_{(G,\mu)}\left(\overline{H}^p(G,L^2(G))\right).$$

If X is a geometric model of G, then

$$\beta_p(G,\mu) = \dim_{(G,\mu)}\left(\overline{H}^p\left(\mathrm{hom}_{\mathbb{C}}(C_*^{\mathrm{cw}}(X),L^2(G))^G\right)\right).$$

Many properties of L^2-Betti numbers of discrete groups possess analogues for locally compact groups. We refer to [25] for more information and discuss only the Euler–Poincaré formula:

Let X be a cocompact proper G-CW complex. Let $K_1,\dots,K_n < G$ be the stabilisers of G-orbits of p-cells of X. The **weighted number of equivariant p-cells** of X is then defined as

$$c_p(X;G,\mu) = \mu(K_1)^{-1} + \cdots + \mu(K_n)^{-1}.$$

Definition 14.44   The **equivariant Euler characteristic** of X is defined as

$$\chi(X;G,\mu) := \sum_{p\geq 0}(-1)^p c_p(X;G).$$

Theorem 14.45 (Euler–Poincaré formula)   *Let G be totally disconnected, and let X be a cocompact geometric model of G. Then*

$$\sum_{p\geq 0}(-1)^p \beta_p(G,\mu) = \chi(X;G,\mu).$$

Proof Let $C^* := \hom_{\mathbf{C}}(C_*^{\mathrm{cw}}(X), L^2(G))^G$, and let H^* be the cohomology of C^*. Let Z^p be the cocycles in C^p and B^p be the coboundaries in C^p. Note that $c_p(X;G,\mu) = \dim_{(G,\mu)}(C^p)$. We have exact sequences $0 \to Z^p \to C^p \to B^{p+1} \to 0$ and $0 \to B^p \to Z^p \to H^p \to 0$ of $L(G)$-modules. By additivity of $\dim_{(G,\mu)}$ we conclude that

$$
\begin{aligned}
\chi(X;G,\mu) &= \sum_p (-1)^p \dim_{(G,\mu)}(C^p) \\
&= \sum_p (-1)^p (\dim_{(G,\mu)}(Z^p) + \dim_{(G,\mu)}(B^{p+1})) \\
&= \sum_p (-1)^p (\dim_{(G,\mu)}(B^p) + \dim_{(G,\mu)}(H^p) + \dim_{(G,\mu)}(B^{p+1})) \\
&= \sum_p (-1)^p \beta_p(G,\mu). \qquad\qquad\qquad \Box
\end{aligned}
$$

Remark 14.46 It turns out that

$$
\beta_p(G,\mu) > 0 \Leftrightarrow \overline{H}^p\big(\hom_{\mathbf{C}}(C_*^{\mathrm{cw}}(X), L^2(G))^G\big) \neq 0.
$$

In general, this is false for non-reduced continuous cohomology. So constructing non-vanishing harmonic cocycles is a way to show non-vanishing of L^2-Betti numbers (see Subsection 14.2.4). Example 14.18 shows that the first L^2-Betti number of a non-abelian free group is strictly positive.

14.4.2 Quasi-isometric and coarse invariance

Every locally compact, second countable group G possesses a left-invariant proper continuous metric [30], and any two left-invariant proper continuous metrics on G are coarsely equivalent. Thus G comes with a well defined coarse geometry. If G is compactly generated, then the word metric associated to a symmetric compact generating set is coarsely equivalent to any left-invariant proper continuous metric. Finally, any coarse equivalence between compactly generated second countable groups is a quasi-isometry with respect to word metrics of compact symmetric generating sets. A recommended background reference for these notions is Chapter 4 in [6].

As discussed in the introduction, neither the exact values nor the proportionality of L^2-Betti numbers are coarse invariants. But the vanishing of L^2-Betti numbers is a coarse invariant. The following result, which was proved by Pansu [22] for discrete groups admitting a classifying space of finite type, by Mimura–Ozawa–Sako–Suzuki [19, Cor. 6.3] for all countable discrete groups, and by Schrödl and the author in full generality [29], has a long history; we refer to the introduction in [29] for an historical overview.

Theorem 14.47 *The vanishing of the n-th L^2-Betti number of a unimodular, second countable, locally compact group is an invariant of coarse equivalence.*

To give an idea why this is true, let us consider compactly generated, second countable, totally disconnected, locally compact, unimodular groups G and H. We assume, in addition, that G and H admit cocompact (simplicial) geometric models X and Y, respectively. Endowed with the simplicial path metrics induced by the Euclidean metric on standard simplices, X and Y are quasi-isometric to G and H, respectively.

By Proposition 14.14 and Remark 14.46 we obtain that

$$\beta_p(G) > 0 \Leftrightarrow L^2\overline{H}^p(X) \neq 0.$$

Similarly for H and Y. Since X and Y are quasi-isometric and both are uniformly contractible, the connect-the-dots technique (see for example [3, Proposition A.1]) yields a Lipschitz homotopy equivalence $f\colon X \to Y$. Pansu [22] proves that from that we obtain an isomorphism

$$L^2\overline{H}^p(X) \cong L^2\overline{H}^p(Y)$$

in all degrees p, implying the above theorem.

The phenomenon that group homological invariants can be viewed as coarse-geometric invariants is not unique to the theory of L^2-cohomology or L^2-Betti numbers, of course. For instance, $H^*(\Gamma, \ell^\infty(\Gamma))$ is isomorphic to the uniformly finite homology by Block and Weinberger [2]. But unlike for $H^*(\Gamma, \ell^\infty(\Gamma))$ the Hilbert space structure allows to numerically measure the size of the groups $H^*(\Gamma, \ell^2(\Gamma))$, which are in general huge and unwieldy as abelian groups.

14.4.3 Examples and computations

Computations of L^2-Betti numbers of groups are rare, especially the ones where there is a non-zero L^2-Betti number in some degree. But sometimes the computation, at least the non-vanishing result, follows quite formally. We present two such cases.

Example 14.48 Let $\Gamma = F_2$ be the free group of rank 2. A geometric model is the 4-regular tree T. By the explicit cocycle construction in Example 14.18 we already know that $\beta_1(\Gamma) \neq 0$. But since $\beta_0(\Gamma) = 0$ (since Γ is infinite) and $\beta_p(\Gamma) = 0$ for $p > 1$, this also follows from the Euler–Poincaré formula:

$$\beta_1(\Gamma) = -\chi(T;\Gamma) = -\chi(S^1 \vee S^1) = 1.$$

The only relevant information for the previous example was the number of equivariant cells in the geometric model. The same technique helps in the next example (cf. [25, 5.29]).

Example 14.49 Let $G = SL_3(\mathbf{Q}_p)$. For exact computations of L^2-Betti numbers of Chevalley groups over $\mathbf{Q}_p$ or $\mathbf{F}_p((t^{-1}))$ (and their lattices) we refer to [26]. Here we only show $\beta_2(G,\mu) \neq 0$ as easily as possible and then apply this to the deficiency of lattices in G. As geometric model, we take the Bruhat–Tits building X of G, which is 2-dimensional. By the fact that there are no 3-cells and by additivity of dimension, we obtain that

$$\begin{aligned}
\beta_2(G,\mu) &= \dim_{(G,\mu)}(\operatorname{coker}(d^1)) \\
&\geq \dim_{(G,\mu)}\left(\hom_{\mathbf{C}}(C_2^{\mathrm{cw}}(X),L^2(G))^G\right) \\
&\quad - \dim_{(G,\mu)}\left(\hom_{\mathbf{C}}(C_1^{\mathrm{cw}}(X),L^2(G))^G\right).
\end{aligned}$$

In dimension 2 there is only one equivariant cell with stabiliser B, the Iwahori subgroup of G. Hence

$$\hom_{\mathbf{C}}(C_2^{\mathrm{cw}}(X),L^2(G))^G \cong \operatorname{map}(G/B,L^2(G))^G \cong L^2(G)^B$$

and with Example 14.39 it follows that $\dim_{(G,\mu)}\left(\hom_{\mathbf{C}}(C_2^{\mathrm{cw}}(X),L^2(G))^G\right) = 1/\mu(B)$. We normalise μ such that $\mu(B) = 1$. There are three equivariant 1-cells corresponding to the 1-dimensional faces of the 2-dimensional fundamental chamber. The stabiliser of each splits into $p+1$ many cosets of B. Therefore the μ-measure of each stabiliser is $(p+1)$. Similarly as above, this yields

$$\dim_{(G,\mu)}\left(\hom_{\mathbf{C}}(C_1^{\mathrm{cw}}(X),L^2(G))^G\right) = \frac{3}{p+1}.$$

Let $p \geq 3$. Then we obtain that $\beta_2(G,\mu) \geq 1 - \frac{3}{p+1} > 0$. Let us consider a lattice $\Gamma < G$. By Theorem 14.3,

$$\beta_2(\Gamma) = \mu(\Gamma\backslash G)\beta_p(G,\mu) \geq (1 - \frac{3}{p+1})\mu(\Gamma\backslash G).$$

Let R be a finite presentation of Γ, and let g be the number of generators and r be the number of relations in R. Let $X(R)$ be the universal covering of the presentation complex of R. One can regard $X(R)$ as the 2-skeleton of a geometric model Y from which we can compute the L^2-Betti numbers of Γ. By the Euler–Poincaré formula,

$$g - r = 1 - \chi(X(R);\Gamma) \leq 1 - \beta_0(\Gamma) + \beta_1(\Gamma) - \beta_2(\Gamma) \leq 1 - (1 - \frac{3}{p+1})\mu(\Gamma\backslash G).$$

We also used that Γ has property (T) which implies that $\beta_1(\Gamma) = 0$. Hence the deficiency of Γ, which is defined as the maximal value $g - r$ over all finite presentations, is bounded from above by $1 - (1 - 3/(p+1))\mu(\Gamma\backslash G)$.

Remark 14.50 The computation of L^2-Betti numbers of locally compact groups reduces to the case of totally disconnected groups. Let G be a (second countable, unimodular) locally compact group. If its amenable radical K, its largest normal amenable (closed) subgroup, is non-compact, then $\beta_p(G,\mu) = 0$ for all $p \geq 0$ by [15, Theorem C], which generalizes a result of Cheeger and Gromov for discrete groups [5]. So let us assume that K is compact. Endowing G/K with the pushforward v of μ, one obtains that $\beta_p(G/K,v) = \beta_p(G,\mu)$ [25, Theorem 3.14]. So we may and will assume that the amenable radical of G is trivial. Upon replacing G by a subgroup of finite index, G splits then as a product of a centrefree non-compact semisimple Lie group H and a totally disconnected group D. This is an observation of Burger and Monod [20, Theorem 11.3.4], based on the positive solution of Hilbert's 5th problem. Since H possesses lattices, one can use Borel's computations of L^2-Betti numbers of such lattices [17, Chapter 5] and Theorem 14.3 to obtain a computation for H. A Künneth formula [25, Theorem 6.7] then yields the L^2-Betti numbers of $G = H \times D$ provided one is able to compute the L^2-Betti numbers of D.

Acknowledgements. The author is grateful to Henrik Petersen for corrections and helpful comments.

References for this chapter

[1] Michael F. Atiyah. Elliptic operators, discrete groups and von Neumann algebras. In *Colloque "Analyse et Topologie" en l'Honneur de Henri Cartan*, Astérisque, No. 32–33, Soc. Math. France, Paris (1976), 43–72.

[2] Jonathan Block and Shmuel Weinberger. Large scale homology theories and geometry. In *Geometric topology* , AMS/IP Stud. Adv. Math. Vol. 2 Amer. Math. Soc., Providence, RI (1997), 522–569.

[3] Noel Brady and Benson Farb. Filling-invariants at infinity for manifolds of non-positive curvature. *Trans. Amer. Math. Soc.*, 350 (1998) no. 8, 3393–3405.

[4] Kenneth S. Brown. *Cohomology of groups*, volume 87 of *Graduate Texts in Mathematics*. Springer-Verlag, New York (1994).

[5] Jeff Cheeger and Mikhail Gromov. L_2-cohomology and group cohomology. *Topology*, 25 (1986) no. 2, 189–215.

[6] Yves Cornulier and Pierre de la Harpe. *Metric Geometry of Locally Compact Groups*. EMS Tracts in Mathematics Vol. 25. European Mathematical Society, Zürich (2016).

[7] Michael W. Davis, Jan Dymara, Tadeusz Januszkiewicz, John Meier and Boris Okun. Compactly supported cohomology of buildings. *Comment. Math. Helv.*, 85 (2010) no. 3, 551–582.

[8] Pierre de la Harpe. *Topics in geometric group theory*. Chicago Lectures in Mathematics. University of Chicago Press, Chicago, IL (2000).

[9] Jan Dymara. Thin buildings. *Geom. Topol.*, 10 (2006) 667–694.

[10] Michael Farber. von Neumann categories and extended *L²*-cohomology. *K-Theory*, 15 (1998) no. 4, 347–405.

[11] Damien Gaboriau. Invariant percolation and harmonic Dirichlet functions. *Geom. Funct. Anal.*, 15 (2005) no. 5 1004–1051.

[12] Damien Gaboriau. Invariants *L²* de relations d'équivalence et de groupes. *Publ. Math. Inst. Hautes Études Sci.*, 95 (2002) 93–150.

[13] Damien Gaboriau. On orbit equivalence of measure preserving actions. In: *Rigidity in dynamics and geometry*, 167–186, Springer, Berlin (2002).

[14] Alain Guichardet. *Cohomologie des groupes topologiques et des algèbres de Lie*, volume 2 of *Textes Mathématiques*. CEDIC, Paris (1980).

[15] David Kyed, Henrik D. Petersen and Stefaan Vaes. *L²*-Betti numbers of locally compact groups and their cross section equivalence relations. *ArXiv e-prints 1302.6753* (2013).

[16] Wolfgang Lück. Dimension theory of arbitrary modules over finite von Neumann algebras and *L²*-Betti numbers. I. Foundations. *J. reine angew. Math.*, 495 (1998) 135–162.

[17] Wolfgang Lück. *L²-invariants: theory and applications to geometry and K-theory*, volume 44 of *Ergebnisse der Mathematik und ihrer Grenzgebiete*. Springer-Verlag, Berlin (2002).

[18] Wolfgang Lück. Survey on classifying spaces for families of subgroups. In *Infinite groups: geometric, combinatorial and dynamical aspects*, volume 248 of *Progr. Math.*, page 269–322. Birkhäuser, Basel (2005).

[19] Masato Mimura, Narutaka Ozawa, Hiroki Sako and Yuhei Suzuki. Group approximation in Cayley topology and coarse geometry, III: Geometric property (T). *Algebr. Geom. Topol.*, 15 (2015) no. 2, 1067–1091.

[20] Nicolas Monod. *Continuous bounded cohomology of locally compact groups*, volume 1758 of *Lecture Notes in Mathematics*. Springer-Verlag, Berlin (2001).

[21] Francis J. Murray and John von Neumann. On rings of operators. *Ann. of Math. (2)*, 37 (1936) no. 1, 116–229.

[22] Pierre Pansu. Cohomologie *Lᵖ*: invariance sous quasiisometries. *Preprint* (1995).

[23] Gert K. Pedersen. *C*-algebras and their automorphism groups*, volume 14 of *London Mathematical Society Monographs*. Academic Press, Inc., London-New York (1979).

[24] Henrik D. Petersen. *L²*-Betti numbers of locally compact groups. *C. R. Math. Acad. Sci. Paris*, 351 (2013) no. 9–10, 339–342.

[25] Henrik D. Petersen. *L²*-Betti Numbers of Locally Compact Groups. *ArXiv e-prints 1104.3294* (2011).

[26] Henrik D. Petersen, Roman Sauer and Andreas Thom. *L²*-Betti numbers of totally disconnected groups and their approximation by Betti numbers of lattices. *ArXiv e-prints 1612.04559* (2016).

[27] Jesse Peterson and Andreas Thom Group cocycles and the ring of affiliated operators. *Invent. Math.*, 185 (2011) no. 3, 561–592.

[28] Roman Sauer. *L²*-Betti numbers of discrete measured groupoids. *Internat. J. Algebra Comput.*, 15 (2005) no. 5–6, 1169–1188.

226 Sauer

[29] Roman Sauer and Michael Schrödl. Vanishing of ℓ^2-Betti numbers of locally compact groups as an invariant of coarse equivalence. *ArXiv e-prints 1702.01685* (2017).

[30] Raimond A. Struble. Metrics in locally compact groups. *Compositio Math.*, 28 (1974), 217–222.

[31] Masamichi Takesaki. *Theory of operator algebras. I*, volume 124 of *Encyclopaedia of Mathematical Sciences*. Springer-Verlag, Berlin (2002).

[32] Tammo tom Dieck. *Transformation groups*, volume 8 of *de Gruyter Studies in Mathematics*. Walter de Gruyter & Co., Berlin (1987).

15

Minimal normal closed subgroups in compactly generated tdlc groups

Thibaut Dumont and Dennis Gulko

Abstract

We present results of P.-E. Caprace and N. Monod on the existence of minimal closed normal subgroups in compactly generated totally disconnected locally compact groups. Under certain conditions, such subgroups exist and at most finitely many of them are non-abelian.

15.1	Introduction	227
15.2	Cayley–Abels graphs	230
15.3	Minimal closed normal subgroups	232
	References for this chapter	235

15.1 Introduction

In the paper [3], P.-E. Caprace and N. Monod discuss the structure of locally compact groups focusing on compactly generated groups. They investigate the existence of minimal closed normal subgroups or, equivalently, of maximal Hausdorff quotients. By first treating the totally disconnected case, an existence result is produced using the solution to Hilbert's fifth problem under the assumption that the ambient group does not contain:

(a) infinite discrete normal subgroup, nor
(b) non-trivial closed normal subgroup which is compact-by-{connected soluble}.[1]

---

[1] *i.e.* fitting in a short exact sequence $1 \to K \to G \to Q \to 1$ with K compact and Q connected and soluble.

In this note, we focus on totally disconnected groups and thus let G be a compactly generated tdlc group. The result holds if we relax (a) and (b) by merely assuming the absence of non-trivial, discrete or compact, closed normal subgroups in G. In order to obtain minimal closed normal subgroups, we invoke Zorn's lemma, but in fact Proposition 15.3 below shows that the intersection of any filtering family of non-trivial closed normal subgroups is non-trivial; even though checking the same condition for chains would suffice. Assuming G not to have non-trivial compact normal subgroups guarantees the group to act faithfully on its Cayley–Abels graphs as a consequence of Proposition 15.7. Those are connected, locally finite, regular graphs on which G acts transitively. Thanks to the faithfulness, a filtering family of closed normal subgroups has non-trivial intersection as soon as all these subgroups are non-discrete, which is our relaxed assumption, see Proposition 15.8.

Away from inevitable abelian phenomena like in Example 15.4, there cannot be too many minimal closed normal subgroups. Indeed, in the complementary note [4] by the same authors, they prove that only finitely many minimal closed normal subgroups can be non-commutative. However the abelian ones remain controlled by the locally elliptic radical, see Proposition 15.3. Assuming only (a) and (b), the Theorem 15.5 below shows the result to hold for compactly generated locally compact groups that are non-necessarily totally disconnected.

15.1.1 Preliminaries

Throughout the paper, we adopt the following conventions. Every locally compact group is assumed to be Hausdorff. A subgroup of a topological group G is called **characteristic** if it is invariant under all topological automorphisms of G. Accordingly, a topological group is called **characteristically simple** if it has exactly two characteristic subgroups. We use the script font to denote certain families of subgroups of a given group: $\mathscr{E}, \mathscr{F}, \mathscr{M}, \mathscr{N}$, etc. On the other hand, $\mathscr{N}_G(H)$ denotes the normaliser in G of a subset H, $\mathscr{Z}_G(H)$ the centraliser of H in G and $\mathscr{Z}(H) = \mathscr{Z}_H(H)$ denotes the centre of H. The identity element is denoted by 1 and the trivial group by **1**. Finally, whenever a *minimal element M of a family $\mathscr{N}$* of subgroups is mentioned, *e.g.* minimal among the collection of closed normal subgroups, it is implicit that M is minimal in the subfamily $\mathscr{N} \setminus \{1\}$, hence M is non-trivial as well.

We recall the notion of locally elliptic subgroups first introduced by V. P. Platonov [6].

Definition 15.1   A subgroup H of a topological group G is called **locally**

elliptic if every finitely generated subgroup of H has compact closure in G. Compact subgroups are obvious and important examples.

Proposition 15.2 *Any locally compact group G possesses a unique maximal normal locally elliptic subgroup* $\mathrm{Rad}_{\mathscr{LE}}(G)$, *called the* **locally elliptic radical** *of G. Moreover,* $\mathrm{Rad}_{\mathscr{LE}}(G)$ *is a characteristic closed subgroup and* $\mathrm{Rad}_{\mathscr{LE}}(G/\mathrm{Rad}_{\mathscr{LE}}(G)) = 1$. □

15.1.2 The statements

We now state the results mentioned in the introduction.

Proposition 15.3 (Proposition 2.6, [4]) *Let G be a compactly generated tdlc group without non-trivial compact or discrete normal subgroup. Then:*

(i) *Every non-trivial closed normal subgroup contains a minimal one.*

(ii) *Let $\mathcal{M}$ be the set of minimal closed normal subgroups and $\mathcal{M}_{na}$ be the subset of non-abelian ones. Then $\mathcal{M}$ might be infinite but $\mathcal{M}_{na}$ is finite.*

(iii) *Each abelian $M \in \mathcal{M}$ is locally elliptic, hence contained in $\mathrm{Rad}_{\mathscr{LE}}(G)$. In particular, if $\mathrm{Rad}_{\mathscr{LE}}(G) = 1$, then $\mathcal{M} = \mathcal{M}_{na}$ is finite.*

(iv) *For any proper $\mathscr{E} \subset \mathcal{M}$, the subgroup $N_{\mathscr{E}} = \overline{\langle M \mid M \in \mathscr{E} \rangle}$ is properly contained in G.*

Example 15.4 Consider the semi-direct product $G = \mathbf{Q}_p^n \rtimes_p \mathbf{Z}$ where the generator 1 of $\mathbf{Z}$ acts on the n-dimensional $\mathbf{Q}_p$-vector space $\mathbf{Q}_p^n$ by multiplication by p. Then G is a tdlc group generated by the compact subset $\mathbf{Z}_p^n \times \{1\}$ and has no compact or discrete normal subgroup other than $\mathbf{1}$. Any one dimensional subspace of $\mathbf{Q}_p^n$ is a minimal closed normal subgroup. If $n \geq 2$, there are uncountably many such abelian subgroups.

Using the solution to Hilbert's fifth problem and some more machinery, one obtains the following result for arbitrary compactly generated locally compact groups. The proof is omitted.

Theorem 15.5 (Theorem B, [4]) *Let G be a compactly generated locally compact group. Then at least one of the following holds.*

(i) *G has an infinite discrete normal subgroup.*

(ii) *G has a non-trivial closed normal subgroup which is compact-by-{soluble connected}.*

(iii) *There exist non-trivial minimal closed normal subgroups, of which only finitely many are non-abelian.*

15.2 Cayley–Abels graphs

Compactly generated tdlc groups naturally act on certain connected locally finite regular graphs known as **Cayley–Abels graphs**, see §1.3 in Chapter 1 of this volume. The starting point of P.-E. Caprace and N. Monod was to observe that whenever one of the latter actions is faithful, any filtering family of non-discrete closed normal subgroups has non-trivial intersection. We shall see that this is always the case modulo a compact subgroup. The method was inspired by the work [2] of M. Burger and S. Mozes.

Let G be a compactly generated tdlc group and let $\mathscr{B}(G)$ denote the set of compact open subgroups of G. The latter forms a neighbourhood base of the identity thanks to van Dantzig's theorem [5].

Definition 15.6 (Abels, [1])   Let $U \in \mathscr{B}(G)$ and C be a compact generating set. Assume C to be bi-U-invariant by replacing it by UCU if necessary. By definition the **Cayley–Abels graph** $\mathfrak{g} = (V_{\mathfrak{g}}, E_{\mathfrak{g}})$ associated to these data has vertex set $V_{\mathfrak{g}} = G/U$ and one connects gU to gcU with an edge for each $g \in G$ and $c \in C$.

Proposition 15.7   *The Cayley–Abels graph $\mathfrak{g}$ is connected, locally finite and regular. Moreover, G acts transitively and continuously on $V_{\mathfrak{g}}$ by left multiplication on cosets. The kernel of the action is $Q = \cap_{g \in G} gUg^{-1}$, hence G/Q acts faithfully on $\mathfrak{g}$.*

Proof   The connectedness of $\mathfrak{g}$ is clear since C generates G. Transitivity is obvious. Since C is compact and U is open, the set $C/U = \{cU \mid c \in C\}$ is finite, thus any vertex has $|C/U|$ neighbours. Here $\mathfrak{g}$ has the discrete topology. Nevertheless, the action is continuous if and only if each vertex stabiliser is open. But clearly the stabiliser of $v = gU$ is $G_v = gUg^{-1}$, hence compact and open, implying continuity. The last statement is now clear.   □

Proposition 15.8   *Let U and C be as above and suppose G acts faithfully on $\mathfrak{g}$. Then any filtering family of non-discrete closed normal subgroups of G has a non-trivial intersection.*

Proof   Let $\mathscr{N}$ be a filtering family of non-discrete closed normal subgroups of G. For every vertex $v_0 \in V_{\mathfrak{g}}$ and every $N \in \mathscr{N}$, the stabiliser of v_0 in N, namely $N_{v_0} = N \cap G_{v_0}$, acts on the set of neighbours of v_0. Let $v_0^{\perp}$ denote this set and write $F_N < \mathrm{Sym}(v_0^{\perp})$ for the permutation subgroup defined by N in this way. We claim that the finite group F_N is independent of the choice of the vertex v_0. For every $v_0, v_1 \in V_{\mathfrak{g}}$, the subgroups N_{v_1} and N_{v_0} are conjugate.

Indeed, transitivity implies that $v_1 = gv_0$ for some $g \in G$, thus

$$N_{v_1} = N \cap G_{gv_0} = gNg^{-1} \cap gG_{v_0}g^{-1} = gN_{v_0}g^{-1},$$

where we used that N is a normal subgroup. Thus, N_{v_0} and N_{v_1} yield isomorphic groups of permutations as claimed.

We now show that F_N is non-trivial. For if N_{v_0} acts trivially on $v_0^{\perp}$, then $N_{v_0} \subset N_{v_1}$ for all $v_1 \in v_0^{\perp}$. Recall that N_{v_1} permutes its own neighbours via F_N; this means that N_{v_0} fixes every vertex at distance 2 of v_0. Inductively, N_{v_0} fixes all vertices in $\mathfrak{g}$. By faithfulness, $\mathbf{1} = N_{v_0} = N \cap G_{v_0}$ which proves N to be discrete, a contradiction.

It is clear that $\{F_N \mid N \in \mathcal{N}\}$ is a filtering family of subgroups sitting inside the finite group $\mathrm{Sym}(v_0^{\perp})$. In other words, it is a filtering family of finitely many non-trivial subgroups. Therefore, there is a unique minimal non-trivial subgroup F_{N_0} for some $N_0 \in \mathcal{N}$ and

$$\bigcap_{N \in \mathcal{N}} F_N = F_{N_0} \neq \mathbf{1}.$$

Fix $g \neq 1$ in F_{N_0} and let $N_g \subset N_{v_0}$ be the subset of elements acting as g on $v_0^{\perp}$. The family $\{N_g \mid N \in \mathcal{N}\}$ consists of closed non-empty subsets of the compact subgroup G_{v_0}. This family is filtering; by compactness, it must have non-empty intersection,

$$\varnothing \neq \bigcap_{N \in \mathcal{N}} N_g \subset \bigcap_{N \in \mathcal{N}} N \setminus \mathbf{1}.$$

$\square$

The two previous propositions together show that for compactly generated tdlc groups the conclusion of Proposition 15.8 holds up to modding out a compact kernel. The next result shows that the latter can be chosen as small as one wishes.

Proposition 15.9 (Proposition 2.5, [3]) *Let G be a compactly generated tdlc group and V an identity neighbourhood. Then there is a compact normal subgroup $Q_V \subset V$ such that any filtering family of non-discrete closed normal subgroups of G/Q_V has a non-trivial intersection.*

Proof Given a neighbourhood V of the identity, there is $U \in \mathcal{B}(G)$ such that $U \subset V$ and let Q_V be the intersection of all conjugates of U. Then Q_V is compact normal and the group G/Q_V acts faithfully on the Cayley–Abels graph $\mathfrak{g}$ defined with respect to U thanks to Proposition 15.7. One uses Proposition 15.8 to conclude. $\square$

Remarks 15.10   *(i) Readily, if G has no compact normal subgroup other than 1, then $Q_V = 1$, hence the result holds for G itself.*
(ii) If, in addition, the only non-trivial closed normal subgroups of G are the non-discrete ones, then the conclusion of Proposition 15.9 holds for any filtering family of closed normal subgroups. In this situation we can finally apply Zorn's lemma to obtain that any non-trivial closed normal subgroup must contain a minimal one.

15.3 Minimal closed normal subgroups

In this section we would like to prove the existence of finitely many non-abelian minimal closed normal subgroups as stated in Proposition 15.3. We will start by proving a corollary to the last proposition, which will play an important role later on.

Corollary 15.11 (Corollary to Proposition 2.5, [4])   *Let G be a compactly generated totally disconnected locally compact group and let $\mathcal{N}$ be a filtering family of closed normal subgroups of G. Then there exist $N \in \mathcal{N}$ and a closed subgroup $Q \trianglelefteq G$ such that $B = \bigcap \mathcal{N}$ is contained in Q, Q/B is compact and $N/N \cap Q$ is discrete.*

Proof   Apply Proposition 15.9 to $\overline{G} := G/B$. We get a compact subgroup $\overline{Q} \trianglelefteq \overline{G}$ which lifts to a closed subgroup $Q \trianglelefteq G$ containing B as a cocompact subgroup. Furthermore, the isomorphism $\overline{G}/\overline{Q} \cong G/Q$ implies that any filtering family of closed normal subgroups of G/Q has non-trivial intersection, see (ii) of Remarks 15.10. Recall that a continuous map of locally compact Hausdorff spaces is proper if and only if it is closed and the inverse image of a point is compact. Hence any continuous quotient homomorphism of locally compact groups is closed as soon as its kernel is compact. Therefore, the quotient map $\pi : G/B \to G/Q$ is closed and $\{NQ/Q\}_{N \in \mathcal{N}}$ is a filtering family of closed normal subgroups of G/Q. By the definition of B, $\bigcap\{N/B\}_{N \in \mathcal{N}}$ is trivial in G/B, but this implies that $\bigcap\{NQ/Q\}_{N \in \mathcal{N}}$ is trivial in G/Q (as the map $\bigcap\{NQ/Q\}_{N \in \mathcal{N}} \to \bigcap\{N/B\}_{N \in \mathcal{N}}$, defined by $nQ \mapsto nB$ is injective). This, by Proposition 15.9, implies in turn that for at least one $N \in \mathcal{N}$, $NQ/Q \cong N/N \cap Q$ is discrete, as required.   □

We will also use the following lemma.

Lemma 15.12 (Lemma, [4])   *A profinite group which is characteristically simple is isomorphic to the direct product of copies of a given finite simple group.*

We now complete the proof of Proposition 15.3.

Proof of Proposition 15.3 As part (ii) is the most complicated, we will prove it last:

(i) It's an immediate consequence of Proposition 15.9 and Zorn's lemma, as mentioned in the Remarks 15.10.

(iii) Let $M \in \mathcal{M}$ be abelian. Since M is non-discrete, it must contain a non-trivial compact open subgroup. Hence $\mathrm{Rad}_{\mathscr{LE}}(M) \neq \mathbf{1}$. As $\mathrm{Rad}_{\mathscr{LE}}(M)$ is a characteristic subgroup of M, we have $\mathrm{Rad}_{\mathscr{LE}}(M) = M$. In particular, if $\mathrm{Rad}_{\mathscr{LE}}(G) = \mathbf{1}$ then $\mathcal{M} = \mathcal{M}_{\mathrm{na}}$.

(iv) Let $\mathscr{E} \subsetneq \mathcal{M}$ be a proper subset. Denote $M_{\mathscr{E}} = \langle M \mid M \in \mathscr{E} \rangle$ and let $N_{\mathscr{E}} = \overline{M_{\mathscr{E}}}$ be its closure. Observe that for any $M' \in \mathcal{M} \setminus \mathscr{E}$ and $M \in \mathcal{M}$, we have $[M', M] \subseteq M' \cap M$. By the minimality assumption on both M', M we have $M \cap M' = \mathbf{1}$. Hence $[M', M] = \mathbf{1}$ and so $[M', M_{\mathscr{E}}] = \mathbf{1}$, which implies that $[M', N_{\mathscr{E}}] = \mathbf{1}$ as well.

Assume for contradiction that $N_{\mathscr{E}} = G$. Then $M' \leq \mathscr{Z}(G)$ and hence M' is abelian. As any subgroup of M' will be normal in G, this implies that M' has to be finite of prime order, but this contradicts our assumption on G.

(ii) As we have seen in (iv), any distinct $M, M' \in \mathcal{M}$ commute. Hence, if $\mathscr{E}_1 \cap \mathscr{E}_2 = \varnothing$ then $N_{\mathscr{E}_1}, N_{\mathscr{E}_2}$ commute.

Consider the family $\mathcal{N} = \left\{ N_{\mathcal{M}_{\mathrm{na}} \setminus \mathscr{F}} \mid \mathscr{F} \subseteq \mathcal{M}_{\mathrm{na}}, \; |\mathscr{F}| < \infty \right\}$. It is a filtering family of closed normal subgroups of G, which are non-discrete. Denote $B = \bigcap \mathcal{N}$. By Proposition 15.8, if B is trivial then one of the elements of $\mathcal{N}$ is trivial, *i.e.* $\mathcal{M}_{\mathrm{na}}$ is finite.

Assume from now on that B is non-trivial. By Corollary 15.11, we can find a closed subgroup $Q \trianglelefteq G$ containing B and $N = N_{\mathcal{M}_{\mathrm{na}} \setminus \mathscr{F}} \in \mathcal{N}$ such that Q/B is compact and $N/N \cap Q$ is discrete.

Claim 15.13 *Under our notations and assumptions, there exists an $M \in \mathcal{M}_{\mathrm{na}} \setminus \mathscr{F}$ which admits a discrete quotient with a non-trivial centre.*

Proof Firstly, we would like to show that $N \leq Q$. Indeed, if $M \in \mathcal{M}_{\mathrm{na}} \setminus \mathscr{F}$ and $M \not\leq Q$ then $M \cap Q = \mathbf{1}$ (by the minimality of M). Hence the map $N \to N/N \cap Q$ restricted to M is injective, with discrete image, so M is discrete, which is a contradiction. Hence $N \leq Q$.

Using the remark above, we see that B commutes with any $M \in \mathcal{M}_{\mathrm{na}}$, and hence $B \leq \mathscr{Z}(N) \leq N \leq Q$. As Q/B is compact and N is closed in Q, so N/B is compact, hence so is $N/\mathscr{Z}(N)$.

Take any compact open subgroup $U \leq N$. Then $\mathcal{N}_N(U)$ is open since it contains U. And since $\mathscr{Z}(N) \leq \mathcal{N}_N(U)$, it is also cocompact in N. This implies that $N/\mathcal{N}_N(U)$ is compact and discrete, hence finite, and in particular, U has only finitely many conjugates in N. Hence $V = \bigcap_{n \in N} nUn^{-1}$ is a compact open normal subgroup of N. As N/V is discrete, we have $N/V = \langle MV/V \mid M \in \mathcal{M}_{na} \setminus \mathcal{F} \rangle$. If $\mathcal{Z}(N/V)$ is trivial then $B \leq V$ and hence B is compact. This contradicts the assumption on G. Thus $\mathcal{Z}(N/V)$ is non-trivial and we can find $M_0, \ldots, M_n \in \mathcal{M}_{na} \setminus \mathcal{F}$ and $m_i \in M_i$ such that $m_0 \cdot \ldots \cdot m_n$ is projected to a non-trivial central element in N/V. Assume that m_0 has non-trivial image. Then, as all elements of $\mathcal{M}_{na} \setminus \mathcal{F}$ commute pairwise, m_0 has to project to a non-trivial central element of $M_0 V/V$, *i.e.* $M_0 V/V$ is discrete with non-trivial centre. $\square$

Together with the next claim, we obtain the desired contradiction.

Claim 15.14   *Under our notations and assumptions, for all $M \in \mathcal{M}_{na} \setminus \mathcal{F}$, any discrete quotient of M is centrefree.*

Proof   Fix any $M \in \mathcal{M}_{na} \setminus \mathcal{F}$. As Q/B is compact, so is N/B and hence so is $\overline{BM}/B$. Since $B \leq \mathcal{Z}_G(M)$, we have $\overline{M\mathcal{Z}_G(M)}/\mathcal{Z}_G(M)$ is compact. Consider now $\pi : G \to G/\mathcal{Z}_G(M)$. We should have $\pi|_M$ injective, as if $\ker(\pi|_M)$ is non-trivial, then it has to be equal to M (by minimality) but this implies that $M \leq \mathcal{Z}_G(M)$, *i.e.* M is abelian, contradicting the choice of M. Now we would like to show that $\overline{\pi(M)}$ is a minimal closed normal subgroup of $G/\mathcal{Z}_G(M)$.

Take any $\overline{P} \leq G/\mathcal{Z}_G(M)$ non-trivial normal closed subgroup. It lifts to a closed subgroup $P \trianglelefteq G$ such that $\mathcal{Z}_G(M) \subsetneq P$. Thus $1 \neq [P,M] \leq P \cap M$, and, by minimality, $M \leq P$. Hence $\overline{\pi(M)} \leq \overline{P}$.

As $\overline{\pi(M)}$ is a minimal closed normal subgroup, it has to be characteristically simple. By Lemma 15.12, $\overline{\pi(M)}$ is isomorphic to a product of a finite simple group S, which is not abelian (as M is not abelian and $\pi|_M$ is injective). Let S denote such a factor. As $\overline{\pi(M)}$ is not abelian, we have $1 \neq [S, \pi(M)] \leq S \cap \pi(M)$. As S is simple, we must have $[S, \pi(M)] = S$. Hence $\pi(M)$ contains the direct sum of all the simple factors of $\overline{\pi(M)}$, call it D. Now, $D' = \pi^{-1}(D) \cap M$, is a normal subgroup of G and contained in M, isomorphic to D and dense in M, by minimality. D' is a direct sum of copies of a simple group, hence any discrete quotient of D' is centrefree. Since D' is dense, it implies that any discrete quotient of M is centrefree. $\square$

$\square$

Acknowledgements. We would like to thank the MFO institute for its hospitality during the Arbeitsgemeinschaft. The authors wish to warmly thank P.-E.

Caprace and N. Monod for taking the time to answer our questions, as well as the anonymous referee for his/her careful reading of this note. Once again, all results presented here are due to the former. We admittedly followed closely the original papers [3], [4] as the formulation suited the purpose of the present volume.

References for this chapter

[1] H. Abels, *Specker-Kompaktifizierungen von lokal kompakten topologischen Gruppen*, Math. Z. **135**, 325–361 (1973/74).

[2] M. Burger, S. Mozes, *Groups acting on trees: from local to global structure*, Inst. Hautes Etudes Sci. Publ. Math. **92** (2000), 113–150.

[3] P.-E. Caprace, N. Monod, *Decomposition locally compact groups into simple pieces*, Math Proc. Cambridge Philos. Soc. **150** (2011), no. 1, 97–128.

[4] P.-E. Caprace, N. Monod, *Correction to: "Decomposition locally compact groups into simple pieces"*, to appear.

[5] D. van Dantzig, *Studien over topologische algebra* (proefschrift). Ph.D. thesis (Groningen, 1931).

[6] V. P. Platonov, *Locally projectively nilpotent radical in topological groups* (Russian), Dokl. Akad. Nauk BSSR **9** (1965), 573–577.

16

Elementary totally disconnected locally compact groups, after Wesolek

Morgan Cesa and François Le Maître

Abstract

We present the main properties of the class of elementary totally disconnected locally compact (tdlc) groups, recently introduced by Wesolek, along with a decomposition result by the same author which shows that these groups along with topologically characteristically simple groups may be seen as building blocks of totally disconnected locally compact groups.

16.1	Introduction	236
16.2	Prerequisites	238
16.3	Examples and non-examples	240
16.4	Closure properties	243
16.5	The construction rank on elementary groups	244
16.6	More permanence properties	250
16.7	The decomposition theorem	254
	References for this chapter	257

16.1 Introduction

The class $\mathscr{E}$ of elementary tdlc (totally disconnected locally compact) Polish[1] groups was recently introduced by Wesolek in [12] and is defined below.

Definition 16.1 The class $\mathscr{E}$ of **elementary groups** is the smallest class of tdlc Polish groups such that

(E1) $\mathscr{E}$ contains all Polish groups which are either profinite or discrete;

---

[1] Or equivalently second-countable, see Sec. 16.2.

(E2) whenever $N \leqslant G$ is a closed normal subgroup of a tdlc Polish group G, if $N \in \mathscr{E}$ and G/N is profinite or discrete then $G \in \mathscr{E}$;

(E3) if a Polish tdlc group G can be written as a countable increasing union of open subgroups belonging to $\mathscr{E}$, then $G \in \mathscr{E}$.

Much like in the case of amenable elementary groups, the class of elementary groups enjoys strong closure properties: it is closed under group extension, taking closed subgroups, Hausdorff quotients, and inverse limits.

Examples of elementary groups include solvable tdlc Polish groups. It is not known whether every tdlc Polish amenable group is an elementary group. A wealth of non-elementary groups is provided by compactly generated, topologically simple, non-discrete groups. As a consequence, for all $n \geqslant 3$, neither the group of automorphisms of the n-regular tree nor the special linear group of degree n over $\mathbf{Q}_p$ are elementary.

The most remarkable feature of elementary groups is that they (along with topologically characteristically simple non-elementary groups) may be seen as building blocks for general tdlc Polish groups. To be more precise, using results of Caprace and Monod [2], Wesolek proved the following structure theorem.

Theorem 16.2 ([12, Thm. 1.6]) *Let G be a compactly generated tdlc Polish group. Then there exists a finite increasing sequence*

$$H_0 = \{e\} \leqslant \cdots \leqslant H_n$$

of closed characteristic subgroups of G such that

(i) G/H_n is an elementary group and

(ii) for all $i = 0, \ldots, n-1$, the group $(H_{i+1}/H_i)/\mathrm{Rad}_{\mathscr{E}}(H_{i+1}/H_i)$ is a finite quasi-product[2] of topologically characteristically simple non-elementary subgroups, where $\mathrm{Rad}_{\mathscr{E}}(H)$ denotes the elementary radical[3] of H.

Our main goal here is to present the proof of the above result in details, which requires us to study the aforementioned closure properties of the class of elementary groups closely. To prove these closure properties, Wesolek makes heavy use of the ordinal-valued construction rank, which is basically a tool for using induction on elementary groups. Here, we try to avoid the construction rank as much as possible so as to make the proofs simpler and to highlight where this rank is really needed. In particular, we prove the following results directly.

---

[2] For the definition of quasi-products, see Def. 16.40.

[3] The elementary radical is the greatest elementary closed normal subgroup; see Thm. 16.38.

- Every topologically simple compactly generated elementary group is discrete (Prop. 16.11).
- $\mathscr{E}$ is closed under extension (Prop. 16.15).
- $\mathscr{E}$ is closed under taking closed subgroups (Prop. 16.16).

However, in order to prove that the class of elementary groups is closed under taking Hausdorff quotients, we could not avoid the use of ordinals. We thus took the opportunity to give a gentle introduction to ordinals and define the construction rank in Section 16.5. Moreover, we use this construction rank only once, so as to get a general scheme for proving results by induction on elementary groups (Thm. 16.29). It is worth noting here that Wesolek defines a second rank on elementary groups, called the decomposition rank, which turns out to be more useful for studying elementary groups once their closure properties have been established (see [12, Sec. 4.3]). We will not need to use it here and so we will not define it.

With this induction scheme in hand, we prove the technical Lemma 16.33 which has the following two consequences.

- Every Hausdorff quotient of an elementary group is elementary (Thm. 16.34).
- If $N_1, \dots, N_k$ are elementary closed normal subgroups of a tdlc group G, then $\overline{N_1 \cdots N_k}$ is elementary (Cor. 16.36).

The latter result is fundamental, since it allows us to define the elementary radical of a tdlc Polish group G as the biggest closed normal elementary subgroup of G (Thm. 16.38). This radical is then used along with a fundamental result of Caprace and Monod to prove Theorem 16.2 in Section 16.7.

Section 16.2 contains some basic definitions and gives an introduction to Cayley–Abels graphs. In Section 16.3, we prove several propositions which lead to easy examples and non-examples of elementary tdlc Polish groups. Sections 16.4 and 16.6 cover closure properties of $\mathscr{E}$, with a discussion of ordinals and some background on the construction rank contained in Section 16.5. In Section 16.7, we discuss the elementary radical and prove the decomposition theorem for tdlc Polish groups.

16.2 Prerequisites

A topological group is **Polish** if it is separable and its topology admits a compatible complete metric. Polish groups should be regarded as nice groups: the completeness of a compatible metric allows for Baire category arguments, while the separability allows for constructive arguments. Polish locally compact groups are characterized as follows.

Theorem 16.3 (see [7, Thm. 5.3]) *For a locally compact group G, the following are equivalent.*

(i) *G is Polish.*

(ii) *G is second-countable.*

(iii) *G is σ-compact and the topology of G is metrizable.*

For general topological groups, being Polish is the strongest of the three properties above, so the totally disconnected locally compact groups satisfying these conditions will be called **tdlc Polish groups**.

We will also need Cayley–Abels graphs, and we first recall a lemma which gives an easy way of understanding them (see [12, Prop. 2.4]).

Lemma 16.4 *Let G be a compactly generated tdlc group, and let U be a compact open subgroup of G. Then there exists a finite symmetric[4] set $A \subseteq G$ such that $AU = UAU$ and*

$$G = \langle A \rangle U.$$

Moreover, if D is a dense subgroup of G, one can choose A as a subset of D.

Proof Let S be a compact symmetric generating set of G, and let D be a dense subgroup of G. Then $\{xU : x \in D\}$ is an open cover of S, so we may find a finite symmetric subset $B \subseteq D$ such that $S \subseteq BU$. Now, UB is also compact, and $UBU \cap D$ is dense in UBU, which contains UB. So we may find a finite symmetric subset A of $UBU \cap D$ such that $UB \subseteq AU$, and we may assume that A contains B. Now, since $A \subseteq UBU$ and U is a group, we actually have

$$UAU = UBU \subseteq AUU = AU,$$

and we conclude by induction that for all $n \geqslant 1$, $(UAU)^n = A^n U$. Since S is symmetric and $S \subseteq BU \subseteq UAU$ we deduce that $G = \langle A \rangle U$. $\square$

Now, given a compact open subgroup U of a tdlc Polish group G and a finite symmetric set $A \subseteq G$ such that $G = \langle A \rangle U$ and $UAU = AU$, the Cayley–Abels graph $\mathscr{C}_{A,U}(G)$ is defined the following way:

- its set of vertices is G/U and
- its set of edges is $\{(gU, gaU) : a \in A, g \in G\}$.

---

[4] A subset A of G is **symmetric** if $A^{-1} = A$.

Then the left action of G on G/U extends to a continuous transitive action of G on $\mathscr{C}_{A,U}(G)$ by graph automorphisms. In particular, all the vertices in $\mathscr{C}_{A,U}(G)$ have the same degree, and we will call that fixed number the **degree** of $\mathscr{C}_{A,U}(G)$. The fact that $UAU = AU$ ensures that the set of neighbours of U is exactly the set of aU for $a \in A$. But A is finite, so the degree of the Cayley–Abels graph $\mathscr{C}_{A,U}(G)$ is finite.

Suppose in addition that N is a closed normal subgroup of G, and let $\pi : G \to G/N$ be the natural projection. Then $G/N = \langle \pi(A) \rangle \, \pi(U)$ and $\pi(U)\pi(A)\pi(U) = \pi(A)\pi(U)$, so we may form the Cayley–Abels graph $\mathscr{C}_{\pi(A),\pi(U)}(G/N)$.

Observe that $\mathscr{C}_{\pi(A),\pi(U)}(G/N)$ is just the quotient of $\mathscr{C}_{A,U}(G)$ by the action of N by graph automorphisms, hence its degree is smaller than or equal to the degree of $\mathscr{C}_{A,U}(G)$. Moreover, by definition of the quotient graph, if two distinct neighbours of a vertex $v \in \mathscr{C}_{A,U}(G)$ are in the same N-orbit, then the degree of $\mathscr{C}_{\pi(A),\pi(U)}(G/N)$ is strictly smaller than the degree of $\mathscr{C}_{A,U}(G)$. This observation will be crucial to the proof of the decomposition theorem (see Lem. 16.43).

16.3 Examples and non-examples

Since it is central, we recall here the definition of the class of elementary groups.

Definition 16.5　　The class $\mathscr{E}$ of elementary groups is the smallest class of tdlc Polish groups satisfying the following properties.

(E1) The class $\mathscr{E}$ contains all Polish groups which are either profinite or discrete.[5]

(E2) Whenever $N \leqslant G$ is a closed normal subgroup of a tdlc Polish group G, if $N \in \mathscr{E}$ and G/N is profinite or discrete[6] then $G \in \mathscr{E}$.

(E3) If a Polish tdlc group G can be written as a countable increasing union of open subgroups belonging to $\mathscr{E}$, then $G \in \mathscr{E}$.

By definition, a topological group is **SIN**[7] if it admits a basis of open neighbourhoods of the identity, each of which is invariant under conjugacy. Note that this condition is automatically satisfied in an abelian topological group. Now, if G is a tdlc Polish SIN group, then by van Dantzig's theorem (see Chapter 1

---

[5]　A profinite group is Polish if and only if it is metrizable, while a discrete group is Polish if and only if it is countable.

[6]　Since G/N is automatically Polish (see [5, Thm. 2.2.10]), we see that equivalently, one could ask that G/N is profinite metrizable, or countable discrete.

[7]　SIN stands for "small invariant neighbourhoods".

in this volume) we may find a compact open subgroup $K \leqslant G$. The SIN condition ensures that the intersection U of the conjugates of K is still open. Then G/U must be a discrete group, so G is elementary by (E2). We have proved the following proposition.

Proposition 16.6   *Every SIN tdlc Polish group is elementary. In particular, any abelian tdlc Polish group is elementary.*

As we will see in the next section (Prop. 16.15), the class of elementary groups is closed under extension, which yields that every solvable tdlc Polish group is elementary by the previous proposition.

Question 16.7 (Wesolek)   Is every amenable[8] tdlc Polish group elementary?

Remark 16.8   Since amenability passes to closed subgroups and to quotients, the question may be reformulated by first defining the class of "amenable elementary groups" to be the smallest class containing profinite amenable groups and discrete amenable groups, stable under extension and exhaustive open unions. The question then becomes: Must every amenable tdlc Polish group be elementary?

A tdlc Polish group is **residually discrete** if for every finite subset $F \subseteq G \setminus \{1\}$, there is an open normal subgroup disjoint from F. It is **residually elementary** if for every finite subset $F \subseteq G \setminus \{1\}$, there is a closed normal subgroup N, disjoint from F, such that G/N is elementary.

Proposition 16.9   *Every residually discrete tdlc Polish group is elementary.*

Proof   Every tdlc Polish group can be written as an increasing union of compactly generated open subgroups (see Lem. 16.30). Moreover, being residually discrete passes to open subgroups, so by (E3) we only have to prove the proposition in the case when G is compactly generated. It is a theorem of Caprace and Monod that every locally compact, compactly generated, residually discrete group is SIN [2, Cor. 4.1], so we can conclude that G is elementary by Proposition 16.6.   □

In fact, adapting the proof of Caprace and Monod's aforementioned result, Wesolek was able to show the following remarkable result, which we state without proof.

Theorem 16.10 (see [12, Thm. 3.14])   *Every residually elementary group is elementary. In particular, any tdlc Polish group which can be written as the inverse limit of elementary groups is elementary.*

---

[8]   A good reference on amenability for locally compact groups is Appendix G in [1].

Note that the above result fails completely for the class of elementarily amenable discrete groups, since free groups are residually finite.

Let us now turn to non-examples of elementary groups. The following fact is closely related to Lemma 16.28, but as a warm-up to the next section we give here a detailed "ordinal-free" proof.

Proposition 16.11 *Every topologically simple compactly generated elementary group has to be discrete.*

Remark 16.12 As a consequence, the existence of a topologically simple compactly generated amenable non-discrete tdlc Polish group[9] would provide a negative answer to Question 16.7.

Proof of Proposition 16.11 Let G be a topologically simple compactly generated non-discrete group, and let $\mathscr{F}$ be the class of elementary groups which are not isomorphic to G. We want to show that $\mathscr{F}$ contains the class $\mathscr{E}$ of elementary groups, and so it suffices to check that $\mathscr{F}$ satisfies the same defining properties as $\mathscr{E}$ (see Def. 16.5) in order to conclude that $\mathscr{E} \subseteq \mathscr{F}$ by minimality.

Because G is neither discrete nor profinite,[10] $\mathscr{F}$ contains profinite and discrete groups, so (E1) is satisfied. The class $\mathscr{F}$ satisfies (E2) because G is topologically simple. Suppose that (E3) is not satisfied. Then we can write $G = \bigcup_{i \in \mathbf{N}} G_i$, where $(G_i)_{i \in \mathbf{N}}$ is an increasing chain of open subgroups of G and every G_i belongs to $\mathscr{F}$, hence is different from G. Let S be a compact generating set for G. Then $(G_i)_{i \in \mathbf{N}}$ is an open cover of S, so by compactness and the fact that $(G_i)_{i \in \mathbf{N}}$ is increasing, there exists $i \in \mathbf{N}$ such that G_i contains S. But then $G_i = G$, which is a contradiction. So (E3) is also satisfied by $\mathscr{F}$, which ends the proof.   □

Let us now apply Proposition 16.11 and give some non-examples of elementary groups.

Proposition 16.13 *Let $n \geqslant 3$. Neither the group $\mathrm{Aut}^+(T_n)$ generated by the vertex stabilizers in the automorphism group of the n-regular tree, nor the projective linear group $\mathrm{PSL}_n(\mathbf{Q}_p)$ are elementary.*

Proof By Proposition 16.11, it suffices to show that these are topologically simple non-discrete compactly generated groups. It is well known that both groups are simple: for $\mathrm{Aut}^+(T_n)$ this is a result of Tits [11], while for $PSl_n(\mathbf{Q}_p)$ a proof may be found in [3, Ch. II, 2]. That these groups are not discrete is

---

[9] In other words, we seek a non-discrete analogue of the derived groups of topological full groups of minimal subshifts (these are finitely generated infinite amenable simple groups, see [6, 9]).

[10] A non-discrete profinite group cannot be topologically simple!

clear. $\mathrm{Aut}^+(T_n)$ acts properly and cocompactly on the tree T_n, and $\mathrm{PSL}_n(\mathbf{Q}_p)$ acts properly and cocompactly on its Bruhat–Tits building, and hence both are compactly generated (a more direct proof for $\mathrm{PSL}_n(\mathbf{Q}_p)$ may be found in [4]). □

Remark 16.14 Since $\mathrm{Aut}^+(T_n)$ is an index 2 subgroup of $\mathrm{Aut}(T_n)$, we deduce from the above proposition and Corollary 16.17 that $\mathrm{Aut}(T_n)$ is non-elementary as well. And since every Hausdorff quotient of an elementary group is elementary, the special linear group $\mathrm{SL}_n(\mathbf{Q}_p)$ is also non-elementary.

16.4 Closure properties

Let us first present and prove some of the closure properties of the class of elementary groups which do not require the use of ordinals.

Proposition 16.15 *$\mathscr{E}$ is closed under extension: whenever G is a tdlc Polish group, and $H \lhd G$ is a closed normal subgroup such that both H and G/H are elementary, then G is elementary.*

Proof Fix an elementary group H, and consider the class $\mathscr{F}_H$ of tdlc Polish groups Q such that whenever G is a tdlc Polish group with $Q = G/H$, then G is elementary. By property (E2), $\mathscr{F}_H$ contains all profinite Polish groups and discrete Polish groups, so we have to show that $\mathscr{F}_H$ has the other two defining properties of $\mathscr{E}$.

- Suppose $N \in \mathscr{F}_H$ is a closed normal subgroup of some tdlc Polish group Q, and that Q/N is a profinite or discrete group. Moreover, suppose that G is a tdlc Polish group such that $Q = G/H$. We want to show that $G \in \mathscr{E}$, and thus that $Q \in \mathscr{F}_H$. To this aim we let $\pi : G \to Q$ denote the quotient map, then $\pi^{-1}(N)$ is a closed normal subgroup of H, and $\pi^{-1}(N)/H = N$. Because $N \in \mathscr{F}_H$, the group $\pi^{-1}(N)$ is elementary. But $G/\pi^{-1}(N) = Q/N$ is profinite or discrete Polish, so G is elementary.
- Suppose $Q = G/H$ is written as a countable increasing union of open subgroups O_i belonging to $\mathscr{F}_H$. Then $\pi^{-1}(O_i)$ belongs to $\mathscr{E}$ because $O_i \in \mathscr{F}_H$. We deduce that $G = \bigcup_i \pi^{-1}(O_i)$ is elementary. □

Proposition 16.16 *Let G be an elementary group. If H is a tdlc Polish group such that there exists a continuous injective homomorphism $\pi : H \to G$, then H is elementary. In particular, any closed subgroup of an elementary group is elementary.*

Proof Let $\mathscr{F}$ be the class of elementary groups G such that if H is a tdlc Polish group with a continuous injective homomorphism $\pi : H \to G$, then H is elementary.

First, $\mathscr{F}$ clearly contains discrete Polish groups. Let us show that it contains profinite Polish groups. Suppose G is profinite, and $\pi : H \to G$ is continuous and injective. Then H is residually discrete, hence it is elementary by Proposition 16.9.

Next, suppose that G is a tdlc Polish group, and $N \in \mathscr{F}$ is a closed normal subgroup of G such that G/N is profinite or discrete. Let $\pi : H \to G$ be a continuous injective morphism. Then π induces a continuous injective homomorphism $\tilde{\pi} : H/\pi^{-1}(N) \to G/N$. Because $\mathscr{F}$ contains discrete and profinite groups, we deduce that $H/\pi^{-1}(N)$ is elementary. But $\pi^{-1}(N)$ injects continuously into N, which belongs to $\mathscr{F}$, hence $\pi^{-1}(N)$ is elementary. We deduce from Proposition 16.15 that H is elementary.

Finally, suppose that G is a countable increasing union of open subgroups O_i which belong to $\mathscr{F}$. We must show that G also belongs to $\mathscr{F}$. Let $\pi : H \to G$ be a continuous injective homomorphism. Then the restriction of π to $\pi^{-1}(O_i)$ is also a continuous injective homomorphism, so the open subgroups $\pi^{-1}(O_i)$ are elementary and thus $H = \bigcup_{i \in \mathbf{N}} \pi^{-1}(O_i)$ is elementary. $\square$

Corollary 16.17 *Let G be a tdlc Polish group, and let $H \leqslant G$ be a closed subgroup of finite index. Then G is elementary if and only if H is elementary.*

Proof First, if G is elementary then H is elementary by the previous proposition. Conversely, if H is elementary, let N be the kernel of the action of G on G/H by left translation. The set G/H is finite, so N has finite index in G. Moreover, N is a closed subgroup of H, hence is elementary by the previous proposition. The group G/N is finite, so by property (E2) of the class of elementary groups, G also has to be elementary. $\square$

The next closure property will require the use of an inductive argument based on the construction rank, which we now introduce in detail. The reader who is allergic to ordinals may just take Theorem 16.29 for granted and move directly to Section 16.6. However, he or she might also try to cure this allergy by reading what follows.

16.5 The construction rank on elementary groups

In order to motivate this section, let us digress a bit and discuss an easy example of a rank which may already be familiar. Suppose we have a finitely generated group Γ and a fixed finite symmetric generating set S containing the identity element. Then we can reconstruct every element of Γ as follows: start with $\Gamma_0 = S$, and then define by induction $\Gamma_{n+1} = \Gamma_n S$. Because Γ is generated by S, which is symmetric and contains the identity element, we have that $\Gamma = \bigcup_{n \in \mathbf{N}} \Gamma_n$. The rank of an element $\gamma \in \Gamma$ is then defined to be the smallest $n \in \mathbf{N}$ such that $\gamma \in \Gamma_n$. Of course, the rank of $\gamma \in \Gamma$ is just $d_S(\gamma, e) + 1$ where d_S is the word metric defined by S. It measures how hard it is to construct γ using S and the group multiplication.

Now, let us try to define a rank on elementary groups the same way: we know that the class of elementary groups is generated by profinite and discrete Polish groups, so we let $\mathscr{E}_0$ be the class of Polish groups which are either profinite or discrete. For a Polish tdlc group G and $n \in \mathbf{N}$, we then say that $G \in \mathscr{E}_{n+1}$ if either

- there exists $N \in \mathscr{E}_n$ such that G/N is either profinite or discrete, or
- there exists a countable increasing family $(G_i)_{i \in \mathbf{N}}$ of open subgroups of G such that for all $i \in \mathbf{N}$, we have $G_i \in \mathscr{E}_n$, and moreover $G = \bigcup_{i \in \mathbf{N}} G_i$.

The problem is that, as opposed to the group case where every element is a product of finitely many elements of S, here the second operation needs countably many elementary groups in order to build a new one. Thus, there is no reason to have $\bigcup_{n \in \mathbf{N}} \mathscr{E}_n = \mathscr{E}$, and indeed one can show that the inclusion $\bigcup_{n \in \mathbf{N}} \mathscr{E}_n \subseteq \mathscr{E}$ is strict [12, Sec. 6.2] (see also the recent preprint [10] for compactly generated examples). So we need a rank taking values in something bigger than the integers: in particular, we want a rank which takes values into something "stable under countable supremums", and that is exactly what the set of countable ordinals will do for us.

16.5.1 The well ordered set of countable ordinals

We will proceed with a crash course on (countable) ordinals. But first, let us define the fundamental property of the set into which a rank takes values, allowing for inductive arguments.

A strictly ordered set $(A, <)$ is **well ordered** if every non-empty subset of A has a minimum. Note that every subset of A is a well ordered set for the induced order.

Example 16.18 The set of rational numbers with the usual order $(\mathbf{Q}, <)$ is not well ordered. The set $(\mathbf{N}^2, <_{lex})$, where $<_{lex}$ denotes the lexicographic order, is well ordered.

Given a well ordered set A and a property $P(x)$, to prove that $P(x)$ is true for all $x \in A$, one may use a proof by induction. Such a proof boils down to showing that the following holds:

($*$) For all $x \in A$, if $P(y)$ is true for all $y < x$, then $P(x)$ is true.

Let us see why ($*$) implies that $P(x)$ is true for all $x \in A$. Consider the set $B = \{x \in A : P(x)$ is not true$\}$. If B were non-empty, then it would have a minimum x_1, and by definition, for all $y < x_1$, $P(y)$ would be true. But because $x_1 \in B$, this is a contradiction.

Remark 16.19 Let x_0 be the minimum of A. Then the assertion "$P(y)$ is true for all $y < x_0$" is necessarily verified, so if we want to show that ($*$) holds, we will have to show that $P(x_0)$ is true.

Example 16.20 For the well ordered set **N** of non-negative integers, we recover what is sometimes called the "principle of generalized recurrence".

The following proposition should be regarded by the non-set theory inclined as an axiom. Its refinements form the foundations of the theory of ordinals, a nice exposition of which can be found in [8].

Proposition 16.21 *There exists a unique (up to order isomorphism) uncountable well ordered set $(\omega_1, <)$ such that for all $\alpha \in \omega_1$, the well ordered set*

$$\{\beta \in \omega_1 : \beta < \alpha\}$$

is countable. Moreover, ω_1 is stable under countable supremums, meaning that every countable family $(\alpha_i)_{i \in \mathbf{N}}$ of elements of ω_1 has a smallest upper bound in ω_1, denoted by $\sup\{\alpha_i : i \in \mathbf{N}\}$.

We will call ω_1 the **set of countable ordinals**. It will allow us to define a *rank* on elementary groups, and this rank will be useful to prove some of the permanence properties of this class of groups by induction.

Every countable ordinal $\alpha \in \omega_1$ is the strict supremum[11] of the set $\{\beta \in \omega_1 : \beta < \alpha\}$ of its predecessors. In this way, every element of ω_1 can actually be thought of as a well ordered set. Moreover, for every countable well ordered set C, there exists a unique $\alpha \in \omega_1$ such that C is order-isomorphic to the set of predecessors of α.

So the elements of ω_1 are precisely the isomorphism classes of countable well orders. In particular, ω_1 contains the isomorphism class of the set of integers **N** with its usual order, which is denoted by ω. Identifying every $n \in \mathbf{N}$ with its set of predecessors $\mathbf{n} := \{0, ..., n-1\}$ equipped with the induced order, we see that the set of predecessors of ω in ω_1 is $\{\mathbf{n} : n \in \mathbf{N}\}$.

---

[11] The strict supremum of a subset $A \subseteq B$ of an ordered set $(B, <)$ is, if it exists, the smallest $x \in B$ such that $x > y$ for all $y \in A$.

16.5.2 The construction rank

Now that we have our well ordered set $(\omega_1, <)$ of countable ordinals at hand, let us see how to define functions by induction on ω_1. In general, one builds functions by induction on a well ordered set $(A, <)$ the same way one proves assertions by induction: if we suppose that $f(y)$ is defined for all $y < x$, we have to prescribe a way of building $f(x)$. When A is the set of integers, this is the usual construction by induction of a function f, and we often only need to know $f(n)$ in order to define $f(n+1)$. For ω_1, the situation is complicated by the fact that not every element is of the form $\alpha + 1$. Let us make sense of this last sentence and do a tiny bit of ordinal arithmetic.

Given a countable ordinal $\alpha \in \omega_1$, the non-empty set $\{\beta \in \omega_1 : \beta > \alpha\}$ has a minimum, which we denote by $\alpha + 1$. The reader may check that, when seeing $\alpha + 1$ as a well ordered set, it is obtained by adding to the well ordered set α an element $+\infty$ bigger than every element in α. Moreover, such a notation is consistent with the addition on **N**.

Elements of the form $\alpha + 1$ are called **successors**. But not every element is a successor: for instance ω is not a successor.

Proposition 16.22 *A countable ordinal $\alpha \in \omega_1$ is not a successor if and only if it is the supremum of its set of predecessors: $\alpha = \sup_{\beta < \alpha} \beta$.*

Proof Suppose that $\alpha \in \omega_1$ is a successor, and write $\alpha = \beta + 1$. Then the supremum of the set of predecessors of α is equal to β, hence it is not equal to α.

Conversely, if $\alpha \in \omega_1$ is not a successor, recall that α is the strict supremum of its set of predecessors. So the supremum of the set of predecessors of α has to be either α or a predecessor β of α. Let us see why the latter case cannot happen. If the supremum of the set of predecessors of α were some $\beta < \alpha$, because α is not a successor, we would have $\alpha \neq \beta + 1$. Then $\alpha > \beta + 1 > \beta$, contradicting the fact that β was the supremum of the set of predecessors of α. □

Definition 16.23 A *non-zero* countable ordinal $\alpha \in \omega_1$ which is not a successor is called a **limit**.

Let us now go back to our initial problem and define by induction the class of elementary groups. Here our function f will map a countable ordinal α to a class of tdlc Polish groups $\mathscr{E}_\alpha$.

Definition 16.24 Let $\mathscr{E}_0$ be the class of Polish groups which are either profinite or discrete. Then, if $\alpha \in \omega_1$ is given and $\mathscr{E}_\beta$ is defined for all $\beta < \alpha$, define $\mathscr{E}_\alpha$ as follows.

- If $\alpha = \beta + 1$ is a successor, we then say that $G \in \mathscr{E}_{\beta+1}$ if either

  - there exists $N \in \mathscr{E}_\beta$ such that G/N is either profinite or discrete, or
  - there exists a countable increasing family $(G_i)_{i \in \mathbf{N}}$ of open subgroups of G such that for all $i \in \mathbf{N}$, we have $G_i \in \mathscr{E}_\beta$, and moreover $G = \bigcup_{i \in \mathbf{N}} G_i$.

- If $\alpha = \sup_{\beta < \alpha} \beta$ is a limit, we let $\mathscr{E}_\alpha = \bigcup_{\beta < \alpha} \mathscr{E}_\beta$.

Note that by construction, for all $\beta < \alpha$ we have $\mathscr{E}_\beta \subseteq \mathscr{E}_\alpha$. Let us check that we have exhausted the class $\mathscr{E}$ of elementary groups. The proof is easy, but we give full details for the reader who is unacquainted with ordinals.

Proposition 16.25   *We have $\mathscr{E} = \bigcup_{\alpha \in \omega_1} \mathscr{E}_\alpha$.*

Proof   First, let us prove by induction that for all $\alpha < \omega_1$, we have $\mathscr{E}_\alpha \subseteq \mathscr{E}$. So let $\alpha \in \omega_1$, and suppose that for all $\beta < \alpha$, $\mathscr{E}_\beta \subseteq \mathscr{E}$.

- If $\alpha = 0$, because $\mathscr{E}_0$ is the class of profinite or discrete Polish groups, it is contained in $\mathscr{E}$.
- If $\alpha = \beta + 1$ is a successor, by construction of $\mathscr{E}_{\beta+1}$ and the stability properties of $\mathscr{E}$, the assumption that $\mathscr{E}_\beta \subseteq \mathscr{E}$ implies that $\mathscr{E}_{\beta+1} \subseteq \mathscr{E}$.
- If $\alpha = \sup_{\beta < \alpha} \beta$ is a limit, we have $\mathscr{E}_\alpha = \bigcup_{\beta < \alpha} \mathscr{E}_\beta \subseteq \mathscr{E}$.

This concludes our proof by induction, so we have $\bigcup_{\alpha \in \omega_1} \mathscr{E}_\alpha \subseteq \mathscr{E}$. By minimality of the class $\mathscr{E}$, we now only have to check that the class $\mathscr{F} := \bigcup_{\alpha \in \omega_1} \mathscr{E}_\alpha$ shares the defining properties of $\mathscr{E}$ in order to conclude that $\mathscr{E} = \mathscr{F}$ (see Def. 16.1).

- The class $\mathscr{F} = \bigcup_{\alpha \in \omega_1} \mathscr{E}_\alpha$ contains $\mathscr{E}_0$, which is the class of profinite or discrete Polish groups.
- Let G be a tdlc Polish group, suppose that $N \in \mathscr{F}$ is a closed normal subgroup of G such that G/N is profinite or discrete Polish. Then, let $\alpha \in \omega_1$ such that $N \in \mathscr{F}$. By definition of $\mathscr{E}_{\alpha+1}$, we have $G \in \mathscr{E}_{\alpha+1} \subseteq \mathscr{F}$.
- Suppose that G is written as a countable increasing union of open subgroups $(G_i)_{i \in \mathbf{N}}$ belonging to $\mathscr{F}$, and for all $i \in \mathbf{N}$, pick $\alpha_i \in \omega_1$ such that $G_i \in \mathscr{E}_{\alpha_i}$. Using Proposition 16.21, we may define $\alpha = \sup_{i \in \mathbf{N}} \alpha_i \in \omega_1$, and again by definition of $\mathscr{E}_{\alpha+1}$, we have $G \in \mathscr{E}_{\alpha+1} \subseteq \mathscr{F}$. $\qquad\qquad\square$

Definition 16.26   Let G be an elementary group. Its **construction rank** is the smallest ordinal $\alpha \in \omega_1$ such that $G \in \mathscr{E}_\alpha$.

Since whenever α is a limit ordinal we have $\mathscr{E}_\alpha = \bigcup_{\beta < \alpha} \mathscr{E}_\beta$, the rank of an elementary group is always a successor ordinal, except when it is equal to zero.

16.5.3 Using the construction rank

Proposition 16.27 *Let G be an elementary group, and let $H \leqslant G$ be an open subgroup. Then H is elementary and the rank of H is smaller than or equal to the rank of G.*

Proof The proof is by induction on the construction rank, which is either null or a successor ordinal.

- The statement is clearly true for rank 0 elementary groups (recall that these are the Polish groups which are either profinite or discrete).
- Suppose G has rank $\alpha + 1$ and that the proposition is true for every elementary group of rank at most α. Let H be an open subgroup of G. We have two subcases to consider.

  – Either G has a closed normal subgroup N of rank at most α such that G/N is profinite or discrete. Then $N \cap H$ is a closed normal subgroup of H. But $N \cap H$ is also an open subgroup of N, so by our induction hypothesis it has rank at most α. Since $H/N \cap H$ is topologically isomorphic to the open subgroup $HN/N \leqslant G/N$, we deduce that H has rank at most $\alpha + 1$, which is smaller than or equal to the rank of G.

  – Or G may be written as an increasing union of open subgroups $(G_i)_{i \in \mathbf{N}}$ of rank at most α. Then for all $i \in \mathbf{N}$, $G_i \cap H$ is an open subgroup of G_i, so by our induction hypothesis it has rank at most α. Again, we deduce that H has rank at most $\alpha + 1$, hence is smaller than or equal to the rank of G. $\qquad\square$

The theorem that ends this section is the only place in this survey where we need the construction rank. Let us first present one useful lemma on elementary groups from which the result will follow easily.

Lemma 16.28 *Let G be a compactly generated elementary group of rank $\alpha + 1$. Then G has a closed normal subgroup N of rank α such that G/N is profinite or discrete.*

Proof Because G has rank $\alpha + 1$, by definition either G has a closed normal subgroup N of rank α such that G/N is profinite or discrete, or G can be written as an increasing union of open subgroups $(G_i)_{i \in \mathbf{N}}$ of rank at most α. We want to show that the latter case is contradictory. Let S be a compact generating set of G. Then $(G_i)_{i \in \mathbf{N}}$ is an open cover of S. Because $(G_i)_{i \in \mathbf{N}}$ is increasing and S is compact, there exists $i \in \mathbf{N}$ such that G_i contains S. But then $G_i = G$, which contradicts the fact that G_i has rank at most α. $\qquad\square$

Theorem 16.29   *Let $P(G)$ be a property. Then to show that $P(G)$ is true for every elementary group G, it suffices to show that:*

  (i) *$P(G)$ is true whenever G is a Polish group which is either profinite or discrete;*

 (ii) *if G is an elementary group and N is a closed normal subgroup of G such that $P(N)$ is true and G/N is profinite or discrete, then $P(G)$ is true;*

(iii) *if G is an elementary group, then there exists an increasing chain $(G_i)_{i \in \mathbf{N}}$ of open compactly generated subgroups of G such that the implication*

$$(\forall i \in \mathbf{N}, P(G_i)) \Rightarrow P(G)$$

holds.

Proof   The proof is by induction on the construction rank, which is either zero or a successor ordinal.

• By assumption (i), $P(G)$ is true for every rank 0 elementary group.

• Suppose that G has rank $\alpha + 1$ and that $P(H)$ is true for all H of rank at most α. By assumption (iii), we may fix an increasing chain $(G_i)_{i \in \mathbf{N}}$ of open compactly generated subgroups of G such that the implication

$$(\forall i \in \mathbf{N}, P(G_i)) \Rightarrow P(G)$$

holds. Let $i \in \mathbf{N}$. By Proposition 16.27, the group G_i has rank at most $\alpha + 1$. The previous lemma provides a closed normal subgroup N_i of G_i which has rank at most α, and such that G_i/N_i is profinite or discrete. By our induction hypothesis, we have that $P(N_i)$ is true, and by assumption (ii), this implies that $P(G_i)$ is true. So $P(G_i)$ is true for all $i \in \mathbf{N}$, and by assumption (iii) we conclude that $P(G)$ is true.   □

16.6 More permanence properties

The main goal of this section is to show that every Hausdorff quotient of an elementary group is elementary (Thm. 16.34). Along the way, we will also prove some results needed for the decomposition theorem (Thm. 16.2).

The following lemma echoes item (iii) in the previous theorem.

Lemma 16.30   *Let G be a tdlc Polish group. Then there exists an increasing sequence $(G_i)_{i \in \mathbf{N}}$ of open compactly generated subgroups of G such that $G = \bigcup_{i \in \mathbf{N}} G_i$. Moreover, if K is a compact group acting continuously on G by automorphisms, then one can choose the G_i so that each of them is setwise fixed by K.*

Proof Let U be an open compact subgroup of G, and let $(g_i)_{i \in \mathbf{N}}$ enumerate a countable dense family of elements of G. For every $i \in \mathbf{N}$, we let $S_i = U \cup \{g_0, ..., g_i\}$. Then $K \cdot S_i$ is compact and open, and we let G_i be the group generated by $K \cdot S_i$. Clearly the sequence $(G_i)_{i \in \mathbf{N}}$ is increasing, and each G_i is compactly generated. The sequence $(G_i)_{i \in \mathbf{N}}$ is also exhaustive by density of (g_i) and the fact that U is an open subgroup of G. Furthermore, the fact that K acts by automorphisms on G guarantees that each G_i is setwise fixed by K. $\square$

Definition 16.31 A tdlc group G is called **quasi-discrete** if it has a dense subgroup whose elements have an open centraliser.

Caprace and Monod have shown that every compactly generated quasi-discrete group is SIN [2, Prop. 4.3]. The following lemma is very close to their result.

Lemma 16.32 *Let G be a quasi-discrete tdlc Polish group. Then G is elementary.*

Proof Let U be a compact open subgroup of G, let $(g_i)_{i \in \mathbf{N}}$ be a countable dense set such that for all $i \in \mathbf{N}$, the centraliser of g_i is open. For all $n \in \mathbf{N}$, let $V_n \lhd U$ be an open subgroup such that every element of V_i commutes with $g_1, ..., g_n$. Then V_n is a compact open normal subgroup of $G_n = \langle U, g_1, ..., g_n \rangle$. Because it has such a compact open normal subgroup, G_n is an elementary open subgroup of G, hence $G = \bigcup_{n \in \mathbf{N}} G_n$ is elementary. $\square$

We now need a technical lemma which will be superseded later by Corollary 16.35.

Lemma 16.33 *Let G be a tdlc Polish group, and let M, L be two closed normal subgroups of G intersecting trivially. If M is elementary, then $\overline{ML}/L$ also is.*

Proof This is done using the inductive scheme provided by Theorem 16.29, where the property $P(M)$ we want to prove is "whenever M arises as a closed normal subgroup of some tdlc Polish group G, and L is another closed normal subgroup of G intersecting M trivially, then $\overline{ML}/L$ is elementary". To make the proof lighter, we won't make any reference to the ambient group G.

Note that the fact that $M \cap L$ is trivial implies $[M, L] = \{e\}$. So the conjugacy of L on M is trivial, in particular L normalises any subgroup of M.

(i) First suppose that M is profinite. Then ML is closed, so $\overline{ML}/L$ is a continuous quotient of M, hence elementary. Next, if M is discrete, each element of M has an open centraliser in $\overline{ML}$, and so we may apply Lemma 16.32 to $\overline{ML}/L$ and deduce that $\overline{ML}/L$ is elementary.

(ii) Suppose that there exists a closed normal subgroup N of M such that $P(N)$ holds and M/N is either profinite or discrete. Then $\overline{NL}/L$ is elementary, moreover it is a closed normal subgroup of $\overline{ML}/L$, so because the class of elementary groups is stable under group extension (Proposition 16.15), we only need to show that

$$(\overline{ML}/L)/(\overline{NL}/L) \simeq \overline{ML}/\overline{NL}$$

is elementary. But M/N is profinite or discrete, so by case (i) applied to $M' = M/N$ and $L' = \overline{NL}/N$, we have that $\overline{M'L'}/L'$ is elementary. By the isomorphism theorem, the latter is isomorphic to $\overline{ML}/\overline{NL}$, which is thus elementary.

(iii) Let K be a compact open subgroup of $\overline{ML}$. Then K acts continuously on M by conjugation, so Lemma 16.30 provides an exhaustive increasing chain $(M_i)_{i \in \mathbf{N}}$ of open compactly generated subgroups of M such that each M_i is normalised by K. Suppose that $P(M_i)$ is true for every $i \in \mathbf{N}$, and fix $i \in \mathbf{N}$. Since the closed subgroup M_i is normalised by L, it is a normal subgroup of $\overline{M_iL}$. So we can apply $P(M_i)$ and deduce that the group $\overline{M_iL}/L$ is elementary. Recall that K normalises M_i, so K normalises M_iL. Because $(KM_iL/L)/(\overline{M_iL}/L)$ is a quotient of K, hence profinite, we deduce that KM_iL/L is elementary.

Now, $\overline{ML}/L = KML/L$ arises as the increasing union of the elementary open subgroups KM_iL/L, hence it is elementary.   □

Theorem 16.34 *Let G be an elementary tdlc Polish group, and let N be a closed normal subgroup of G. Then G/N is elementary.*

Proof Consider the smallest class $\mathscr{F}$ of elementary tdlc Polish groups G such that for all closed $N \lhd G$, the group G/N is elementary. We will show that $\mathscr{F}$ satisfies the same properties as the class of elementary groups, hence coincides with it. First, $\mathscr{F}$ clearly contains profinite and discrete groups.

Then, let G be an elementary group and assume that $M \in \mathscr{F}$ is a normal subgroup of G such that G/M is either profinite or discrete. Let N be a normal subgroup in G, consider the group $\tilde{G} = G/M \cap N$. Then if we let $\tilde{M} = M/M \cap N$ and $\tilde{N} = N/M \cap N$, the groups $\tilde{M}$ and $\tilde{N}$ are closed normal subgroups of $\tilde{G}$ intersecting trivially. Moreover, since $M \in \mathscr{F}$ the group $\tilde{M}$ is elementary. Thus, we may apply Lemma 16.33 to them and deduce that $\overline{\tilde{M}\tilde{N}}/\tilde{N}$ is elementary. But then, $\overline{MN}/N$ is isomorphic to $\overline{\tilde{M}\tilde{N}}/\tilde{N}$, hence elementary. Since $(G/M)/(\overline{MN}/N)$ is isomorphic to the group $G/\overline{MN}$ which is a quotient of the profinite or discrete group G/N, we deduce that $(G/N)/(\overline{MN}/N)$ is elementary. But elementariness is stable under extensions (Prop. 16.15), so G/N is elementary.

In order to conclude the proof, we need to deal with increasing unions: let

G be an elementary group, written as an increasing union $G = \bigcup_{i \in \mathbf{N}} G_i$ of open subgroups belonging to $\mathscr{F}$. Let N be a closed normal subgroup in G. Then for all $i \in \mathbf{N}$, $N \cap G_i$ is a closed normal subgroup of G_i, hence $G_i/(N \cap G_i)$ is elementary, but now G/N may be written as an increasing union of the projections of the G_is onto G/N. Moreover, each projection of G_i onto G/N is isomorphic to $G_i/(N \cap G_i)$, so G/N is an increasing union of open elementary subgroups, hence elementary. $\square$

Corollary 16.35 *Let G be a tdlc Polish group, let M, L be two closed normal subgroups of G. If M is elementary, then $\overline{ML}/L$ also is.*

Proof Consider as in the proof of Theorem 16.34 the quotient group $\tilde{G} = G/(M \cap L)$, inside which $\tilde{M} := M/(M \cap L)$ and $\tilde{L} := L/(M \cap L)$ are closed, normal and have trivial intersection. By Theorem 16.34 the group $\tilde{M}$ is elementary, so we may apply Lemma 16.33 and deduce that $\overline{\tilde{M}\tilde{L}}/\tilde{L}$ is elementary. But the latter is isomorphic to $\overline{ML}/L$, which concludes the proof. $\square$

Corollary 16.36 *Let G be a tdlc Polish group, and let L, M be two elementary closed normal subgroups of G. Then $\overline{ML}$ is elementary.*

Proof By the previous corollary, $\overline{ML}/L$ is elementary. But because L is elementary and the class of elementary groups is stable under extension (Proposition 16.15), the group $\overline{LM}$ is also elementary. $\square$

Here is the last permanence property that we will need in order to prove the decomposition theorem. Actually, we only need it for the much easier case where all the C_is are normalised by some fixed open subgroup of G.

Theorem 16.37 *Let G be a Polish tdlc group. Suppose that there exists an increasing sequence of elementary subgroups $(C_i)_{i \in \mathbf{N}}$ of G such that each C_i has an open normaliser in G, and that $G = \bigcup_{i \in \mathbf{N}} C_i$. Then G is elementary.*

Proof By Lemma 16.30, we can write G as an increasing countable union of open compactly generated subgroups. Since these subgroups will satisfy the same assumption as G, and since an increasing union of open elementary subgroups is elementary, we only have to show that the theorem holds for G compactly generated.

So assume that G is compactly generated and fix a compact open subgroup U of G. By Lemma 16.4 applied to the dense subgroup $D = \bigcup_{i \in \mathbf{N}} C_i$, there exists $i \in \mathbf{N}$ and a finite subset $A \subseteq C_i$ such that $G = \langle A \rangle U$. Let V be an open subgroup of U that normalises C_i, and let B be the compact reunion of the V-conjugates of A. Then $\langle \overline{B} \rangle$ is normalised by V, and it is a closed subgroup of C_i, hence elementary by Proposition 16.16. Because $\overline{\langle B \rangle} V/\overline{\langle B \rangle}$ is a quotient of V, hence profinite, we deduce that $\overline{\langle B \rangle}V$ is elementary. But V has finite index in U and B contains A, so $\overline{\langle B \rangle}V$ has finite index in $G = \langle A \rangle U$. Having an elementary closed subgroup of finite index, G has to be elementary by Corollary 16.17.   □

16.7 The decomposition theorem

We are now almost ready to understand how a compactly generated tdlc Polish group can be decomposed into elementary and topologically characteristically simple non-elementary pieces. The main tool for doing this is the existence, inside any tdlc Polish group, of a maximum elementary closed normal subgroup.

Theorem 16.38 ([12, Thm 1.5])   *Let G be a tdlc Polish group. Then the family of closed elementary normal subgroups of G has a unique maximum with respect to inclusion.*

This largest elementary normal subgroup is called the **elementary radical** of G, and denoted by $\mathrm{Rad}_{\mathscr{E}}(G)$. Note that it is a topologically characteristic subgroup of G, meaning that every continuous automorphism of G fixes $\mathrm{Rad}_{\mathscr{E}}(G)$ setwise.[12]

Moreover, by Proposition 16.15, the quotient $G/\mathrm{Rad}_{\mathscr{E}}(G)$ must have trivial elementary radical.

Proof of Theorem 16.38   Let $(U_n)_{n \in \mathbf{N}}$ be a countable basis of open subsets of G. Let $\mathscr{F}$ be the set of $n \in \mathbf{N}$ such that there exists a closed elementary normal subgroup intersecting U_n. For each $n \in \mathscr{F}$, choose such a closed elementary normal subgroup N_n intersecting U_n. Enumerate[13] $\mathscr{F} = \{n_k : k \in \mathbf{N}\}$, and let

$$N = \overline{\bigcup_{k \in \mathbf{N}} N_{n_1} \cdots N_{n_k}}.$$

By Corollary 16.36, each $\overline{N_{n_1} \cdots N_{n_k}}$ is elementary, so by Theorem 16.37, N is an elementary closed normal subgroup of G. By definition, N intersects every basic open set U_n which intersects some elementary closed normal subgroup. So every U_n that does not intersect N must intersect no elementary closed normal subgroup. But because N is closed, its complement may be written as a reunion of such U_ns. This implies that N is the unique maximum of the class of closed elementary normal subgroups of G.   □

The second tool was developed by Caprace and Monod, and provides a way

---

[12] Continuous automorphisms of G are also homeomorphisms, since G is Polish.
[13] If $\mathscr{F}$ is finite, we allow for repetitions in the enumeration.

to decompose the Polish tdlc compactly generated groups which have a trivial elementary radical.

Definition 16.39 A tdlc Polish group G is **locally elliptic** if every finite subset of G generates a group with compact closure.

A result of Platonov asserts that every locally elliptic tdlc Polish group is an increasing union of open compact subgroups, so that in particular it is elementary (see [12, Sec. 2.4]).

Definition 16.40 Let G be a tdlc Polish group. One says that G is the **quasi-product** of the closed normal subgroups $N_1, ..., N_k \leqslant G$ if the product map $N_1 \times \cdots \times N_k \to G$ which maps $(n_1, ..., n_k)$ to $n_1 \cdots n_k$ is injective and has dense image.

We recall the following result due to Caprace–Monod [2]; see Proposition 15.3 in Chapter 15 of this volume for a proof.[14]

Theorem 16.41 (Caprace–Monod) *Let G be a compactly generated tdlc group. Then one of the following holds:*

(1) G has an infinite discrete normal subgroup;

(2) G has a non-trivial locally elliptic closed normal subgroup;

(3) G has exactly $0 < n < \infty$ minimal non-trivial closed normal subgroups.

Corollary 16.42 (Wesolek) *Let G be a non-trivial compactly generated tdlc Polish group with trivial elementary radical. Then G contains a topologically characteristic closed subgroup which decomposes as a quasi-product of $0 < n < \infty$ non-elementary closed normal subgroups.*

Proof We apply the previous theorem to G. Because G has a trivial elementary radical, cases (1) and (2) cannot hold, so we deduce that G has exactly $0 < n < \infty$ minimal closed normal subgroups $N_1, ..., N_n$. Let $H = \overline{N_1 \cdots N_n}$. The subgroup H is topologically characteristic for its definition is clearly invariant under continuous group automorphisms.

Observe that the N_is pairwise commute since for all $i \neq j$, $[N_i, N_j]$ is contained in $N_i \cap N_j$, which is trivial by minimality.

We now prove that for all $1 \leqslant m < n$, $\overline{N_1 \cdots N_m} \cap N_{m+1}$ is trivial. Indeed, if it is not trivial, it must contain N_{m+1} by minimality. But N_{m+1} commutes with every N_i for $i \leqslant m$, so in particular $\overline{N_1 \cdots N_m}$ has a non-trivial centre C. Now such a centre C is a closed characteristic subgroup of $\overline{N_1 \cdots N_m}$, hence C is a non-trivial abelian closed normal subgroup of G. Since abelian tdlc Polish

---

[14] The original statement in [2] contains an error; a corrected version is included in Chapter 15.

groups are elementary, this contradicts the fact that G has a trivial elementary radical.

So for all $1 \leqslant m < n$, the groups $\overline{N_1 \cdots N_m}$ and N_{m+1} intersect trivially, which easily yields by induction that the product map $N_1 \times \cdots \times N_n \to H$ is injective. In other words, $H = \overline{N_1 \cdots N_n}$ is the quasi-product we seek. □

The next lemma is where Cayley–Abels graphs show up (see Lem. 16.4 and the paragraph thereafter for a reminder about these).

Lemma 16.43 *Let G be a non-trivial tdlc Polish group with trivial elementary radical, and let N be a non-trivial closed normal subgroup of G. Denote by $\pi : G \to G/N$ the natural projection. Let U be a compact open subgroup of G, and A be a finite subset of G such that $G = \langle A \rangle U$. Then the Cayley–Abels graph $\mathscr{C}_{A,U}(G)$ has a degree strictly greater than the degree of $\mathscr{C}_{\pi(A),\pi(U)}(G/N)$.*

Proof First, the action on the Cayley–Abels graph $\mathscr{C}_{A,U}(G)$ has compact kernel, so since G has trivial elementary radical, this action must be faithful. Now consider the action of N on $\mathscr{C}_{A,U}(G)$. Because G has trivial elementary radical, N is non-discrete, hence there exists $g \in N \cap U \setminus \{1\}$. Such a g fixes the vertex $v := U$. Now, since the N-action is faithful, there exists another vertex $w \in \mathscr{C}_{A,U}(G)$ which is not fixed by g. In particular, because $\mathscr{C}_{A,U}(G)$ is connected, the boundary[15] of the set of g-fixed vertices has to be non-empty, implying that there exists a vertex in $\mathscr{C}_{A,U}(G)$ with two distinct neighbours belonging to the same N-orbit. As observed at the end of section 16.2, this yields that the degree of $\mathscr{C}_{\pi(A),\pi(U)}(G/N)$ is strictly less than the degree of $\mathscr{C}_{A,U}(G)$. □

Now the path to a decomposition theorem is clear. If G is a compactly generated tdlc Polish group, we apply the following algorithm to it until we get a Cayley–Abels graph of degree 0, that is, a compact (hence elementary) group:

(A) quotient G by its elementary radical to obtain a group G' with trivial elementary radical. If $G = G'$ then stop. If $G \neq G'$, then

(B) quotient G' by the non-trivial quasi-product provided by Corollary 16.42, obtaining a new group H. If H is not compact, proceed to step (A), replacing G by H.

Indeed, the previous lemma guarantees that the degree of the associated Cayley–Abels graphs drops by at least one each time we apply both steps (in particular, the length of the characteristic series obtained by running this algorithm is bounded above by the minimal degree of the Cayley–Abels graphs of G). By

---

[15] The boundary of a set F of vertices is the set of vertices not belonging to F, but connected to an element of F.

lifting the obtained quasi-products back to G, we obtain a proof of the following theorem, also known as Theorem 16.2.

Theorem 16.44 *Let G be a compactly generated tdlc Polish group. Then there exists a finite increasing sequence*

$$H_0 = \{e\} \leqslant \cdots \leqslant H_n$$

of closed characteristic subgroups of G such that

(i) G/H_n is an elementary group and
(ii) for all $i = 0,...,n-1$, the group $(H_{i+1}/H_i)/\mathrm{Rad}_{\mathscr{E}}(H_{i+1}/H_i)$ is a finite quasi-product of topologically characteristically simple non-elementary subgroups.

References for this chapter

[1] Bachir Bekka, Pierre de la Harpe and Alain Valette. *Kazhdan's property (T)*, volume 11 of *New Mathematical Monographs*. Cambridge University Press, Cambridge, 2008.

[2] Pierre-Emmanuel Caprace and Nicolas Monod. Decomposing locally compact groups into simple pieces. *Math. Proc. Camb. Philos. Soc.*, 150(1):97–128, 2011.

[3] Jean A. Dieudonné. *La géométrie des groupes classiques*. Springer-Verlag, Berlin-New York, 1971. Troisième édition, Ergebnisse der Mathematik und ihrer Grenzgebiete, Band 5.

[4] Pierre de la Harpe and Yves de Cornulier. *Metric geometry of locally compact groups*. Book in preparation, available online.

[5] Su Gao. *Invariant descriptive set theory*, volume 293 of *Pure and Applied Mathematics (Boca Raton)*. CRC Press, Boca Raton, FL, 2009.

[6] Kate Juschenko and Nicolas Monod. Cantor systems, piecewise translations and simple amenable groups. *Ann. Math. (2)*, 178(2):775–787, 2013.

[7] Alexander S. Kechris. *Classical descriptive set theory*, volume 156 of *Graduate Texts in Mathematics*. Springer-Verlag, New York, 1995.

[8] Jean-Louis Krivine. *Introduction to axiomatic set theory*. Translated from the French by David Miller. D. Reidel Publishing Co., Dordrecht; Humanities Press, New York, 1971.

[9] Hiroki Matui. Some remarks on topological full groups of Cantor minimal systems. *Internat. J. Math.*, 17(2):231–251, 2006.

[10] Colin D. Reid and Phillip R. Wesolek. Homomorphisms into totally disconnected, locally compact groups with dense image. *ArXiv:1509.00156*.

[11] Jacques Tits. Sur le groupe des automorphismes d'un arbre. In *Essays on topology and related topics (Mémoires dédiés à Georges de Rham)*, pages 188–211. Springer, New York, 1970.

[12] Phillip Wesolek. Elementary totally disconnected locally compact groups. *Proc. Lond. Math. Soc. (3)*, 110(6):1387–1434, 2015.

17

The structure lattice of a totally disconnected locally compact group

John S. Wilson

Abstract

We explain how the collection of locally normal subgroups of an arbitrary tdlc group gives rise to the structure lattice of that group, following [2, 3].

17.1	Definition of the lattice	258
17.2	Quasi-centralisers	261
17.3	Technical lemmas	263
17.4	Groups with no abelian subgroups in $\mathscr{L}$	264
References for this chapter		266

17.1 Definition of the lattice

Throughout this chapter, G is always a tdlc group, and $\mathscr{B} = \mathscr{B}(G)$ is the set of compact open subgroups; this is a base of neighbourhoods of 1 by van Dantzig's theorem; see Chapter 1 in this volume. Any two elements B_1, B_2 of $\mathscr{B}$ are commensurable since the cover of each by the open cosets of the intersection $B_1 \cap B_2$ must be finite.

We shall be working with the family $\mathscr{L}$ of *locally normal* subgroups: these are the closed subgroups with open normaliser. By van Dantzig's theorem, a closed subgroup is in $\mathscr{L}$ if and only if it is normalised by some $U \in \mathscr{B}$. It is very easy to see that the compact subgroups in $\mathscr{L}$ are precisely the compact normal subgroups of members of $\mathscr{B}$. We write $K \leq_O H$ (resp. $K \leq_f H$) to indicate that K is an open subgroup (resp. a subgroup of finite index) in H.

We note the following facts:

Lemma 17.1 (a) *If* $H_1,\ldots,H_n \in \mathscr{L}$ *then there are subgroups* $K_1,\ldots,K_n \in \mathscr{L}$ *that normalise each other and satisfy* $K_i \leq_o H_i$ *for each* i.

(b) *If* $H_1,H_2 \in \mathscr{L}$ *then* $H_1 \cap H_2 \in \mathscr{L}$; *if in addition* H_1,H_2 *are compact and normalise each other then* $H_1 H_2 \in \mathscr{L}$.

(c) *If* $H,K \in \mathscr{L}$ *then* $\mathrm{C}_H(K) \in \mathscr{L}$.

Proof (a) Choose for each i a subgroup $U_i \in \mathscr{B}$ that normalises K_i, and let $K_i = H_i \cap \bigcap_{j=1}^{n} U_j$.

(b) If $U \in \mathscr{B}$ normalises H_1,H_2 then it normalises $H_1 \cap H_2$ and $\langle H_1,H_2 \rangle$; the latter equals $H_1 H_2$ and is closed if H_1, H_2 are compact and normalise each other.

(c) We have $\mathrm{C}_H(K) = H \cap \mathrm{C}_G(K)$, and $\mathrm{N}_G(K)$ is open and normalises $\mathrm{C}_G(K)$. $\square$

We define an equivalence relation on $\mathscr{L}$ by writing $H_1 \sim H_2$ if and only if $H_1 \cap U = H_2 \cap U$ for some $U \in \mathscr{B}$. We write $\widehat{\mathscr{L}}$ for the quotient set $\mathscr{L}/\sim$.

Lemma 17.2 *Let* $H_1,H_2,L_1,L_2 \in \mathscr{L}$ *with* $H_1 \sim H_2$ *and* $L_1 \sim L_2$. *Then*

$$H_1 \cap L_1 \sim H_2 \cap L_2.$$

If in addition H_i, L_i *are compact and* $H_i,L_i \lhd \langle H_i,L_i \rangle$ *for* $i = 1,2$ *then*

$$H_1 L_1 \sim H_2 L_2.$$

Proof The first statement is clear. To prove the second, choose $U \in \mathscr{B}$ such that $H_1 \cap U = H_2 \cap U$ and $L_1 \cap U = L_2 \cap U$. Moreover, write P for the subgroup $(H_1 \cap U)(L_1 \cap U)$. Since $H_1 \cap U \leq_o H_1$ we have $H_1 = (H_1 \cap U)X$ for a finite subset X. Since $V := \bigcap_{x \in X} x^{-1}Ux$ is in $\mathscr{B}$ we have $L_1 = (L_1 \cap V)Y$ for a finite subset Y. If $x \in X$ then $x(L_1 \cap V) \subseteq xL_1 \cap x(x^{-1}Ux) = (L_1 \cap U)x$. Thus $H_1 L_1 \subseteq PXY$ and so $P \leq_o H_1 L_1$. Therefore $(H_1 L_1) \cap W_1 = P$ for some $W_1 \in \mathscr{B}$ (see for example [8, 0.3.3]). Similarly $(H_2 L_2) \cap W_2 = P$ for some $W_2 \in \mathscr{B}$ and so $(H_1 L_1) \cap W = (H_2 L_2) \cap W$ with $W = W_1 \cap W_2 \in \mathscr{B}$. $\square$

Given equivalence classes α, β, choose compact representatives H, L that normalise each other and define

$$\alpha \wedge \beta = [H \cap L] \quad \text{and} \quad \alpha \vee \beta = [HL].$$

Lemma 17.2 shows that these operations of meet and join are well defined. The operations $\wedge$, $\vee$ are clearly commutative and associative. Moreover, if α, β as above are represented by H,L then $\alpha \wedge (\alpha \vee \beta)$, $\alpha \vee (\alpha \wedge \beta)$ and α are represented by $H \cap (HL)$, $H(H \cap L)$ and H, and so are all equal. This shows that $\widehat{\mathscr{L}}$ satisfies the lattice axioms. The lattice $\widehat{\mathscr{L}}$ is called the *structure lattice* of G.

We make $\widehat{\mathscr{L}}$ into a partially ordered set in the usual way by writing $\alpha \leq \beta$ if $\alpha \wedge \beta = \alpha$. This coincides with the order induced by subgroup inclusion in $\mathscr{L}$: for $\alpha, \beta \in \widehat{\mathscr{L}}$ we have $\alpha \leq \beta$ if and only if α, β have representatives H, L satisfying $H \leq L$. The largest element of $\widehat{\mathscr{L}}$ is $\mathscr{B}$ and the smallest is the class of all discrete subgroups with open normalisers.

Proposition 17.3 *The structure lattice $\widehat{\mathscr{L}}$ of G is a modular lattice.*

Proof Let $\alpha, \beta, \gamma \in \widehat{\mathscr{L}}$ with $\alpha \leq \beta$; we need $(\alpha \vee \gamma) \wedge \beta = \alpha \vee (\gamma \wedge \beta)$. Choose compact representatives K, L, M of α, β, γ that normalise each other and satisfy $K \leq L$. Thus $(KM) \cap L = K(M \cap L)$, and the desired conclusion follows. $\square$

Example. Let G be a finite-dimensional vector space over $\mathbf{Q}_p$ for some prime p. Then $\mathrm{N}_G(H) = G$ for each subgroup H, and two closed subgroups are equivalent if and only if they span the same $\mathbf{Q}_p$-subspace. Thus $\widehat{\mathscr{L}}$ is isomorphic to the lattice of subspaces of G. This lattice is not distributive if $\dim V > 1$.

The following lemma will be needed in the next section.

Lemma 17.4 *Suppose that G is topologically simple and compactly generated.*

(a) *For each compact subgroup $K \in \mathscr{L}$ there is a finite family $\{L_1, \ldots, L_n\}$ of open subgroups of conjugates of K, normal in their join L, with L commensurated by G and $K \cap L \sim K$.*

(b) *If $\mathscr{L}$ contains an infinite compact nilpotent subgroup K, then it contains an infinite compact nilpotent subgroup that is commensurated by G.*

Proof (a) Choose $U \in \mathscr{B}$ with $U \leq \mathrm{N}_G(K)$. Since G is topologically simple we have $G = \langle U, \Omega \rangle$ where $\Omega = \{gKg^{-1} \mid g \in G\}$. For $g \in G$ the element gKg^{-1} is fixed by $U \cap gUg^{-1}$ in the conjugation action of U on Ω, and so U has finite orbits on Ω. Since K is compact, $|M : M \cap U|$ is finite for each $M \in \Omega$. Since G is compactly generated, it is generated by U and finitely many elements of Ω, and hence by U and a finite subset $\{K_1, \ldots, K_n\} \subseteq \Omega$ that is a union of U-orbits; we can assume that $K_1 = K$. Set $V = \bigcap_{i=1}^n \mathrm{N}_U(K_i)$ and $L_i = V \cap K_i$ for each i. Thus the subgroups L_i normalise each other. Since $\{K_1, \ldots, K_n\}$ is a union of U-orbits we also have $L := \prod L_i \lhd U$. Clearly $K \cap L \geq L_1$ and so $K \cap L \sim K$. For each i the index $|K_i L : L|$ is finite and so $K_i L$ commensurates L. Since $U \leq \mathrm{N}_G(L)$ and $G = \langle U, K_1, \ldots, K_n \rangle$ it follows that G commensurates L.

(b) Define L as above. Then L is nilpotent by Fitting's theorem (see [5, (5.2.8)]). $\square$

17.2 Quasi-centralisers

The definition of the structure lattice $\widehat{\mathscr{L}}$ of G is inspired by the definition of the structure lattice of a just infinite group, introduced in [6]. In that case, the lattice turned out to have unique complementation, essentially because there are no non-trivial abelian normal subgroups and because centralisers are plentiful. Complemented modular lattices are Boolean lattices, and so the action induced by conjugation also gave rise to actions on the Stone space of the Boolean lattice and on geometric structures associated with the structure lattice. This provided an effective tool for analysing the structure of just infinite groups.

One would like to do something similar for tdlc groups. One possibility is to study the subfamily $\widehat{\mathscr{C}}$ of $\widehat{\mathscr{L}}$ consisting of classes containing elements $C_G(H)$ with $H \in \mathscr{L}$. This turns out often to be a Boolean lattice with complementation coming from the operation $C_G(H) \mapsto C_G C_G(H)$ and using the general fact that $C_G C_G C_G(H) = C_G(H)$. To verify that this works, one needs to study centralisers in a context that is invariant under passage to commensurable subgroups. The details are described in Sections 3, 4 below. They are somewhat technical but not difficult.

For subgroups H, K of G write

$$D_H(K) = \bigcup_{U \in \mathscr{B}} C_H(K \cap U).$$

Clearly this depends only on $[K]$, and we may denote it by $D_H([K])$. If $[K]$ is fixed by H in the action induced by conjugation then for $h \in H$ we have $(D_H([K]))^h = (D_H[K]^h) = D_H([K])$ and hence $D_H([K]) \lhd H$.

The subgroup $D_H(K)$ is the *quasi-centraliser* of K in H. The *quasi-centre* $Y(H)$ of H is $D_H(H)$, the set of elements of H with open centralisers. We note the following easy facts.

Lemma 17.5

(a) $Y(H)$ *is a subgroup and is invariant under all topological automorphisms of H. It is not necessarily closed in H.*
(b) $Y(H) \cap U = Y(H \cap U)$ *for each $U \in \mathscr{B}$. If $H \leq_o G$ then $Y(H) \leq Y(G)$.*
(c) *If $H \in \mathscr{L}$ and $H \cap V = 1$ for some $V \in \mathscr{B}$ then $H \leq Y(G)$.*

Proof (a) The first assertion is clear. To see the second we may take G to be a Cartesian product of infinitely many non-trivial finite groups with trivial centre: $Y(G)$ is the dense subgroup of elements with only finitely many non-trivial co-ordinates. (A similar example shows that the FC-centre of a profinite subgroup group need not be closed.)

(b) If $x \in Y(H \cap U)$ then $x \in D_H(H \cap U) \cap U \leq Y(H) \cap U$. If on the other hand $x \in Y(H) \cap U$ then for some $V \in \mathscr{B}$ we have

$$x \in C_H(H \cap V) \cap U \leq C_{H \cap U}((H \cap U) \cap V) \leq Y(H \cap U).$$

If $H \leq_O G$ and $C_H(g)$ is open then so is $C_G(g)$, and the conclusion follows.

(c) Set $U = N_V(H)$; then $U \in \mathscr{B}$. If $h \in H$ then $h^U = \{uhu^{-1} \mid u \in U\}$ is a compact subset of H (being a continuous image of U) and so finite since H inherits the discrete topology from G. Thus $C_U(h)$ has finite index in U. But $C_U(H)$ is also closed in U, and so $C_G(h)$ contains a subgroup in $\mathscr{B}$. Therefore $h \in Y(G)$, as required. $\qquad\qquad\square$

Lemma 17.6 *Suppose that G is topologically simple, compactly generated and non-discrete. Then $D_G(H) = 1$ for every infinite subgroup $H \in \mathscr{L}$ that is commensurated by G. In particular, $Y(G) = D_G(G) = 1$.*

Proof Choose $U \in \mathscr{B}$ with $U \leq N_G(H)$. Let $Q = D_G(H)$ and assume that $Q \neq 1$. For each $g \in G$ we have $Q \sim Q^g$, so that $D_G(H) = D_G(H^g) = D_G(H)^g$. Hence $Q \lhd G$ and $\overline{Q} = G$ since G is topologically simple. Therefore $UQ = G$. Since G is compactly generated and $Q = \bigcup_{V \in \mathscr{B}} C_G(H \cap V)$ it follows that $G = UC_G(H \cap V)$ for some $V \in \mathscr{B}$; replacing V by a smaller subgroup if necessary we can assume that $V \lhd U$. Thus the closed subgroup $N_G(H \cap V)$ contains both $C_G(H \cap V)$ and U, so is equal to G. Hence $H \cap V$ is a closed normal subgroup of G. If $H \cap V = G$ then G is a profinite simple group and hence finite, and this is a contradiction since H is infinite. Therefore $H \cap V = 1$ and so $H \leq Y(G) = D_G(G)$ by Lemma 17.5 (c). Since G is compactly generated it follows that $G = UC_G(W)$ for some subgroup $W \in \mathscr{B}$ with $W \leq U$. But then all conjugates of W lie in U, and so the closure of the group that they generate is normal in G and contained in U. Since G is topologically simple we conclude that the subgroups in $\mathscr{B}$ are finite and that G is discrete, and the result follows from this contradiction. $\qquad\qquad\square$

Theorem 17.7 *Suppose that G is topologically simple, compactly generated and non-discrete. Then $Y(G) = 1$ and $\mathscr{L}$ contains no non-trivial virtually soluble subgroups.*

Proof The first assertion follows from Lemma 17.6. Suppose that $\mathscr{L}$ contains a non-trivial virtually soluble subgroup H.

First suppose that H is finite. Then $N_G(H)$ is open, and $C_G(H)$ is a closed subgroup of finite index in $N_G(H)$ and so again open in G. Choose $U \in \mathscr{B}$ with $U \leq C_G(H)$. Then $H \leq D_G(U)$, and since U is commensurated by G this is a contradiction to Lemma 17.6.

Now we consider the general case. For each $U \in \mathscr{B}$ we have $H \cap U \neq 1$ by

Lemmas 17.5 (c) and 17.6. From above each $H \cap U$ is infinite, and since H has an open soluble subgroup, some $H_1 := H \cap U$ is also soluble. Let A be the closure of the last non-trivial term of the series of iterated abstract commutator subgroups of H_1. Then A is normalised by $N_U(H_1)$ and so is in $\mathscr{L}$. Moreover A is abelian and closed in the compact subgroup U, hence compact. Therefore $\mathscr{L}$ has an infinite compact nilpotent subgroup K commensurated by G by Lemma 17.4. Since $D_G(K) \geq Z(K)$ we have another contradiction to Lemma 17.6.

$\square$

17.3 Technical lemmas

Lemma 17.8 (cf. [1, Proposition 2.6]) *Let* $U, V \leq G$, *let* $K \leq U \cap V$ *and suppose that* $C_G(K \cap g^{-1}Kg) = 1$ *for all* $g \in U$. *If* $\theta : U \to V$ *is an isomorphism fixing K pointwise then* $\theta = \mathrm{id}_U$ *and so* $U = V$.

Proof Let $g \in U$. For $x \in K \cap g^{-1}Kg$ we have

$$gxg^{-1} = \theta(gxg^{-1}) = \theta(g)\theta(x)\theta(g^{-1}) = \theta(g)x\theta(g^{-1}).$$

Hence $g^{-1}\theta(g) \in C_G(K \cap g^{-1}Kg) = 1$. Thus $g = \theta(g)$. $\square$

Lemma 17.9 *Let* $H \leq G$ *with* $Y(H) = 1$. *The following assertions hold.*

(a) $D_G(H) \cap N_G(H) = C_G(H)$.
(b) *If* K *is a finite subgroup of* G *and* $|H : N_H(K)|$ *is finite then* $N_K(H) = C_K(H)$.
(c) *If* $K \in \mathscr{L}$ *and* K *is finite then* $H \cap K = 1$.

Proof (a) Let $t \in D_G(H) \cap N_G(H)$. Then t acts by conjugation as an automorphism of H that fixes pointwise some subgroup $K = H \cap U$ with $U \in \mathscr{B}$. For $g \in H$ we have $K \cap g^{-1}Kg = H \cap (U \cap g^{-1}Ug)$; since $U \cap g^{-1}Ug \in \mathscr{B}$ and $Y(H) = 1$ it follows that $C_H(K \cap g^{-1}Kg) = 1$. Hence $t \in C_G(H)$ by Lemma 17.8. This proves that $D_G(H) \cap N_G(H) \leq C_G(H)$ and the reverse inclusion is clear.

(b) We can replace K by $N_K(H)$ and assume that K normalises H. Then K also normalises $L = N_H(K)$. So $C_L(L \cap K) \leq_f L \leq_f H$. Thus $L \cap K$ centralises an open subgroup of H and $L \cap K \leq Y(H) = 1$. Therefore $[K, L] = 1$. It follows that $C_H(K) \leq_o H$ and that $K \leq D_G(H)$. But also $K \leq N_G(H)$, and so $K \leq C_G(H)$ by (a).

(c) Find $V \in \mathscr{B}$ with $K \lhd V$. Let $H_1 = H \cap V$ and $K_1 = H \cap K$. Then $K_1 \lhd H_1$ and H_1, K_1 satisfy the hypotheses on H, K in (b). Therefore

$$K_1 = N_{K_1}(H_1) = C_{K_1}(H_1) \leq Y(H) = 1. \qquad \square$$

Lemma 17.10 *Suppose that* $Y(G) = 1$. *Let* $L \in \mathscr{L}$ *and suppose that* $C_U(L)$ *is finite for some* $U \in \mathscr{B}$.

(a) *If* θ *is an automorphism of* G *that fixes* L *pointwise then* $\theta = \mathrm{id}_G$.
(b) $C_G(L) = 1$.

Proof (a) We may replace U by the subgroup $N_U(L) \in \mathscr{B}$ and assume that U normalises L.

Let $g \in U$ and let $x \in L$. Then

$$gxg^{-1} = \theta(gxg^{-1}) = \theta(g)x\theta(g^{-1}),$$

and so $g^{-1}\theta(g)$ centralises x. Hence $g^{-1}\theta(g) \in C_U(L) = 1$. Thus θ fixes U pointwise. Since $C_G(U \cap g^{-1}Ug) \leq Y(G) = 1$ for all $g \in G$, Lemma 17.8 gives $\theta = \mathrm{id}_G$.

(b) Applying (a) to the conjugation action of $C_G(L)$ on G we conclude that $C_G(L) \leq Y(G) = 1$, as required. $\qquad\square$

Lemma 17.11 *Suppose that* $Y(G) = 1$ *and that* $\mathscr{L}$ *contains no non-trivial compact abelian subgroups. Then* $Y(H) = 1$ *for each compact subgroup* H *in* $\mathscr{L}$.

Proof Choose $U \in \mathscr{B}$ with $H \triangleleft U$. Let $g \in Y(H)$; so $g \in C_H(H \cap V)$ for some $V \in \mathscr{B}$. The subgroup $C_H(H \cap V)$ is in $\mathscr{L}$ and since $Y(G) = 1$ it must be trivial or infinite by Lemma 17.9 (c). In the latter case, it must intersect $H \cap V$ non-trivially as $|U : U \cap V|$ is finite, so $Z(H \cap V)$ is a non-trivial compact abelian subgroup in $\mathscr{L}$, a contradiction. Therefore $C_H(H \cap V) = 1$ and $g = 1$. $\qquad\square$

17.4 Groups with no abelian subgroups in $\mathscr{L}$

It is now easy to develop some properties of the structure lattice $\widehat{\mathscr{L}}$ and the subset $\widehat{\mathscr{C}}$ consisting of the elements $[C_G(H)]$ with $H \in \mathscr{L}$. We shall see that in many important cases $\widehat{\mathscr{C}}$ is a Boolean lattice with respect to suitably defined operations.

Theorem 17.12 *If* $Y(G) = 1$ *and* $\mathscr{L}$ *contains no non-trivial compact abelian subgroups then every subgroup* $H \in \mathscr{L}$ *has the following properties.*

(a) $Y(H) = 1$.
(b) $D_G(H) = C_G(H)$. *In particular,* $C_G(H)$ *depends only on* $[H]$.
(c) *If also* $K \in \mathscr{L}$, *then* H, K *commute if and only if* $H \cap K = 1$.

Proof　Choose $U \in \mathscr{B}$ with $U \leq N_G(H)$.

(a) By Lemmas 17.5 (b) and 17.11 we have $Y(H) \cap U = Y(H \cap U) = 1$, and so the result follows from Lemma 17.5 (c).

(b) Let $H_1 = D_G(H)$. Since $Y(H) = Y(H_1) = 1$ from (a), Lemma 17.9 (a) implies that

$$C_G(H) = D_G(H) \cap N_G(H) = N_{H_1}(H) \quad \text{and} \quad D_G(H_1) \cap N_G(H_1) = C_G(H_1).$$

Since U normalises H we have $H_1 \cap U \leq N_{H_1}(H)$ and so the first equality above shows that $H \leq C_G(H_1 \cap U) \leq D_G(H_1)$. Clearly $H \leq N_G(H_1)$ and so the second equality gives $H \leq C_G(H_1)$. Thus $D_G(H) \leq C_G(H)$, and the reverse equality is clear.

(c) If H, K commute then $H \cap K = 1$ by (a).

Conversely, suppose that $H \cap K = 1$. There is a subgroup $W \in \mathscr{B}$ that normalises H and K. Thus $H \cap W$ and $K \cap W$ are disjoint and normal in their join. From (b) we have $H \cap W \leq D_G(K) = C_G(K)$ and $K \leq D_G(H) = C_G(H)$. Hence H, K commute, as required.　□

Proposition 17.13　*Suppose that* $Y(G) = 1$ *and that* $\mathscr{L}$ *contains no non-trivial compact abelian subgroups. Define a map* $^\circ \colon \widehat{\mathscr{L}} \to \widehat{\mathscr{C}}$ *by* $[H] \mapsto [C_G(H)]$. *For elements* $\alpha, \beta \in \widehat{\mathscr{C}}$ *the following are equivalent:*

(i) $\alpha \wedge \beta = [1]$;

(ii) $\alpha \leq \beta^\circ$.

Proof　The map $^\circ$ is well defined by Theorem 17.12 (a). Let $\alpha = [C_G(L)]$, $\beta = [C_G(M)]$ with $L, M \in \mathscr{L}$. Replacing L, M by their intersections with a group in $\mathscr{B}$ that normalises both, we can assume that L, M are compact and normalise each other.

Suppose that (i) holds: then $[C_G(L) \cap C_G(M)] = [1]$ and so for some $V \in \mathscr{B}$ that normalises L, M we have $C_V(L) \cap C_V(M) = 1$. Thus $[C_V(L), C_V(M)] = 1$ and $C_V(L) \leq C_G C_V(M)$. But $C_V(M) \sim C_G(M)$, and so $C_G C_V(M) = C_G C_G(M)$. Hence $\alpha = [C_V(L)] \leq \beta^\circ$.

Now suppose instead that (ii) holds. Then for some $U \in \mathscr{B}$ we have $C_U(L) \leq C_U C_G(M)$ and hence $[C_U(L), C_G(M)] = 1$. This implies that $C_U(L) \cap C_G(M)$ has trivial derived group and so is itself trivial. Thus $C_G(L) \cap C_G(M) \cap U = 1$, and so $[1] = [C_G(L) \cap C_G(M)] = \alpha \wedge \beta$.　□

We need the following result about partially ordered sets due to Frink [4].

Lemma 17.14　*Let M be a partially ordered set with least element 0 in which every two elements have a greatest lower bound. Suppose that* $^\circ \colon M \to M$ *is an order-reversing bijection such that for all* $\alpha, \beta \in M$ *we have* $\alpha \wedge \beta = 0$ *if and* *only if* $\alpha \leq \beta°$. *Then M is a Boolean lattice, with minimum* 0 *and maximum* 0°, *complementation given by* ° *and join operation given by* $\alpha \vee \beta = (\alpha° \wedge \beta°)°$.

Theorem 17.15 *Suppose that* $Y(G) = 1$ *and that* $\mathscr{L}$ *contains no non-trivial compact abelian subgroups. Let* $\mathscr{C}$ *be the subfamily of* $\mathscr{L}$ *consisting of elements* $[C_G(H)]$ *with* $H \in \mathscr{L}$. *Then* $\mathscr{C}$ *is a Boolean lattice.*

Proof Since $C_G(H) \cap C_G(K) = C_G(HK)$ for subgroups $H, K \in \mathscr{L}$ that normalise each other, $\mathscr{C}$ is closed with respect to taking meets in $\mathscr{L}$. The map $° \colon \mathscr{L} \to \mathscr{C}$ discussed in Proposition 17.13 is order-reversing since $H_1 \leq H_2$ implies $C_G(H_2) \leq C_G(H_1)$, and its restriction to $\mathscr{C}$ is an involution since the equality $C_G C_G C_G(H) = C_G(H)$ holds for each subgroup H. The conclusion now comes from Proposition 17.13 and Lemma 17.14. The latter of these also explains how the join operation in $\mathscr{C}$ is defined. □

Corollary 17.16 *Let G be non-discrete, compactly generated and topologically simple. Then the conclusions of Theorems* 17.12 *and* 17.15 *hold.*

Proof The hypotheses of these theorems hold by Theorem 17.7. □

Acknowledgements. This account was written while the author was the Leibniz Professor at the University of Leipzig. He thanks the university for its generous hospitality.

References for this chapter

[1] Y. Barnea, M. Ershov and T. Weigel. Abstract commensurators of profinite groups. *Trans. Amer. Math. Soc.* **363** (2011), 5381–5417.

[2] P.-E. Caprace, C. D. Reid and G. A. Willis. Locally normal subgroups of totally disconnected groups. Part I: General theory. Forum Math. Sigma 5 (2017), e11, 76 pp.

[3] P.-E. Caprace, C. D. Reid and G. A. Willis. Locally normal subgroups of totally disconnected groups. Part II: Compactly generated simple groups. Forum Math. Sigma 5 (2017), e12, 89 pp.

[4] O. Frink. Pseudo-complements in semi-lattices. *Duke Math. J.* **29** (1962), 505–514.

[5] D. J. S. Robinson. *A course in the theory of groups,* 2nd edn. (Springer, 1996).

[6] J. S. Wilson. Groups with every proper quotient finite. *Proc. Cambridge Philos. Soc.* **69** (1971), 373–391.

[7] J. S. Wilson. Structure theory for branch groups. In *Geometric and Homological Topics in Group Theory,* London Math. Soc. Lecture Note Ser. 358 (Cambridge University Press, 2009), 306–320.

[8] J. S. Wilson. *Profinite groups* (Clarendon Press, Oxford, 1998).

18

The centraliser lattice

David Hume and Thierry Stulemeijer

Abstract

In this chapter, we review the definition of the centraliser lattice $\mathscr{L}\mathscr{C}(G)$ of a totally disconnected locally compact (tdlc) group, and explain why it is a Boolean algebra. We then study the action of G on Ω, the associated Stone space of $\mathscr{L}\mathscr{C}(G)$, and prove that under certain hypotheses on G, the action is continuous, minimal, strongly proximal and micro-supported. Moreover, Ω satisfies a universal property in the category of profinite G-spaces that are micro-supported.

18.1	The centraliser lattice as a Boolean algebra	267
18.2	Topological dynamics	271
	References for this chapter	274

18.1 The centraliser lattice as a Boolean algebra

The centraliser lattice is a subset of the structure lattice defined in Chapter 17 in this volume. Let us just record its definition here.

Definition 18.1

(i) A subgroup $K \leq G$ is called **locally normal** if it is compact and normalised by an open subgroup of G.

(ii) Two subgroups H, K of G are **locally equivalent** if there exists a compact open subgroup U of G such that $H \cap U = K \cap U$, or equivalently if $H \cap K$ has finite index in both H and K.

(iii) The set of all local equivalence classes having a locally normal representative is called the **structure lattice** of G, and is denoted by $\mathscr{L}\mathscr{N}(G)$ in
the present chapter.

So, informally, $\mathscr{L}\mathscr{N}(G)$ is the lattice of subgroups that are compact and
normal in an open subgroup, everything happening 'up to finite index'. As
explained in Chapter 17, this is a lattice in a natural way. The aim of this
first part is to find lattices living inside $\mathscr{L}\mathscr{N}(G)$ that are Boolean algebras.
The following example shows that $\mathscr{L}\mathscr{N}(G)$ itself is not a Boolean algebra in
general.

Example 18.2 Let $G = \mathbf{Q}_p^n$ (where $n \geq 2$), and consider three distinct copla-
nar lines V_1, V_2 and V_3. Each V_i is isomorphic to $\mathbf{Q}_p$. Using this identification,
define $K_i \leq V_i$ to be the p-adic integers inside this line. Now, $[K_i] \in \mathscr{L}\mathscr{N}(G)$,
but we have

$$[\{e\}] = [(K_1 \wedge K_2) \vee (K_1 \wedge K_3)] \neq [K_1 \wedge (K_2 \vee K_3)] = [K_1].$$

This gives an example where $\mathscr{L}\mathscr{N}(G)$ is not even distributive.

Before defining the centraliser lattice, it will be more intuitive to first define
local decomposition lattices.

Definition 18.3

(i) Given a topological group H and subgroup K, say K is an **almost direct
factor** of H if there is a closed subgroup L of H of finite index such that
K is a direct factor of L.

(ii) Let $\alpha \in \mathscr{L}\mathscr{N}(G)$, say $\alpha = [H]$, where H is locally normal. We define
the **local decomposition lattice** $\mathscr{L}\mathscr{D}(G;H)$ of G at H to be the subset of
$\mathscr{L}\mathscr{N}(G)$ consisting of elements $[K]$ where K is locally normal in G and
K is an almost direct factor of H.

We think about this lattice simply as the lattice of direct factors, up to finite
index. Now, let us pretend for a minute that we are not working up to finite
index, but instead that we truly consider the lattice of direct factors of H. In
this situation, it is really straightforward to prove that we obtain a Boolean
algebra, under the obviously necessary condition that H has a trivial centre.

Lemma 18.4 *Let H be a group such that $Z(H) = \{1\}$.*

*(i) If H decomposes as a direct product $H = K \times L$, then $L = C_H(K)$, and
hence $K = C_H(C_H(K))$.*

*(ii) If K and L are direct factors of H, then $K = (K \cap L) \times (K \cap C_H(L))$, and
hence $LK = L(K \cap C_H(L))$.*

(iii) If K and L are direct factors of H, then $H = KL \times (C_H(L) \cap C_H(K))$.

Proof

(i) L is clearly contained in $C_H(K)$. Now, if $kl \in H \simeq K \times L$ and kl centralises K, then in particular $k \in C_H(K) \cap C_H(L) = Z(H) = \{1\}$. Hence $C_H(K) \leq L$.

(ii) Writing $H = L \times C_H(L)$, let π_1 (resp. π_2) denote the projection onto L (resp. $C_H(L)$). Then $K \leq \pi_1(K) \times \pi_2(K)$. But $C_H(K)$ centralises $\pi_i(K)$, i.e. $\pi_i(K) \leq C_H(C_H(K)) = K$ (the last equality uses (1)). Hence, $K = \pi_1(K) \times \pi_2(K)$, $\pi_1(K) = K \cap L$, and $\pi_2(K) = K \cap C_H(L)$. The last assertion follows from the equalities :

$$LK = L(K \cap L)(K \cap C_H(L)) = L(K \cap C_H(L)).$$

(iii) Applying (ii) to $C_H(K)$, we find

$$H = K \times C_H(K) = K \times (C_H(K) \cap L) \times (C_H(K) \cap C_H(L)).$$

So that by the last part of (ii), we obtain

$$H = KL \times (C_H(K) \cap C_H(L)).$$

$\square$

Lemma 18.4 shows that the set of direct factors of H is a Boolean algebra, where the complementation map is 'taking the centraliser'. The major technical input in [1] is to show that, under a somewhat intuitive assumption on H (namely **C-stability**, see definition 18.5), it is possible to define this complementation map in the local setting, i.e. when considering subgroups 'up to finite index'. The crucial result in that direction is Lemma 18.6.

Definition 18.5 Let G be a topological group, and $H \leq G$ be a subgroup.

(i) We define the **quasi-centraliser** of H in G, denoted by $QC_G(H)$, to be the subgroup of G consisting of those elements that centralise an open subgroup of H. Obviously, $QC_G(H)$ only depends on H up to finite index. So that for $\alpha \in \mathscr{LN}(G)$, we can also define $QC_G(\alpha)$ to be equal to $QC_G(H)$, where H is any representative of α.

(ii) H is called **C-stable** if $QC_G(H) \cap QC_G(C_G(H)) \cap U$ is trivial, for all open compact subgroups U of G.

Lemma 18.6 (Corollary 3.13 in [1]) *Let G be a tdlc group and let K be a locally normal subgroup which is C-stable. Then for all open compact subgroups U of G, we have*

$$[C_U(K \cap U)] = [QC_G(K)].$$

With just a bit more work, Lemma 18.6 enables us to transpose to the local setting the observation we made about direct factors in Lemma 18.4.

Lemma 18.7 (Theorem 4.5 in [1]) *Let G be a tdlc group and let $[H] \in \mathscr{LN}(G)$ have a C-stable representative. Then $\mathscr{LD}(G;H)$ is a sublattice of $\mathscr{LN}(G)$ and every $\beta \in \mathscr{LD}(G;H)$ has a locally normal C-stable representative. Furthermore, $\mathscr{LD}(G;H)$ is internally a Boolean algebra (relative to the maximum $[H]$), with complementation map*

$$\perp : [K] \to [QC_G(K)] \wedge [H].$$

Remark 18.8   Note that the C-stability condition can be rephrased in the following intuitive way:

$$H^{\perp} \wedge (H^{\perp})^{\perp} = 0.$$

We finally arrive at the definition of the centraliser lattice, which is a Boolean algebra whenever G is **locally C-stable**.

Definition 18.9   A tdlc group G is called **locally C-stable** if $QC_G(G) = \{1\}$ and if all locally normal subgroups are C-stable. Or equivalently, if $QC_G(G) = \{1\}$ and the only abelian locally normal subgroup is the trivial one (see Proposition 3.13 in [1] for a proof that those properties are equivalent).

Definition 18.10   Let G be a locally C-stable tdlc group. As above, define the map $\perp : \mathscr{LN}(G) \to \mathscr{LN}(G) : \alpha \mapsto [QC_G(\alpha)]$. This is well defined by Lemma 18.6. The **centraliser lattice** $\mathscr{LC}(G)$ is defined to be the set $\{\alpha^{\perp} | \alpha \in \mathscr{LN}(G)\}$ together with the map $\perp$ restricted to $\mathscr{LC}(G)$, the partial order inherited from $\mathscr{LN}(G)$ and the binary operations $\wedge_c$ and $\vee_c$ given by:

$$\alpha \wedge_c \beta = \alpha \wedge \beta$$

$$\alpha \vee_c \beta = (\alpha^{\perp} \wedge \beta^{\perp})^{\perp}.$$

Theorem 18.11 (Theorem 5.2 in [1]) *Let G be a locally C-stable tdlc group. The poset $\mathscr{LC}(G)$ is a Boolean algebra and $\perp^2 : \mathscr{LN}(G) \to \mathscr{LC}(G)$ is a surjective lattice homomorphism.*

Remark 18.12   Note that in order to get a Boolean algebra, the local C-stability is obviously necessary. Indeed, as mentioned in Definition 18.9, G being locally C-stable is equivalent to $QC_G(G)$ being trivial and G having no non-trivial abelian locally normal subgroup. But both conditions are necessary to allow $\perp$ to be a complementation.

We end up with an observation that might be of some interest. In this section, we have just constructed a Boolean algebra in $\mathscr{L}\mathscr{N}(G)$. The aim was to study the topological dynamics of the action of G on the associated Stone space. But in fact, one could more generally look for distributive lattices, and then study the dynamics of the action on the associated Priestley space. It is possible to define a quotient of $\mathscr{L}\mathscr{N}(G)$ which is distributive, and which is maximal for this property (one has just to kill all derived subgroups by identifying $[K]$ with $[K']$). A nice feature of this maximal distributive quotient is that it naturally contains a copy of the centraliser lattice. The hope is that the associated Priestley space might supply an interesting tool to study G when its centraliser lattice is trivial.

18.2 Topological dynamics

In this section we concentrate on the dynamics of the action of a group G in the class $\mathscr{S}$ — compactly generated, topologically simple, non-discrete, totally disconnected locally compact groups — on the Stone space Ω of its centraliser lattice $\mathscr{L}\mathscr{C}(G)$.

We recall that the Stone representation theorem assigns a profinite space $\Omega(\mathscr{B})$ to any Boolean algebra $\mathscr{B}$ where elements of $\mathscr{B}$ are in $1-1$ correspondence with clopen subsets of $\Omega(\mathscr{B})$. Since every $G \in \mathscr{S}$ is locally C-stable, the centraliser lattice $\mathscr{L}\mathscr{C}(G)$ is a Boolean algebra, see Chapter 17 in this volume.

18.2.1 Automorphism group of a regular tree

Let T be a regular tree and let $G = \mathrm{Aut}(T)^+$ be the group of all automorphisms of T which act **without edge inversion**. This group satisfies Tits' property P and therefore it is an element of $\mathscr{S}$, see Chapter 6 in this volume.

Theorem 18.13 *Let T be a tree and let $\alpha \in Aut(T)^+$. Define the **translation length** of α to be $\lambda(\alpha) = \min\{d_T(v, \alpha(v)) \,|\, v \in VT\}$.*

*If $\lambda(\alpha) > 0$ then the **axis** of α: $\{v \in VT \,|\, d(v, \alpha(v) = \lambda(\alpha)\}$ is a geodesic line ℓ and for every vertex $w \in VT$ we have $d(w, \alpha^n(w)) = 2d(w, \ell) + n\lambda(\alpha)$.*

An automorphism with translation length 0 (a fixed vertex) is said to be **elliptic**, all other elements are said to be **hyperbolic**.

Let T be the regular tree of degree 3. The **boundary** of T, $\partial_\infty T$, is defined to be the set of geodesic rays beginning at a fixed vertex v. Notice that given any two geodesic rays γ_1, γ_2, there is a unique geodesic line γ and a natural number

k such that

$$\gamma(n) = \begin{cases} \gamma_1(n+k) & \text{if } n \geq 0 \\ \gamma_2(-n+k) & \text{if } n \leq 0. \end{cases}$$

To see this notice that the union of the geodesic rays $\gamma_1 \cup \gamma_2$ is a subtree of T with at most one point of valence 3. The value k is sometimes called the **Gromov product** of γ_1 and γ_2 — denoted $(\gamma_1 . \gamma_2)_v$.

We define a topology on $\partial_\infty T$ with basic open sets

$$V(\gamma, r) = \left\{ \gamma' \mid (\gamma . \gamma')_v \geq r \right\}.$$

With respect to this topology (which does not depend on the choice of the vertex v), ∂T is a Cantor set.

The group $G = \operatorname{Aut}(T)^+$ lies in $\mathscr{S}$ and its centraliser lattice contains all almost direct factors of the compact open subgroup U consisting of all automorphisms which fix a vertex v, as a result the pointwise stabiliser of any clopen subset of ∂T defines an element of the centraliser lattice.

In this case, the Stone representation theorem gives a G-equivariant homeomorphism $\partial T \to \Omega(\mathscr{LC}(G))$, so it is sufficient to study the dynamics of the G-action on ∂T. In particular, continuity follows immediately from the definition of the topology on ∂T.

Now let T' be a component of $T \setminus v$. The boundary of T' is a clopen subset $Y \subset \partial T$; denote by H the subgroup of G which fixes $\partial T \setminus Y$ pointwise. Let $g \in \operatorname{Aut}(T)^+$ be a hyperbolic isometry such that the repelling fixed point ℓ_- of g in ∂T lies in Y.

Notice that for each n, $g^{-n} H g^n$ defines the subgroup of G which fixes $\partial T \setminus g^{-n} Y$ pointwise. As $g^{-n} Y$ is strictly contained in Y and $Y \setminus g^{-n} Y$ contains a non-trivial clopen subset of ∂T, we see that $[g^{-n} H g^n]$ a strictly smaller element of the centraliser lattice and in the boundary the clopen subsets converge in the weak-$*$ topology to the singleton $\{\ell_-\}$.

Definition 18.14   An action of a group G on a topological space X by homeomorphisms is said to be **strongly proximal** if the closure of every G-orbit in the space of probability measures on X (under the weak-$*$ topology) contains a Dirac mass.

A closed subset $V \subset X$ is said to be **compressible** if the closure of the orbit of V in the space of closed subsets of X contains a singleton.

In the above discussion we prove that there is a clopen compressible subset for the action of G on ∂T and that this action is strongly proximal.

Moreover, the action of G on ∂T is transitive. A map which sends one geodesic ray starting at v to another can be extended to an automorphism of the tree.

Definition 18.15 An action of a group on a topological space X is **minimal** if every orbit is dense.

The final condition we will introduce is a triviality for this action.

Definition 18.16 An action of a topological group G on a totally disconnected space X by homeomorphisms is said to be **micro-supported** if it is continuous (in the sense given above) and the pointwise stabiliser of every proper clopen subset of X is non-trivial.

18.2.2 Groups in $\mathscr{S}$

Theorem 18.17 ([2, Theorem F]) *Let* $G \in \mathscr{S}$ *and let* Ω *be the Stone space associated to the Boolean algebra* $\mathscr{LC}(G)$.

(i) *The G-action on* Ω *is continuous, minimal, strongly proximal and micro-supported; moreover,* Ω *contains a compressible clopen subset.*

(ii) *Given a profinite space* X *with a continuous G-action, the G-action on* X *is micro-supported if and only if there is a continuous G-equivariant surjective map* $\Omega \to X$. *In particular, every micro-supported G-action is minimal and strongly proximal.*

The first part of the above theorem states that the phenomena described above for the action of $\mathrm{Aut}(T)^+$ on ∂T hold for the action of G on Ω whenever $G \in \mathscr{S}$. This is trivial if Ω is a singleton, in other words, when $\mathscr{LC}(G) = \{0, \infty\}$. When $\mathscr{LC}(G) \neq \{0, \infty\}$, Theorem 18.17 implies that the action of G on Ω is faithful.

The second part states that any micro-supported action is a 'quotient' of the action of G on Ω, so in particular, implies the following.

Corollary 18.18 *Let* $G \in \mathscr{S}$. *Then the centraliser lattice* $\mathscr{LC}(G) \neq \{0, \infty\}$ *if and only if* G *admits a continuous micro-supported action on a space containing at least* 3 *points.*

$\mathrm{Aut}(T)^+$ has North-South dynamics — hyperbolic isometries have attracting and repelling fixed points — so a standard ping-pong argument finds non-Abelian free subgroups. By contrast, for a general $G \in \mathscr{S}$ it is currently only possible to find isometries which are attracting, hence we may obtain free subsemigroups.

Corollary 18.19 ([2, Corollary H]) *Let $G \in \mathscr{S}$ and suppose $\mathscr{LC}(G) \neq \{0,\infty\}$. Then G contains a non-Abelian discrete free subsemigroup.*

The question of whether every $G \in \mathscr{S}$ with non-trivial centraliser lattice contains a discrete non-Abelian free subgroup remains open.

It is therefore appropriate — as pointed out to us by Rémi Coulon — to say that the topological dynamics of such groups may resemble the action of a Baumslag–Solitar group (on its Bass–Serre tree) more closely than the action of $\mathrm{Aut}(T)^+$. However, one should be wary of this analogy as well, since the incompatibility of amenability and strongly proximal actions yields

Corollary 18.20 ([2, Corollary G]) *Let $G \in \mathscr{S}$. Every closed cocompact amenable subgroup of G fixes a point in Ω. In particular, if the centraliser lattice is non-trivial, then G is not amenable.*

References for this chapter

[1] P.-E. Caprace, C.D. Reid and G.A. Willis. Locally normal subgroups of totally disconnected groups. Part I: General theory. Forum Math. Sigma 5 (2017), e11, 76 pp.

[2] P.-E. Caprace, C.D. Reid and G.A. Willis. Locally normal subgroups of totally disconnected groups. Part II: Compactly generated simple groups. Forum Math. Sigma 5 (2017), e12, 89 pp.

19

On the quasi-isometric classification of locally compact groups

Yves de Cornulier

Abstract

This (quasi-)survey addresses the quasi-isometry classification of locally compact groups, with an emphasis on amenable hyperbolic locally compact groups. This encompasses the problem of quasi-isometry classification of homogeneous negatively curved manifolds. A main conjecture provides a general description; an extended discussion reduces this conjecture to more specific statements.

In the course of the paper, we provide statements of quasi-isometric rigidity for general symmetric spaces of non-compact type and also discuss accessibility issues in the realm of compactly generated locally compact groups.

19.1	Introduction	276
19.2	Preliminaries	281
19.3	Quasi-isometric rigidity of symmetric spaces	292
19.4	Quasi-isometric rigidity of trees	298
19.5	Commable groups, and commability classification of amenable hyperbolic groups	310
19.6	Towards a quasi-isometric classification of amenable hyperbolic groups	324
19.7	Beyond the hyperbolic case	335
References for this chapter		339

19.1 Introduction

19.1.1 Locally compact groups as geometric objects

It has long been well understood in harmonic analysis (notably in the study of unitary representations) that locally compact groups are the natural objects unifying the setting of connected Lie groups and discrete groups. In the context of geometric group theory, this is still far from universal. For a long period, notably in the 70s, this unifying point of view was used essentially by Herbert Abels, and, more occasionally, some other people including Behr, Guivarc'h, Houghton. The considerable influence of Gromov's work paradoxically favored the bipolar point of view discrete vs continuous, although Gromov's ideas were applicable to the setting of locally compact groups and were sometimes stated (especially in [25]) in an even greater generality.

If a locally compact group is generated by a compact subset S, it can be endowed with the word length with respect to S, and with the corresponding left-invariant distance. While this distance depends on the choice of S, the metric space (G, d_S) — or the 1-skeleton of the corresponding Cayley graph — is uniquely determined by G up to quasi-isometry, in the sense that if T is another compact generating subset, the identity map $(G, d_S) \to (G, d_T)$ is a quasi-isometry.

We use the usual notion of Gromov-hyperbolicity [25] for geodesic metric spaces, which we simply call "hyperbolic"; this is a quasi-isometry invariant. A locally compact group is called *hyperbolic* if it is compactly generated and if its Cayley graph with respect to some/any compact generating subset is hyperbolic.

This paper is mainly concerned with the quasi-isometric classification and rigidity of amenable hyperbolic locally compact groups among compactly generated locally compact groups. Within discrete groups, this problem is not deep: the answer is that it falls into two classes, finite groups and infinite virtually cyclic groups, both of which are closed under quasi-isometry among finitely generated groups. On the other hand, in the locally compact setting it is still an open problem. In order to tackle it, we need a significant amount of nontrivial preliminaries. Part of this paper (esp. Sections 19.3 and 19.4) appears as a kind of survey of important necessary results about groups quasi-isometric to symmetric spaces of non-compact type and trees, unjustly not previously stated in the literature but whose proofs gather various ingredients from existing work along with minor additional features. Although only the case of rank 1 symmetric spaces and trees is needed for the application to hyperbolic groups, we state the theorems in a greater generality.

19.1.2 From negatively curved Lie groups to focal hyperbolic groups

In 1974, in answer to a question of Milnor, Heintze [27] characterized the connected Lie groups of dimension at least 2 admitting a left-invariant Riemannian metric of negative curvature as those of the form $G = N \rtimes \mathbf{R}$ where $\mathbf{R}$ acts on N as a one-parameter group of contractions; the group N is necessarily a simply connected nilpotent Lie group; such a group G is called a *Heintze group*. He also showed that any negatively curved connected Riemannian manifold with a transitive isometry group admits a simply transitive isometry group; the latter is necessarily a Heintze group (if the dimension is at least 2). The action of a Heintze group $G = N \rtimes \mathbf{R}$ on the sphere at infinity ∂G has exactly two orbits: a certain distinguished point ω and the complement $\partial G \smallsetminus \{\omega\}$, on which the action of N is simply transitive. This sphere admits a visual metric, which depends on several choices and is not canonical; however its quasi-symmetric type is well defined and is a functorial quasi-isometry invariant of G. The study of quasi-symmetric transformations of this sphere was used by Tukia, Pansu and R. Chow [11, 46, 58] to prove the quasi-isometric rigidity of the rank one symmetric spaces of dimension at least 3. Pansu also initiated such a study for other Heintze groups [45, 46].

On the other hand, hyperbolic groups were introduced by Gromov [25] in 1987. The setting was very general, but for many reasons (mainly unrelated to the quasi-isometric classification), their subsequent study was especially focused on discrete groups with a word metric. In particular, a common belief was that amenability is essentially incompatible with hyperbolicity. This is not true in the locally compact setting, since there is a wide variety of amenable hyperbolic locally compact groups, whose quasi-isometry classification is open at the moment. One purpose of this note is to describe the state of the art as regards this problem.

In 2005, the author asked Pierre Pansu whether there was a known characterization of (Gromov-) hyperbolic connected Lie groups, who replied that the answer could be obtained from his L^p-cohomology computations [47] (which were extracted from an unpublished manuscript going back to 1995) combined with a vanishing result later obtained by Tessera [56]; an algebraic characterization of hyperbolic groups among connected Lie groups, based on this approach, was finally given in [15], namely such groups are either compact, Heintze-by-compact, or compact-by-(simple of rank one). This approach consisted in proving, using structural results of connected Lie groups, that any connected Lie group not of this form cannot be hyperbolic, by showing that its L^p-cohomology in degree 1 vanishes for all p.

Cornulier

In [7], using a more global approach, namely by studying the class of *focal* hyperbolic groups, this was extended to a general characterization of all amenable hyperbolic locally compact groups, and more generally of all hyperbolic locally compact groups admitting a cocompact closed amenable subgroup (every connected locally compact group, hyperbolic or not, admits such a subgroup).

Theorem 19.1 ([7]) *Let G be a non-elementary hyperbolic compactly generated locally compact group with a cocompact closed amenable subgroup. Then exactly one of the following holds:*

(a) (focal case) G is amenable and non-unimodular. Then G is isomorphic to a semidirect product $N \rtimes \mathbf{Z}$ or $N \rtimes \mathbf{R}$, where the non-compact subgroup N is compacted by the action of positive elements t of $\mathbf{Z}$ or $\mathbf{R}$, in the sense that there exists a compact subset K of N such that $tKt^{-1} \subset K$ and $\bigcup_{n \geq 1} t^{-n} K t^n = N$.

(b) G is non-amenable. Then G admits a continuous proper isometric boundary-transitive (and hence cocompact) action on a rank 1 symmetric space of non-compact type, or on a finite valency tree with no valency 1 vertex and not reduced to a line (necessarily biregular).

Groups in (a) are precisely the amenable non-elementary hyperbolic locally compact groups, and are called *focal hyperbolic* (or *focal*) locally compact groups.

We assumed for simplicity that G is **non-elementary** in the sense that its boundary has at least 3 points (and is indeed uncountable), ruling out compact groups and 2-ended locally compact groups (which are described in Corollary 19.39). It should be noted that any group as in (b) admits a closed cocompact subgroup of the form in (a), but conversely most groups in (a) are not obtained this way, and are actually generally not quasi-isometric to any group as in (b), see the discussion in Section 19.6.3. This is actually a source of difficulty in the quasi-isometry classification: namely those groups in (a) that embed cocompactly in a non-amenable group bear "hidden symmetries"; see Conjecture 19.104 and the subsequent discussion.

For the discussion below, it is natural to split the class of focal hyperbolic locally compact groups G into three subclasses (see §19.2.3 for more details):

- G is of *connected type* if its boundary is homeomorphic to a positive-dimensional sphere, or equivalently if it admits a continuous proper cocompact isometric action on a complete negatively curved Riemannian manifold of dimension ≥ 2;

- G is of *totally disconnected type* if its identity component is compact, or equivalently if its boundary is a Cantor space, or equivalently if it admits a continuous proper cocompact isometric action on a regular tree of finite valency;
- G is of *mixed type* otherwise; then its boundary is connected but not locally connected.

19.1.3 Synopsis

We define the *commability* equivalence between locally compact groups as the equivalence relation generated by the requirement that any two locally compact groups G_1, G_2 with a continuous proper homomorphism $G_1 \to G_2$ with cocompact image are commable. Thus G_1 and G_2 are commable if and only if there exists an integer k, a family of locally compact groups $G_1 = H_0$, $H_1, \ldots, H_k = G_2$, and continuous proper homomorphisms f_i with cocompact image, either from H_i to H_{i+1} or from H_{i+1} to H_i. Obviously, commable compactly generated locally compact groups are quasi-isometric (the converse is not true: after the author asked about a counterexample within discrete groups, Carette and Tessera checked that some free products of suitable lattices in Lie groups with $\mathbf{Z}$ are indeed quasi-isometric but not commable, see §19.5.2).

A general study of commability is not an easy matter, as it is difficult in general to describe in a satisfactory way, given a locally compact group G (e.g. a Baumslag–Solitar group as given in Remark 19.55), those locally groups H in which G embeds as a cocompact subgroup. A study of commability in the realm of focal hyperbolic locally compact groups is carried out in Section 19.5, with a comprehensive description (except in the totally disconnected case). Actually, the Mostow rigidity theorem [43] was maybe the first time that quasi-isometries were used to solve a commability problem.

A comprehensive study of commability in the realm of focal hyperbolic locally compact groups is carried out in Section 19.5.

Section 19.6 addresses the quasi-isometry classification of focal hyperbolic groups. It discusses, using the results of all the previous sections, the following:

Conjecture 19.2 (Main conjecture) Two hyperbolic locally compact groups G, H with G focal are quasi-isometric if and only if they are commable.

This conjecture can be split between the *internal* case (H focal) and the *external* case (H non-focal), providing more explicit reformulations, which may seem unrelated at first sight. Notably, the external part of Conjecture 19.2 has the following equivalent restatement:

Conjecture 19.3 (slightly restated)   A focal hyperbolic locally compact group is quasi-isometric to a non-focal hyperbolic locally compact group if and only if it is quasi-isometric to a rank 1 symmetric space of non-compact type or a 3-regular tree.

Because of the very special role played by rank 1 symmetric spaces and trees, and because the extensive literature about them is not formulated in the context of locally compact groups, the important results concerning their quasi-isometric rigidity are surveyed (and slightly extended) in Sections 19.3 and 19.4.

In the external part, the totally disconnected type of Conjecture 19.2 is a bit at odds with the other two types because there is a complete understanding of the quasi-isometric classification in this case (Section 19.4), so it is rather a question about commability itself. See §19.5.6 (esp. Question 19.78). After being asked in a first version of this survey, it has been solved by M. Carette [8].

The internal part of Conjecture 19.2 (H focal) can be split into three cases, according to the type of the focal group G (see §19.2.5). Here the totally disconnected case is an easy theorem rather than a conjecture, since all focal hyperbolic locally compact groups of totally disconnected type are in the same commability class (Proposition 19.74). The remaining cases are the mixed type and the connected type.

Recall that a *purely real Heintze group* is a Heintze group $N \rtimes \mathbf{R}$ as above, for which the action of $\mathbf{R}$ on the Lie algebra of N has only real eigenvalues. Heintze groups are focal of connected type, and actually every focal group of connected type is commable to a purely real Heintze group, unique up to isomorphism (see §19.5.3). This gives a reduction of the internal connected type case of Conjecture 19.2 to the following simpler statement:

Conjecture 19.4   Any two purely real Heintze groups are quasi-isometric if and only if they are isomorphic.

The mixed type case can also be reduced to a similar statement (Conjecture 19.92), by finding, in each commability class of focal hyperbolic locally compact group of mixed type, a given group $H[\varpi, q]$, depending on three independent "parameters": a purely real Heintze group H, a non-power integer q, and a positive real number ϖ.

It turns out (see Theorem 19.94 extracted from [13]) that the quasi-isometry class of $H[\varpi, q]$

- determines the quasi-isometry class of H (by a simple argument based on the boundary);

- determines the real number ϖ, using a computation of L^p-cohomology in degree one by the author and Tessera [15], strongly inspired by Pansu [47];
- and finally also determines the non-power integer q, by a recent result of Dymarz about the large-scale geometry of focal groups of mixed type, relying on the study of the fine metrical structure of their boundary.

General remark This paper may seem to reduce, for expository matters, the study of the quasi-isometry equivalence relation between hyperbolic groups to the determination of quasi-isometry classes. It should by no means be the only point of view; the study of quasi-isometry invariants, such as various kinds of dimensions and more refined ones, allows to shed light on the fine geometric structure of many groups. For this reason, the large-scale study of real Heintze groups should not be reduced to aiming at proving Conjecture 19.4 (which reduces the QI-classification to a classification up to isomorphism, which is, in a certain sense, a wild problem), and even a proof of the latter would not supersede the relevance of the study of these invariants.

Here is an outline of the sequel.

- Section 19.2 contains some preliminary material, notably relying on [7].
- In Section 19.3, we give the quasi-isometric rigidity statements for symmetric spaces of non-compact type in the locally compact setting. These results are especially due to Kleiner–Leeb, Tukia, Pansu, R. Chow, Casson–Jungreis and Gabai.
- In Section 19.4, we give the quasi-isometric rigidity statements for trees in the locally compact setting, emphasizing on the notion of accessibility (in its group-theoretic and its graph-theoretic versions). These results are notably due to Stallings, Dunwoody, Abels, Thomassen–Woess, Mosher–Sageev–Whyte, and Krön–Möller.
- Section 19.5 provides a detailed description of commability classes between focal groups, and between focal and non-focal groups, relying on [13];
- The core of this paper is Section 19.6. It contains a discussion about the main conjecture, the link with its specifications, and surveys the main known cases; the first of which being due to P. Pansu, while recent progress has notably been made by X. Xie, T. Dymarz and M. Carrasco.
- The final Section 19.7 notably discusses the nilpotent case.

19.2 Preliminaries

We freely use the shorthand LC-group for locally compact group, and CGLC-group for compactly generated LC-group.

19.2.1 Quasi-isometries

The large-scale language

Recall that a map $f : X \to Y$ between metric spaces is a *large-scale Lipschitz map* if there exist $\mu > 0$ and $\alpha \in \mathbf{R}$ such that

$$d(f(x_1), f(x_2)) \leq \mu d(x_1, x_2) + \alpha \quad \forall x_1, x_2 \in X;$$

we then say f is (μ, α)-Lipschitz. Maps $f : X \to Y$ are at bounded distance, denoted $f \sim f'$ if $\sup_{x \in X} d(f(x), f'(x)) < \infty$; if this supremum is bounded by α we write $f \overset{\alpha}{\sim} f'$.

A quasi-isometry $f : X \to Y$ is a large-scale Lipschitz map such that there exists a large-scale Lipschitz map $f' : Y \to X$ such that both $f \circ f' \sim \mathrm{id}_Y$ and $f' \circ f \sim \mathrm{id}_X$ hold. The map f' is called an *inverse* quasi-isometry to f; it is unique modulo the equivalence $\sim$.

A large-scale Lipschitz map $X \to Y$ is *coarsely proper* if, using the convention $\inf \varnothing = +\infty$, the function $F(r) = \inf\{d(f(x_1), f(x_2)) : d(x_1, x_2) \geq r\}$ satisfies $\lim_{+\infty} F = +\infty$.

Every CGLC-group G can be endowed with the left-invariant distance defined by the word length with respect to a compact generating subset. Given any two such distances, the identity map is a quasi-isometry, and therefore the notions of large-scale Lipschitz map, quasi-isometry, etc. from or into G are independent of the choice of a compact generating set.

Hyperbolicity

A geodesic metric space is *hyperbolic* if there exists $\delta \geq 0$ such that for every triple of geodesic segments $[ab], [bc], [ac]$ and $x \in [bc]$ we have

$$d(x, [ab] \cup [ac]) \leq \delta.$$

To be hyperbolic is a quasi-isometry invariant among geodesic metric spaces. Thus a locally compact group is called *hyperbolic* if it is compactly generated and the 1-skeleton of its Cayley graph with respect to some/any compact generating subset is hyperbolic. By [7, §2], this holds if and only if it admits a continuous proper cocompact isometric action on a proper geodesic hyperbolic metric space.

Metric amenability

A locally compact group with left Haar measure λ is *amenable* (resp. *metrically amenable*) if for every compact subset S and $\varepsilon > 0$ there exists a compact subset F of positive Haar measure such that $\lambda(SF \smallsetminus F)/\lambda(F) \leq \varepsilon$, resp. $\lambda(FS \smallsetminus F)/\lambda(F) \leq \varepsilon$.

Note that for a left-invariant distance, FS is the "1-thickening" of F and justifies the adjective "metric". The following lemma is [55, Theorem 2], see also [14, §4.C].

Lemma 19.5

(a) A locally compact group is metrically amenable if and only if it is both amenable and unimodular.

(b) Metric amenability is a quasi-isometry invariant among CGLC-groups. □

Let us emphasize that being amenable is not a quasi-isometry invariant, in view of the cocompact inclusion $\mathbf{R} \rtimes \mathbf{R} \subset \mathrm{SL}_2(\mathbf{R})$. The problem asking which amenable CGLC-groups are quasi-isometric to non-amenable ones is a very challenging one; it will be addressed in the context of hyperbolic LC-groups in §19.6.3.

19.2.2 Cayley–Abels graph

For a compactly generated locally compact group, the Cayley graph with respect to some compact generating subset is often convenient, but has the drawback, when G is not discrete, to have infinite valency and in addition the action of G on its Cayley graph is not continuous.

Definition 19.6 A *Cayley–Abels* graph for G is a continuous, proper and cocompact action of G on a non-empty finite valency connected graph.

Of course, if G admits a Cayley–Abels graph, then it admits an open compact subgroup, namely the stabilizer of some vertex. The converse is the following elementary fact due to Abels [1, Beispiel 5.2] (see [37, §11.3] or [14, §2.E]).

Proposition 19.7 (Abels) *Let G be a locally compact group with a compact open subgroup (i.e. G is compact-by-(totally disconnected)). Then G admits a Cayley–Abels graph, which can be chosen to be vertex-transitive.* □

19.2.3 Types of hyperbolic groups

Gromov [25, §3.1] splits isometric group actions on hyperbolic spaces into 5 types: bounded, horocyclic, lineal, focal and general type, see [7, §3], from which we borrow the terminology. When specifying this to the action of a CGLC-group G on itself (or any continuous proper cocompact isometric action of G) we get four out of these five types, the first two of which are called elementary and the last two are called non-elementary:

- ∂G is empty: G is compact;
- ∂G has two elements: G admits $\mathbf{Z}$ as a cocompact lattice (see Corollary 19.39 for more characterizations);
- ∂G is uncountable and has a G-fixed point: G is called a focal hyperbolic group;
- ∂G is uncountable and the G-action is minimal: G is called a hyperbolic group of general type.

Among hyperbolic LC-groups, focal groups can be characterized as those that are amenable and non-unimodular, and general type groups can be characterized as those that are not amenable. Most hyperbolic LC-groups of general type (e.g., discrete ones) do not admit cocompact amenable subgroups, the exceptions being listed in Theorem 19.1.

Focal hyperbolic groups soon disappeared from the literature after [25], because the focus was made on proper actions of discrete groups, for which the applications were the most striking.[1] Except in the connected or totally disconnected case (and with another point of view), they reappear in [15], before they were given a structural characterization in [7].

19.2.4 Boundary

Let X be a proper geodesic hyperbolic space and ∂X its boundary. Then $\overline{X} = X \cup \partial X$ has natural compact topology, for which X is open and dense.

If X and Y are proper geodesic hyperbolic spaces, every quasi-isometric embedding $f : X \to Y$ has a unique extension $\hat{f} : \overline{X} \to \overline{Y}$ that is continuous on ∂X. This extension $\hat{f}$ maps ∂X into ∂Y and is functorial in f. Let $\bar{f} : \partial X \to \partial Y$ denote the restriction of $\hat{f}$. Then $\bar{f} = \bar{g}$ whenever f and g are at bounded distance; in particular, for every quasi-isometry, $\bar{f}$ is a homeomorphism $\partial X \to \partial Y$ whose inverse is $\bar{g}$, where g is any inverse quasi-isometry for f.

The boundary carries a so-called visual metric. For such metrics, the homeomorphic embedding $\bar{f}$ above is a *quasi-symmetric* embedding in the sense that there exists an increasing function $F : [0,\infty[\to [0,\infty[$ such that

$$\frac{d(f(x),f(y))}{d(f(y),f(z))} \le F\left(\frac{d(x,y)}{d(y,z)}\right), \quad \forall x \ne y \ne z \in X.$$

[1] There was a semantic shift in the meaning of "elementary", when the terminology from post-Gromov papers, which was only fit for proper actions of discrete groups, was borrowed instead of referring to the general setting duly considered by Gromov.

19.2.5 Focal hyperbolic groups

We say that an automorphism α of a locally compact group N is *compacting* (or is a *compaction*) if there exists a compact subset K of N such that $\alpha(K) \subset K$ and $\bigcup_{n \geq 0} \alpha^{-n}(K) = N$; we say that an action of $\mathbf{Z}$ or $\mathbf{R}$ by automorphisms $(\alpha^n)_{n \in \mathbf{Z}}$ or $(\alpha^t)_{t \in \mathbf{R}}$ on N is *compacting* if α^1 is compacting.

Focal hyperbolic groups of connected type

Recall that Lie groups are not assumed connected and thus include discrete groups; a locally compact group is by definition compact-by-Lie if it has a compact normal subgroup so that the quotient is Lie. This is equivalent to (connected-by-compact)-by-discrete.

We say that a focal hyperbolic group is of *connected type* if it is compact-by-Lie. The following proposition is contained in [7, Theorem 7.3].

Proposition 19.8 *Let G be a focal hyperbolic LC-group. Equivalences:*

- *G is of connected type;*
- *the kernel of its modular function is connected-by-compact;*
- *it admits a continuous, proper cocompact isometric action on a homogeneous simply connected negatively curved manifold of dimension ≥ 2;*
- *G has a maximal compact normal subgroup W and G/W is isomorphic to a semidirect product $N \rtimes \mathbf{Z}$ or $N \rtimes \mathbf{R}$, where N is a virtually connected Lie group and the action of $\mathbf{Z}$ or $\mathbf{R}$ is compacting.* □

Focal hyperbolic groups of totally disconnected type

Similarly we say that a focal hyperbolic group is of *totally disconnected type* if its identity component is compact. From [7, Theorem 7.3] we can also extract the following proposition.

Proposition 19.9 *Let G be a focal hyperbolic LC-group. Equivalences:*

- *G is of totally disconnected type;*
- *the kernel of its modular function has a compact identity component;*
- *it admits a continuous, proper cocompact isometric action on a regular tree of finite valency greater than 2;*
- *it is isomorphic to a strictly ascending HNN-extension over a compact group endowed with an injective continuous endomorphism with open image.*

Focal hyperbolic groups of mixed type

Finally, we say that a focal hyperbolic group is of *mixed type* if it is of neither connected nor of totally disconnected type. For instance, if p is prime and $\lambda \in \;]0,1[$, then the semidirect product $(\mathbf{R} \times \mathbf{Q}_p) \rtimes \mathbf{Z}$, where the positive generator of $\mathbf{Z}$ acts by multiplication by (λ, p) on the ring $\mathbf{R} \times \mathbf{Q}_p$.

Specifying once again [7, Theorem 7.3], we obtain

Proposition 19.10 *Let G be a focal hyperbolic LC-group. Equivalences:*

- *G is of mixed type;*
- *the kernel of its modular function is neither connected-by-compact, nor compact-by-(totally disconnected);*
- *it admits a continuous, proper cocompact isometric action on a* pure mille-feuille space *(see §19.2.6);*
- *it has a maximal compact normal subgroup W such that G/W has an open subgroup of finite index isomorphic to a semidirect product $(N_1 \times N_2) \rtimes \mathbf{Z}$, where $\mathbf{Z}$ acts on $N_1 \times N_2$ by compaction preserving the decomposition, N_1 is a connected Lie group, N_2 is totally disconnected, and N_1 and N_2 are both non-compact.*

As far as the author knows, it seems that focal groups of mixed type (including examples) were not considered before [15].

19.2.6 Millefeuille spaces and amenable hyperbolic groups

We here recall the definition of millefeuille spaces.

Given a metric space X, define a Busemann function $X \to \mathbf{R}$ as a limit, uniform on bounded subsets of X, of functions of the form $x \mapsto d(x, x_0) + c_0$ for $x_0 \in X$ and $c_0 \in \mathbf{R}$. By Busemann metric space, we mean a metric space (X, b) endowed with a Busemann function; a shift-isometry of (X, b) is an isometry f of X preserving b up to adding constants, i.e. such that $x \mapsto b(f(x)) - b(x)$ is constant. A homogeneous Busemann metric space means a Busemann metric space with a transitive group of shift-isometries.

Let (X, b) be a complete CAT(κ) Busemann space ($-\infty \leq \kappa \leq 0$). For k a non-negative integer, let T_k be a $(k+1)$-regular tree (identified with its 1-skeleton), endowed with a Busemann function denoted by b' (taking integer values on vertices). Note that the Busemann space (T_k, b') is uniquely determined by k up to combinatorial shift-isometry. The millefeuille space $X[b, k]$, introduced in [7, §7], is by definition the topological space

$$\{(x, y) \in X \times T_k \mid b(x) = b'(y)\}.$$

Call *vertical geodesic* in T_k, a geodesic in restriction to which b' is an isometry. Call *vertical leaf* in $X[b,k]$ a (closed) subset of the form $X[b,k] \cap (X \times V)$ where V is a vertical geodesic. In [7, §7], it is observed that there is a canonical geodesic distance, defining the topology, and such that in restriction to any vertical leaf, the canonical projection to X is an isometry. Moreover $X[b,k]$ is $\mathrm{CAT}(\kappa)$, and is naturally a Busemann space, the Busemann function mapping (x,y) to $b(x) = b'(y)$. Note that $X[b,0] = X$.

We have the following elementary well known lemma:

Lemma 19.11   *Let X be a homogeneous connected negatively curved Riemannian manifold of dimension ≥ 2. Then exactly one of the following holds:*

(a) $\mathrm{Isom}(X)$ *fixes a unique point in ∂X.*

(b) X *is a rank 1 symmetric space of non-compact type; in particular* $\mathrm{Isom}(X)$ *is transitive on ∂X.*

In particular, $\mathrm{Isom}(X)$ has a unique closed orbit on ∂X, which is either a singleton or the whole ∂X.

Proof   We use that $G = \mathrm{Isom}(X)$ is hyperbolic and the action of G on X is quasi-isometrically conjugate to the left action of G on itself. If G is amenable, it is focal hyperbolic and thus fixes a unique point ω_X on the boundary and G is transitive on $X \smallsetminus \{\omega_X\}$ (see for instance [7, Proposition 5.5(c)]), thus $\{\omega_X\}$ is the unique closed G-orbit.

Otherwise, it is hyperbolic of general type and virtually connected, and thus, by a simple argument (see [7, Proposition 5.10]), is isomorphic to an open subgroup in $\mathrm{Isom}(Y)$ for some rank 1 symmetric space Y of non-compact type. Let $K \subset G$ be the stabilizer in G of one point $x_0 \in X$, by transitivity we have $X \simeq G/K$ as G-spaces. Since X is $\mathrm{CAT}(-1)$ (up to homothety) and complete, K is a maximal compact subgroup of G. Thus K is also the stabilizer of one point in Y, and the identifications $X \simeq G/K \simeq Y$ then exchange G-invariant Riemannian metrics on X and those on Y. Since on Y, G-invariant Riemannian metrics are unique up to scalar multiplication and are symmetric, this is also true on X. So X is symmetric and in particular $\mathrm{Isom}(X)$ is transitive on ∂X.   □

Definition 19.12   Let X be a homogeneous simply connected negatively curved Riemannian manifold. A *distinguished boundary point* is a point in the closed $\mathrm{Isom}(X)$-orbit in ∂X (see Lemma 19.11). A *distinguished Busemann function* is a Busemann function attached to a distinguished boundary point.

Lemma 19.13   *Let X be a homogeneous simply connected negatively curved Riemannian manifold. For any two distinguished Busemann functions b, b' on*

X, there exists an isometry from (X,b) *to* (X,b')*, i.e. there exists an isometry* $f : X \to X'$ *such that* $b = b' \circ f$.

Proof Let ω and ω' be the distinguished points associated to b and b'. Since all distinguished points are in the same $\mathrm{Isom}(X)$-orbit, we can push b' forward by a suitable isometry and assume $\omega' = \omega$. Since the stabilizer in $\mathrm{Isom}(X)$ of ω is transitive on X, it is transitive on the set of Busemann functions attached to ω. Thus there exists f as required.   □

Lemma 19.13 allows to rather write $X[k]$ with no reference to any Busemann function, whenever X is a homogeneous simply connected negatively curved Riemannian manifold.

The relevance of these spaces is due to the following theorem from [7, §7]

Theorem 19.14 *Let G be a non-compact LC-group. Equivalences:*

- *G is amenable hyperbolic;*
- *for some integer $k \geq 1$ and some homogeneous simply connected negatively curved Riemannian manifold of positive dimension d, the group G admits a continuous, proper and cocompact action by isometries on the millefeuille space $X[k]$, fixing a point (or maybe a pair of points if $(k,d) = (1,1)$) on the boundary.*

Let us now describe the topology of the boundary of $\partial X[k]$.

Proposition 19.15 *Let X be homogeneous negatively curved d-dimensional Riemannian manifold ($d \geq 1$) endowed with a surjective Busemann function and $k \geq 1$. Then $\partial X[k]$ is homeomorphic to the one-point compactification of*

- *$\mathbf{R}^{d-1}$ if $k = 1$ and $d \geq 2$;*
- *$\mathbf{R}^{d-1} \times \mathbf{Z} \times C$ if $k \geq 2$, where C is a Cantor space.*

In particular, its topological dimension is $d - 1$.

Note that when $d = 1$ and $k \geq 2$, then $X[k]$ is a $(k+1)$-regular tree and $\partial X[k]$ is homeomorphic to a Cantor space.

Proof of Proposition 19.15 We can pick a group $G_1 = N_1 \rtimes \mathbf{Z}$ acting properly cocompactly on X, shifting the Busemann function by integer values, such that the action of $\mathbf{Z}$ contracts the simply connected nilpotent $(d-1)$-dimensional Lie group N_1, and $G_2 = N_2 \rtimes \mathbf{Z}$ acting continuously transitively on the $(k+1)$-regular tree, where $\mathbf{Z}$ contracts N_2. Then $G = (N_1 \times N_2) \rtimes \mathbf{Z}$ acts properly cocompactly on $X[k]$, and the action of $N_1 \times N_2$ on $\partial X[k] \smallsetminus \{\omega\}$ is simply transitive (see the proof of [7, Proposition 5.5]). So $X[k] \smallsetminus \{\omega\}$ is homeomorphic to $N_1 \times N_2$ and thus $X[k]$ is its one-point compactification. (This proof makes use of homogeneity for the sake of briefness, but the reader can find a more geometric proof in a more general context.) □

Corollary 19.16 *The classification of the spaces $\partial X[k]$ up to homeomorphy is given by the following classes*

- $k = 1$, $d \geq 1$ *is fixed: $\partial X[k]$ is a $(d-1)$-sphere;*
- $k \geq 2$ *is not fixed, $d \geq 1$ is fixed: $\partial X[k]$ is a "Cantor bunch" of d-spheres.*

Proof The homeomorphy type of the space $\partial X[k]$ detects d, since $d - 1$ is its topological dimension. Moreover, it detects whether $k = 1$ or $k \geq 2$, because $\partial X[k]$ is locally connected in the first case and not in the second. □

Corollary 19.17 *Let ω be the distinguished point in $\partial X[k]$ (the origin of its distinguished Busemann function). Consider the action of $\mathrm{Homeo}(\partial X[k])$ on $\partial X[k]$. Then*

- *if $\min(k,d) = 1$ then this action is transitive;*
- *if $\min(k,d) \geq 2$ then this action has two orbits, namely the singleton $\{\omega\}$ and its complement $\partial X[k] \smallsetminus \{\omega\}$.*

Proof The space $\partial X[k] \smallsetminus \{\omega\}$ is homogeneous under its self-homeomorphisms, and thus any transitive group of self-homeomorphisms extends to the one-point compactification. It remains to discuss whether ω belongs to the same orbit.

- If $\min(k,d) = 1$, then $\partial X[k]$ is a Cantor space or a sphere and thus is homogeneous under its group of self-homeomorphisms.
- If $\min(k,d) \geq 2$, and $x \in \partial X[k]$, then $X[k] \smallsetminus \{x\}$ is connected if and only if $x \neq \omega$. In other words, ω is the only cut-point in $\partial X[k]$ and thus is fixed by all self-homeomorphisms. □

Corollary 19.18 *Among focal hyperbolic LC-groups, the three classes of focal groups of connected, totally disconnected and mixed type are closed under quasi-isometry.*

Proof If G is a focal hyperbolic LC-group, it is of connected type if and only if its boundary is locally connected, and of totally disconnected type if and only if its boundary is totally disconnected. □

Remark 19.19 Another natural topological description of the visual boundary $\partial X[k]$ is that it is homeomorphic to the smash product $\partial X \wedge \partial T$ of the boundary of X and the boundary of T, and is thus homeomorphic to the $(d-1)$-fold reduced suspension of ∂T, where $d = \dim(X)$. Actually, this allows to

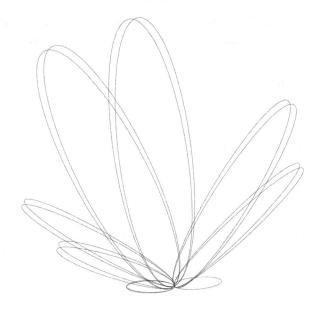

Figure 19.1 The boundary of $\mathbf{H}^2_{\mathbf{R}}[2]$. Except at the singular point, it is locally modeled on the product of a dyadic Cantor space and a line.

describe the whole compactification $\overline{X[k]} = X[k] \cup \partial X[k]$ as the smash product of $\overline{T}$ and a $(d-1)$-sphere.

Remark 19.20 The dimension $\dim(X)$ can also be characterized as the asymptotic dimension of $X[k]$, giving another proof that it is a QI-invariant of $X[k]$.

Let us call *pure millefeuille space* a millefeuille space $X[k]$ with the property that $\min(\dim X, k) \geq 2$, i.e. that is neither a manifold nor a tree; equivalently, its isometry group is focal of mixed type.

Proposition 19.21 *A hyperbolic LC-group G is focal of mixed type if and only if ∂G contains a point fixed by every self-homeomorphism of ∂G.*

Proof The "only if" part follows from Corollary 19.17.

Conversely, assume that the condition is satisfied. Then G cannot be of general type. The condition also implies that the boundary is non-empty and hence rules out compact groups. Otherwise G is 2-ended or focal. But if G is 2-ended, or focal of connected or totally disconnected type, its boundary is a non-empty sphere or a Cantor space, which have at least two elements and transitive homeomorphism groups. □

Corollary 19.22 *A focal hyperbolic LC-group of mixed type (or equivalently a pure millefeuille space) is not quasi-isometric to any hyperbolic LC-group of general type. In particular, it is not quasi-isometric to any vertex-transitive connected graph of finite valency.*

Proof The first statement is a particular case of Proposition 19.21. For the second statement, observe that the isometry group of such a vertex-transitive graph would be focal of totally disconnected type or of general type (by Lemma 19.5), but we have just excluded the general type case and the focal case is excluded by Corollary 19.18. □

19.2.7 Carnot groups

Recall that a *Heintze group* is a Lie group of the form $N \rtimes \mathbf{R}$, where N is a simply connected nilpotent Lie group, and the action of positive reals contracts N. It is *purely real* if it only has real eigenvalues in the adjoint representation, or equivalently in the action of $\mathbf{R}$ on the Lie algebra of N.

Definition 19.23 Let us say that a purely real Heintze group is of *Carnot type* if it admits a semidirect decomposition $N \rtimes \mathbf{R}$ such that, denoting by $(N^i)_{i \geq 1}$ the descending central series, it satisfies one of the three equivalent conditions:

(i) the action of $\mathbf{R}$ on $N/[N,N]$ is scalar;
(ii) the action of $\mathbf{R}$ on N^i/N^{i+1} is scalar for all i;
(iii) there is a linear decomposition of the Lie algebra $\mathfrak{n} = \bigoplus_{j=1}^{\infty} \mathfrak{v}_j$ such that, for some $\lambda \in \mathbf{R} \smallsetminus \{0\}$, the action of every $t \in \mathbf{R}$ on $\mathfrak{v}_j$ is given by multiplication by $\exp(j\lambda t)$ for all $j \geq 1$ and such that $\bigoplus_{j \geq i} \mathfrak{v}_j = \mathfrak{n}^i$.

We need to justify the equivalence between the definitions. The trivial implications are (iii)$\Rightarrow$(ii)$\Rightarrow$(i). To get (ii) from (i), observe that as a module (for the action of the one-parameter subgroup), $\mathfrak{n}^i/\mathfrak{n}^{i+1}$ is a quotient of $(\mathfrak{n}/[\mathfrak{n},\mathfrak{n}])^{\otimes i}$, so if the action on $\mathfrak{n}/[\mathfrak{n},\mathfrak{n}]$ is scalar then so is the action on $\mathfrak{n}^i/\mathfrak{n}^{i+1}$. To get (iii) from (ii), use a characteristic decomposition of the action of the one-parameter subgroup.

Remark 19.24 If a purely real Heintze group $G = N \rtimes \mathbf{R}$ is of Carnot type then N is a Carnot gradable[2] nilpotent group, in the sense that its Lie algebra admits a Carnot grading, i.e. a Lie algebra grading $\mathfrak{n} = \bigoplus_{i \geq 1} \mathfrak{v}_i$ such that $\mathfrak{n}^j = \bigoplus_{i \geq j} \mathfrak{v}_i$ for all j. The converse is not true: for instance most purely real Heintze

---

[2] In the literature, "Carnot gradable" is sometimes referred to as "graded" but this terminology is in practice a source of confusion, inasmuch as nilpotent Lie algebras may admit other relevant gradings.

groups of the form $\mathbf{R}^2 \rtimes \mathbf{R}$ are not of Carnot type. Nevertheless, if N is Carnot gradable, then up to isomorphism it defines a unique purely real Heintze group of Carnot type $\mathrm{Carn}(N)$ of the form $N \rtimes \mathbf{R}$. This is because a simply connected nilpotent Lie group N is gradable if and only if its Lie algebra $\mathfrak{n}$ is isomorphic to the graded Lie algebra $\mathrm{grad}(\mathfrak{n})$ defined as $\bigoplus_{i \geq 1} \mathfrak{n}^i / \mathfrak{n}^{i+1}$ (where the bracket is uniquely defined by factoring the usual bracket $\mathfrak{n}^i \times \mathfrak{n}^j \to \mathfrak{n}^{i+j}$), so that if $\mathfrak{n}$ is Carnot gradable then any two Carnot gradings define the same graded Lie algebra up to graded isomorphism.

19.3 Quasi-isometric rigidity of symmetric spaces

19.3.1 The QI-rigidity statement

The general QI-rigidity statement for symmetric spaces of non-compact type is the following. Recall that a Riemannian symmetric space of non-compact type has a canonical decomposition as a product of irreducible factors; we call the metric *well normalised* if all homothetic irreducible factors are isometric; this can always be ensured by a suitable factor-wise rescaling (for instance, requiring that the infimum of the sectional curvature on each factor is -1).

Theorem 19.25 *Let X be a symmetric space of non-compact type with a well normalised metric. Let G be a compactly generated locally compact group, quasi-isometric to X. Then G has a continuous, proper cocompact action by isometries on X.*

This statement is mostly known in a weaker form, where G is assumed to be discrete and sometimes in an even much weaker form, where one allows to pass to a finite index subgroup. Still, it is part of a stronger result concerning arbitrary cocompact large-scale quasi-actions on X of arbitrary groups (with no properness assumption) which, up to a minor continuity issue, is due to Kleiner and Leeb; see §19.3.2.

Before going into the proof, let us indicate some corollaries of the above Theorem 19.25. They illustrate the interest of having a statement for CGLC-groups instead of only finitely generated groups, even when studying discrete objects such as vertex-transitive graphs.

Corollary 19.26 *Let X be a symmetric space of non-compact type with a well normalised metric. Let Y be a vertex-transitive, finite valency connected graph. Assume that Y is quasi-isometric to X. Then there exists a cocompact lattice $\Gamma \subset \mathrm{Isom}(X)$ and an extension $W \hookrightarrow \tilde{\Gamma} \twoheadrightarrow \Gamma$ with W compact, such that $\tilde{\Gamma}$ admits a proper vertex-transitive action on Y; moreover there exists a* *connected graph Z on which* Γ *acts properly and transitively, with a surjective finite-to-one 1-Lipschitz equivariant quasi-isometry* $Y \to Z$.

Proof Let $\tilde{\Gamma}$ be the automorphism group of Y; this is a totally disconnected LC-group; since Y is vertex-transitive, it is compactly generated and quasi-isometric to Y and hence to X. By Theorem 19.25, there is a proper continuous homomorphism with cocompact image $\tilde{\Gamma} \to \mathrm{Isom}(X)$. Since $\mathrm{Isom}(X)$ is a Lie group, this homomorphism has an open kernel K, compact by properness. So the image is a cocompact lattice Γ.

To obtain Z, observe that the K-orbits in Y have uniformly bounded diameter, so Z is just obtained as the quotient of Y by the K-action. $\qquad\square$

Corollary 19.27 *Let X be a symmetric space of non-compact type with a well normalised metric. Let M be a proper, homogeneous geodesic metric space quasi-isometric to X. Then there exists a closed, connected cocompact subgroup H of* $\mathrm{Isom}(X)$ *and an extension $W \hookrightarrow \tilde{H} \twoheadrightarrow H$ with W compact, and a faithful proper isometric transitive action of $\tilde{H}$ on M. If moreover M is contractible then $W = 1$ and there is a diffeomorphism $M \to X$ intertwining the H-action on M with the original H-action on X.*

Proof The argument is similar to that of Corollary 19.26; here $\tilde{H}$ will be the identity component of the isometry group of M. This proves the first statement.

Fix a point $m_0 \in M$ and let K be its stabilizer in $\tilde{H}$. Then K is a compact subgroup. The image of K in H fixes a point x_0 in X. So the homomorphism $\tilde{H} \to \mathrm{Isom}(X)$ induces a continuous map $j : M = \tilde{H}/K \to X$ mapping m_0 to x_0. Since H is connected and closed cocompact, its action on X is transitive; it follows that j is surjective.

Under the additional assumption, let us check that j is injective; by homogeneity, we have to check that $j^{-1}(\{x_0\}) = \{m_0\}$. This amounts to check that if $g \in \tilde{H}$ fixes x_0 then it fixes m_0. Indeed, the stabilizer of x_0 in $\tilde{H}$ is a compact subgroup containing K; since M is contractible, K is maximal compact (see [3]) and this implies the result. Necessarily $W = 1$, because it fixes a point on X and being normal it fixes all points, so also acts trivially on M; since $\tilde{H}$ acts by definition faithfully on M it follows that $W = 1$. $\qquad\square$

19.3.2 Quasi-actions and proof of Theorem 19.25

We use the large-scale language introduced in §19.2.1.

The following powerful theorem is essentially due to Kleiner and Leeb [32], modulo the formulation, a continuity issue, and the fact we have equivariance instead of quasi-equivariance. It is stronger than Theorem 19.25 because no properness assumption is required.

If G is a locally compact group and acts, not necessarily continuously, by isometries on a metric space X, we say that the action is *locally bounded* if for some/every $x \in X$ and every compact subset $K \subset G$, the subset $Kx \subset X$ is bounded; in particular if the action is continuous, then it is locally bounded. We say that the action is *proper* if for every compact subset B of X, the set $\{g \in G : gB \cap B \neq \varnothing\}$ has a compact closure in G. The action is *cobounded* if there exists $x \in X$ such that $\sup_{y \in X} d(y, Gx) < \infty$.

Theorem 19.28 *Let X be a symmetric space of non-compact type with a well normalised metric (as defined before Theorem 19.25). Let G be a locally compact group with a locally bounded and cobounded isometric action ρ on a metric space Y quasi-isometric to X. Then there exists an isometric action of G on X and a G-equivariant quasi-isometry $Y \to X$. Moreover, the latter action is necessarily continuous, cocompact, and if ρ is proper then the action is proper.*

For the continuity issue, we need the following lemma.

Lemma 19.29 *Let H be a locally compact group with a continuous isometric action on a metric space X. Assume that the only element of H acting on X as an isometry with bounded displacement is the identity. (For instance, X is a minimal proper CAT(0)-space with no Euclidean factor and H is the full isometry group.) Let G be a locally compact group with an action of G on X given by a locally bounded homomorphism $\varphi \colon G \to H$. Assume that the action of G on X is cobounded. Then φ (and hence the action of G on X) is continuous.*

Proof Let $w \in H$ be an accumulation point of $\varphi(\gamma)$ when $\gamma \to 1$. Fix compact neighbourhoods Ω_G, Ω_H of 1 in G and H. Fix $g \in G$. There exists a neighbourhood V_g of 1 in G such that $\gamma \in V_g$ implies $g^{-1} \gamma g \in \Omega_G$. Moreover, if we define

$$C_g = \{\gamma \in G : \varphi(\gamma)^{-1} w \in \varphi(g)\Omega_H \varphi(g)^{-1}\},$$

then 1 belongs to the closure of C_g. Hence $V_g \cap C_g \neq \varnothing$, so if we pick γ in the intersection we get

$$\varphi(g)^{-1}\varphi(\gamma)\varphi(g) \in \varphi(\Omega_G); \quad \varphi(g)^{-1}\varphi(\gamma)^{-1} w \varphi(g) \in \Omega_H,$$

so we deduce, for all $g \in G$

$$\varphi(g)^{-1} w \varphi(g) \in \varphi(\Omega_G)\Omega_H,$$

Since the action is locally bounded, $\varphi(\Omega_G)$ is bounded. This shows that for a given $x \in X$, the distance $d(w\varphi(g)x, \varphi(g)x)$ is bounded independently of g. Since G acts coboundedly, this shows that w has bounded displacement, so

$w = 1$. Since H is locally compact, this implies that φ is continuous at 1 and hence is continuous. □

We need to deduce Theorem 19.25 from [32, Theorem 1.5]. For this we need to introduce all the terminology of quasi-actions. We define a *quasi-action* of a group G on a set X as an arbitrary map $G \to X^X$, written as $g \mapsto (x \mapsto gx)$. If $x \in X$, let $i_x : G \to X$ be the orbital map $g \mapsto gx$.

If G is a group and X a metric space, a *uniformly large-scale Lipschitz* (ULSL) quasi-action of G on X is a map $\rho : G \to X^X$, such that for some $(\mu, \alpha) \in \mathbf{R}_{>0} \times \mathbf{R}$, every map $\rho(g)$, for $g \in G$ is a (μ, α)-Lipschitz map (in the sense defined in §19.2.1), and satisfying $\rho(1) \overset{\alpha}{\sim} \mathrm{id}$ and $\rho(gh) \overset{\alpha}{\sim} \rho(g)\rho(h)$ for all $g, h \in G$. Note that for a ULSL quasi-action, all i_x are pairwise $\sim$-equivalent.

The quasi-action is *cobounded* if the map i_x has a cobounded image for every $x \in X$. If G is a locally compact group, a quasi-action is *locally bounded* if i_x maps compact subsets to bounded subsets for all x (if G is discrete this is an empty condition). The quasi-action is *coarsely proper* if for all $x \in X$, inverse images of bounded subsets by i_x have compact closure.

Given quasi-actions ρ and ρ' of G on X and X', a map $q : X \to X'$ is *quasi-equivariant* if it satisfies

$$\sup_{g \in G, x \in X} d\big(q(\rho(g)x), \rho(g)q(x)\big) < \infty.$$

The ULSL quasi-actions are quasi-isometrically quasi-conjugate if there is a quasi-equivariant quasi-isometry $X \to X'$; this is an equivalence relation.

Lemma 19.30 *Let (X, d) be a metric space and G a group. Then*

(a) *If Y is a metric space with an isometric action of G and with a quasi-isometry $q : Y \to X$, then there is a ULSL quasi-action of G on X so that q is a quasi-isometric quasi-conjugacy.*

(b) *Conversely, if ρ is a ULSL quasi-action of G on X, then there exists a metric space Y with an isometric G-action and a quasi-equivariant quasi-isometry $Y \to X$.*

Proof We begin by the easier (a). Let $s : X \to Y$ be a quasi-isometry inverse to q and define a quasi-action of G on X by $\rho(g)x = q(g \cdot s(x))$; it is straightforward that this is a ULSL quasi-action and that q is a quasi-isometric quasi-conjugacy.

For (b), first, let $Y \subset X$ be a maximal subset for the property that the $\rho(G)y$, for $y \in Y$, are pairwise disjoint. Let us show that $\rho(G)Y$ is cobounded in X. Indeed, denoting, by abuse of notation $a \approx b$ if $|a - b|$ is bounded by a constant depending only on the ULSL constants of ρ, if $x \in X$ by maximality there exists $y \in Y_0$ and $g \in G$ such that $\rho(g)x = \rho(h)y$. Then, denoting by d_X the distance in X

$$d_X(x, \rho(g^{-1}h)y) \approx d_X(x, \rho(g^{-1})\rho(h)y) \approx d_X(\rho(g)x, \rho(h)y) = 0,$$

so $\rho(G)Y$ is cobounded.

Consider the cartesian product $G \times Y$, whose elements we denote by $g \diamond y$ rather than (g, y) for the sake of readability. Define a pseudometric on $G \times Y$ by

$$d(g \diamond y, h \diamond z) = \sup_{k \in G} d_X(\rho(kg)y, \rho(kh)z).$$

If $G \times Y$ is endowed with the G-action $k \cdot (g \diamond y) = (kg \diamond y)$, then this pseudo-metric is obviously G-invariant. Consider the map $j : (g \diamond y) \mapsto \rho(g)y$. Let us check that j is a quasi-equivariant quasi-isometry $(G \times Y, d) \to X$. We already know it has a cobounded image. We obviously have

$$d(g \diamond y, h \diamond z) \geq d_X(\rho(g)y, \rho(h)z) = d_X(j(g \diamond y), j(h \diamond z)), \qquad (19.1)$$

and conversely, writing $a \preceq b$ if $a \leq Cb + C$ where $C > 0$ is a constant depending only on the ULSL constants of ρ

$$
\begin{aligned}
d(g \diamond y, h \diamond z) &= \sup_{k \in G} d_X\big(\rho(kg)y, \rho(kh)z\big) \\
&\approx \sup_{k \in G} d_X\big(\rho(k)\rho(g)y, \rho(k)\rho(h)z\big) \\
&\preceq d_X(\rho(g)y, \rho(h)z) = d_X(j(g \diamond y), j(h \diamond z)).
\end{aligned}
$$

This shows that j is a large-scale bilipschitz embedding. Thus it is a quasi-isometry. It is also quasi-equivariant, as

$$d_X\big(j(h \cdot (g \diamond y)), \rho(h)j(g \diamond y)\big) = d_X\big(\rho(hg)y, \rho(h)\rho(g)y\big) \approx 0.$$

So we obtain the conclusion, except that we have a pseudo-metric; if we identify points in $G \times Y$ at distance zero, the map j factors through the quotient space, as follows from (19.1) and thus we are done. $\qquad\square$

Proof of Theorem 19.28 The statement of [32, Theorem 1.5] is the following: *Let X be a symmetric space of non-compact type with a well normalised metric. Let G be a discrete group with a ULSL quasi-action ρ on X. Then ρ is QI quasi-conjugate to an isometric action.*

By Lemma 19.30, this can be translated into the following: *Let X be as above and let G be a discrete group with an isometric action ρ on a metric space Y quasi-isometric to X. Then there exists an isometric action of G on X and a G-quasi-equivariant quasi-isometry $q : Y \to X$.*

To get the theorem, we need a few improvements. First, we want the quasi-isometry to be equivariant (instead of quasi-equivariant). Starting from q as above, we proceed as follows. Fix a bounded set $Y_0 \subset Y$ containing one point in each G-orbit. For $y_0 \in Y_0$, let G_{y_0} be its stabilizer. Since $G_{y_0}\{y_0\} = \{y_0\}$ and q is quasi-equivariant, we see that $G_{y_0}\{q(y_0)\}$ has its diameter bounded by a constant C depending only on q (and not on y_0). By the centre lemma, G_{y_0} fixes a point at distance at most C of $q(y_0)$, which we define as $q'(y_0)$. For $y \in Y$ arbitrary, we pick g and $y_0 \in Y_0$ (y_0 is uniquely determined) such that $gy_0 = y$ and define $q'(y) = gq'(y_0)$. By the stabilizer hypothesis, this does not depend on the choice of g, and we see that q' is at bounded distance from q. So q' is a quasi-isometry as well, and by construction is G-equivariant.

Now assume that G is locally compact. Applying the previous result to G endowed with the discrete topology, we get all the non-topological conclusions. Now from the additional hypothesis that the action is locally bounded, we obtain that the action on X is locally bounded. We then invoke Lemma 19.29, using that X has no non-trivial bounded displacement isometry, to obtain that the G-action on X is continuous. It is clear that the action on X is cobounded, and that if ρ is metrically proper then the action on X is proper. □

Proof of Theorem 19.25 Fix a left-invariant word metric d_0 on G. Then (G, d) is quasi-isometric to X and the action of G on (G, d) is locally bounded. So by Theorem 19.28, there exists a continuous isometric proper cocompact action on X. □

Theorem 19.25 was initially proved in the case of discrete groups, for symmetric spaces with no factor of rank one by Kleiner–Leeb [31] and in rank one using different arguments by:

- Pansu in the case of quaternionic and octonionic hyperbolic spaces;
- R. Chow in the case of the complex hyperbolic spaces [11];
- Tukia for real hyperbolic spaces of dimension at least three [58];
- In the case of the real hyperbolic plane, Tukia [59], completed by, independently, Casson–Jungreis [10] and Gabai [22].

The synthesis for arbitrary symmetric spaces is due to Kleiner–Leeb [32], where they insightfully consider actions of arbitrary discrete groups, with no properness assumption. This generality is essential because when considering a proper cocompact locally bounded action of a non-discrete LC-group, when viewing it as an action of the underlying discrete group, we lose the properness.

19.4 Quasi-isometric rigidity of trees

19.4.1 The QI-rigidity statement

The next theorem is the locally compact version of the result that a finitely generated group is quasi-isometric to a tree if and only if it is virtually free. The latter is a combination of Bass–Serre theory [50], Stallings' theorem and Dunwoody's result that finitely presented groups are accessible [17], a notion we discuss below.

Theorem 19.31   *Let G be a compactly generated locally compact group. Equivalences:*

(i) *G is quasi-isometric to a tree;*

(ii) *G is quasi-isometric to a bounded valency tree;*

(iii) *G admits a continuous proper cocompact isometric action on a locally finite tree T, or a continuous proper transitive isometric action on the real line;*

(iv) *G is topologically isomorphic to the Bass–Serre fundamental group of a finite connected graph of groups, with compact vertex and edge stabilizers with open inclusions, or has a continuous proper transitive isometric action on the real line.*

Proof (first part)   Let us mention that the equivalence between (iii) and (iv) is just Bass–Serre theory [50]. Besides, the trivial implications are (iii)$\Rightarrow$(ii)$\Rightarrow$(i).

Let us show (i)$\Rightarrow$(ii); let G be compactly generated and let $f : T \to G$ be a quasi-isometry from a tree T (viewed as its set of vertices). We can pick a metric lattice J inside G and assume that f is valued in J; the metric space J has finite balls of cardinality bounded by a constant depending only on their radius. We can modify f to ensure that $f^{-1}(\{j\})$ is convex for every $j \in J$. Since f is a quasi-isometry the convex subsets $f^{-1}(\{j\})$ have uniformly bounded radius. We define a tree T'' by collapsing each convex subset $f^{-1}(\{j\})$ (along with the edges joining them) to a point. The collapsing map $T' \to T''$ is a 1-Lipschitz quasi-isometry and thus f factors through an injective quasi-isometry $T'' \to G$. It follows that T'' has balls of cardinal bounded in terms of the radius; this implies in particular that T'' has finite (indeed bounded) valency.

We postpone the proof of (ii)$\Rightarrow$(iv), which is the deep part of the theorem.
$\square$

Remark 19.32   Another characterization in Theorem 19.31 is the following: G is of type FP_2 and has asymptotic dimension ≤ 1. Here type FP_2 means that the homology of the Cayley graph of G with respect to some compact generating subset is generated by loops of bounded length. Indeed, Fujiwara and Whyte [21] proved that a geodesic metric space satisfying these conditions is quasi-isometric to a tree.

By *essential tree* we mean a non-empty tree with no vertex of degree 1 and not reduced to a line. By *reduced* graph of groups we mean a connected graph of groups in which no vertex group is trivial, and such that for every oriented non-self edge, the target map is not surjective. We call a reduced graph of groups *non-degenerate* if its Bass–Serre tree is not reduced to the empty set, a point or a line, or equivalently if it is not among the following exceptions:

- The empty graph;
- A single vertex with no edge;
- A single vertex and a single self-edge so that both target maps are isomorphisms;
- 2 vertices joined by a single edge, so that both target maps have image of index 2.

Corollary 19.33   *Let G be a compactly generated locally compact group. Equivalences:*

(i) *G is quasi-isometric to an unbounded tree not quasi-isometric to the real line;*

(ii) *G is quasi-isometric to the 3-regular tree;*

(iii) *G admits a continuous proper cocompact action on an essential locally finite non-empty tree T;*

(iii') *G admits a continuous proper cocompact isometric action on non-empty tree T of bounded valency, with only vertices of degree ≥ 3;*

(iv) *G is topologically isomorphic to the Bass–Serre fundamental group of a finite non-degenerate reduced graph of groups, with compact vertex and edge stabilizers, and open inclusions.* $\Box$

19.4.2 Reminders about the space of ends

Recall that the set of ends of a geodesic metric space X is the projective limit of $\pi_0(X - B)$, where B ranges over bounded subsets of X_1 and π_0 denotes the set of connected components. Each $\pi_0(X - B)$ being endowed with the discrete topology, the set of ends is endowed with the projective limit topology and thus is called the space of ends $E(X)$.

If X, Y are geodesic metric spaces, any large-scale Lipschitz, coarsely proper map $X \to Y$ canonically defines a continuous map $E(X) \to E(Y)$. Two maps at bounded distance induce the same map. This construction is functorial. In particular, it maps quasi-isometries to homeomorphisms.

If G is a locally compact group generated by a compact subset S, the space of ends [29, 53] of G is by definition the space of ends of the 1-skeleton of its Cayley graph with respect to S. It is compact. Since the identity map (G, S_1) to (G, S_2) is a quasi-isometry, the space of ends is canonically independent of the choice of S and is functorial with respect to continuous proper group homomorphisms.

In the case when G is hyperbolic, it is not hard to prove that the space of ends is canonically homeomorphic to the space $\pi_0(\partial G)$ of connected components of the visual boundary.

We note for reference the following elementary lemma, due to Houghton. It is a particular case of [29, Th. 4.2 and 4.3].

Lemma 19.34 *Let G be a CGLC-group with G_0 non-compact. Then G is (1 or 2)-ended; if moreover G/G_0 is not compact then G is 1-ended.*

Let us also mention, even if it will be not be used here, the following theorem of Abels [2]: if a compactly generated LC-group has at least 3 ends, the action of G on $E(G)$ is minimal (i.e. all orbits are dense) unless G is focal hyperbolic of totally disconnected type (in the sense of §19.2.5).

19.4.3 Metric accessibility

Definition 19.35 (Thomassen, Woess [57])   Let X be a bounded valency connected graph. Say that X is *accessible* if there exists m such that for every two distinct ends of X, there exists an m-element subset of the 1-skeleton X that separates the two ends, i.e. so that the two ends lie in distinct components of the complement.

Accessibility is a quasi-isometry invariant of connected graphs of bounded valency. Although not needed in view of Theorem 19.31, we introduce the following definition, which is more metric in nature.

Definition 19.36   Let X be a geodesic metric space. Let us say that X is *diameter-accessible* if there exists m such that for every two distinct ends of X, there exists a subset of X of diameter at most m that separates the two ends.

This is obviously a quasi-isometric invariant property among geodesic metric spaces. Obviously for a bounded valency connected graph, the property of diameter-accessibility implies accessibility. The reader can construct, as an exercise, a connected planar graph of valency ≤ 3 that is accessible (with $m = 2$) but not diameter-accessible.

Theorem 19.37 (Dunwoody) *Let X be a connected, locally finite simplicial 2-complex with a cocompact isometry group. Assume that $H^1(X,\mathbf{Z}/2\mathbf{Z}) = 0$ (e.g., X is simply connected). Then X is diameter-accessible.*

On the proof This statement is not explicit, but is the contents of the proof of [17] (it is quoted in [40, Theorem 15] in a closer way). This proof consists in finding, denoting by $G = \mathrm{Aut}(X)$, an equivariant family $(C_i)_{i\in I}$, indexed by a discrete G-set I with finitely many G-orbits, of pairwise disjoint compact subsets of X homeomorphic to graphs and each separating X (each $X \smallsetminus C_i$ is not connected), called tracks, so that each component of $X \smallsetminus \bigcup C_i$ has the property that its stabilizer in G acts coboundedly on it, and is (at most 1)-ended. Thus any two ends of X are separated by one of these components, which have uniformly bounded diameters. □

19.4.4 The locally compact version of the splitting theorem

The following theorem is the locally compact version of Stallings' theorem; it was proved by Abels (up to a minor improvement in the 2-ended case).

Theorem 19.38 (Stallings–Abels) *Let G be a compactly generated, locally compact group. Then G has at least two ends if and only if it splits as a non-trivial HNN-extension or amalgam over a compact open subgroup, unless G is 2-ended and is compact-by-$\mathbf{R}$ or compact-by-$\mathrm{Isom}(\mathbf{R})$.*

On the proof The main case is when G is a closed cocompact isometry group of a vertex-transitive graph; this is [1, Struktursatz 5.7]. He deduces the theorem [1, Korollar 5.8] when G has at least 3 ends in [1, Struktursatz 5.7], the argument also covering the case when G has a compact open subgroup (i.e. G_0 is compact) in the case of at least 2 ends.

Assume now that G_0 is non-compact. By Lemma 19.34, if G/G_0 is non-compact then G is 1-ended, so assume that G/G_0 is compact. Then G has a maximal compact subgroup K so that G/K admits a G-invariant structure of Riemannian manifold diffeomorphic to a Euclidean space ([42, Theorem 3.2] and [38, Theorem 4.6]); the only case where this manifold has at least two ends is when it is one-dimensional, hence isometric to $\mathbf{R}$, whence the conclusion. □

This yields the following corollary, which we state for future reference. The characterization (iv) of 2-ended groups is contained in Houghton [29, Theorem 3.7]; the stronger characterization (v) is due to Abels [1, Satz B]; the stronger versions (vi) or (vii) can be deduced directly without difficulty; they are written in [7, Proposition 5.6].

Corollary 19.39   *Let G be a compactly generated, locally compact group. Equivalences:*

(i) G *has exactly 2 ends;*

(ii) G *is hyperbolic and #$\partial G = 2$;*

(iii) G *is quasi-isometric to* **Z***;*

(iv) G *has a discrete cocompact infinite cyclic subgroup;*

(v) G *has an open subgroup of index* ≤ 2 *admitting a continuous homomorphism with compact kernel onto* **Z** *or* **R***;*

(vi) G *admits a continuous proper cocompact isometric action on the real line;*

(vii) G *has a (necessarily unique) compact open normal subgroup W such that G/W is isomorphic to one of the 4 following groups:* **Z***,* Isom(**Z**)*,* **R***,* Isom(**R**)*.*

Proof   The implications (vii)$\Rightarrow$(vi)$\Rightarrow$(v)$\Rightarrow$(iv)$\Rightarrow$(iii)$\Rightarrow$(ii)$\Rightarrow$(i) are clear. Assume (i). Then, by Theorem 19.38, either (vii) holds explicitly or G splits as a non-trivial HNN-extension or amalgam over a compact open subgroup; the only cases for which this does not yield more than 2 ends is the case of a degenerate HNN extension (given by an automorphism of the full vertex group) or an amalgam over a subgroup of index 2 in both factors. Thus G maps with compact kernel onto **Z** or the infinite dihedral group Isom(**Z**).   □

19.4.5 Accessibility of locally compact groups

It is natural to wonder whether the process of applying iteratively Theorem 19.38 stops. This motivates the following definition.

Definition 19.40   A CGLC-group G is *accessible* if it has a continuous proper transitive isometric action on the real line, or satisfies one of the following two equivalent conditions:

(i) G admits a continuous cocompact action on a locally finite tree T with (at most 1)-ended vertex stabilizers and compact edge stabilizers;

(ii) G is topologically isomorphic to the Bass–Serre fundamental group of a finite graph of groups, with compact edge-stabilizers with open inclusions and (at most 1)-ended vertex groups.

The equivalence between the two conditions is Bass–Serre theory [50]. For either definition, trivial examples of accessible CGLC groups are (at most 1)-ended groups. Also, by Theorem 19.38, 2-ended CGLC groups are accessible, partly using the artifact of the definition. A non-accessible finitely generated group was constructed by Dunwoody in [18], disproving a long-standing conjecture of Wall.

19.4.6 Metric vs group accessibility and proof of Theorem 19.31

The metric notion of accessibility, which, unlike in this paper, was introduced after group accessibility, allowed Thomassen and Woess, using the Dicks–Dunwoody machinery [16], to have a purely geometric characterization of group accessibility (as defined in Theorem 19.31). The following theorem is the natural extension of their method to the locally compact setting, due to Krön and Möller [34, Theorem 15].

Theorem 19.41 *Let X be a bounded valency non-empty connected graph and G a locally compact group acting continuously, properly cocompactly on X by graph automorphisms. Then the following are equivalent:*

(i) G *is accessible (as defined in Definition 19.40);*
(ii) X *is accessible;*
(iii) X *is diameter-accessible.*

On the proof The statement in [34] is the equivalence between (i) and (ii) when G is totally disconnected. But actually the easy implication (i)⇒(ii) yields (i)⇒(iii) without change in the proof, as they check that X is quasi-isometric to a bounded valency graph with a cocompact action, with a G-equivariant family of "cuts", which are finite sets, with finitely many G-orbits of cuts, so that any two distinct ends are separated by one cut. Finally (iii)⇒(ii) is trivial for an arbitrary bounded valency graph.   □

End of the proof of Theorem 19.31 It remains to show the main implication of Theorem 19.31, namely that (ii) implies (iv). Let G be a CGLC-group quasi-isometric to a tree T. We begin with the claim that G has no 1-ended closed subgroup. Indeed, if H were such a group, then, being non-compact, H admits a bi-infinite geodesic, and using the quasi-isometry to T we see that this geodesic has 2 distinct ends in G. Then these two ends are distinct in H, a contradiction. Let us now prove (iv). We begin by three easy cases:

- G is compact. There is nothing to prove.
- G is 1-ended. This has just been ruled out.
- G is 2-ended. In this case Corollary 19.39 gives the result.

So assume that G has at least three ends. By Lemma 19.34, G_0 is compact, so by Proposition 19.7, it admits a continuous proper cocompact isometric action on a connected finite valency graph X. Since X is quasi-isometric to a tree, it is diameter-accessible. By Theorem 19.41, we deduce that G is accessible. This means that it is isomorphic to the Bass–Serre fundamental group of a graph of groups with (at most 1)-ended vertex groups and compact open edge groups. As we have just shown that G has no closed 1-ended subgroup, it follows that vertex groups are compact. So (iv) holds.   □

19.4.7 Cobounded actions

There is a statement implying Theorem 19.31, essentially due to L. Mosher, M. Sageev and K. Whyte, concerning cobounded actions with no properness assumption. In a tree, say that a vertex is an *essential branching vertex* if its complement in the 1-skeleton has at least three unbounded components. Say that a tree is *bushy* if the set of essential branching vertices is cobounded. We here identify any tree to its 1-skeleton.

Theorem 19.42 *Let G be a locally compact group with a locally bounded, isometric, cobounded action on a metric space Y quasi-isometric to a bushy tree T of bounded valency. Then there exists a tree T' of bounded valency with a continuous isometric action of G and an equivariant quasi-isometry $T \to T'$.*

Remark 19.43 It is not hard to check that a bounded valency non-bushy tree T that is quasi-isometric to a metric space with a cobounded isometry group, is either bounded or contains a cobounded bi-infinite geodesic.

Remark 19.44 Theorem 19.42 is not true when T is a linear tree (the Cayley graph of $(\mathbf{Z}, \{\pm 1\})$). Indeed, taking G to be $\mathbf{R}$ acting on $\mathbf{R}$ (which is quasi-isometric to T), there is no isometric cobounded action on any tree quasi-isometric to $\mathbf{Z}$. Indeed this action would preserve a unique axis, while there is no non-trivial homomorphism from $\mathbf{R}$ to $\mathrm{Isom}(\mathbf{Z})$. In this case, we can repair the issue by allowing actions on $\mathbf{R}$-trees. But here is a second more dramatic counterexample: the universal covering $G = \widetilde{\mathrm{SL}}_2(\mathbf{R})$ (endowed with either the discrete or Lie topology) admits a locally bounded isometric action on the Cayley graph of $(\mathbf{R}, [-1, 1])$ (see [7, Example 3.12]), but admits no isometric cobounded action on any $\mathbf{R}$-tree quasi-isometric to $\mathbf{Z}$, because by the same argument, this action would preserve an axis, while there is no non-trivial homomorphism of abstract groups $\widetilde{\mathrm{SL}}_2(\mathbf{R}) \to \mathrm{Isom}(\mathbf{R})$.

On the proof of Theorem 19.42 The statement of [40, Theorem 1] is in terms of quasi-actions, see the conventions in §19.3.2. It reads: *Let G be a discrete group. Suppose that G has a ULSL quasi-action on a bushy tree T of bounded valency. Then there exists a tree T' of bounded valency with a continuous isometric action of G and a quasi-equivariant quasi-isometry $T \to T'$.*

By Lemma 19.30, it can be translated as: *(*) Let G be a discrete group. Consider a locally bounded, isometric, cobounded action of G on a metric space Y quasi-isometric to a bushy tree T of bounded valency. Then there exists a tree T' of bounded valency with an isometric action of G and a quasi-equivariant quasi-isometry $T \to T'$.*

To get the statement of Theorem 19.42, apply (*) to the underlying discrete group; the action on T' we obtain is then locally bounded. Since T' has at least three ends, the only isometry with bounded displacement is the identity, so we deduce by Lemma 19.29 that the action on T' is continuous.

Let us now sketch the proof of (*) (which is the discrete case of Theorem 19.42), as the authors of [40] did not seem to be aware of the Thomassen–Woess approach and repeat a large part of the argument.

Start from the hypotheses of (*). A trivial observation is that we can suppose Y to be a connected graph. Namely, using that T is geodesic, there exists r such that if we endow Y with a graph structure by joining points at distance $\leq r$ by an edge, then the graph is connected and the graph metric and the original metric on Y are quasi-isometric through the identity. The difficulty is that Y need not be of finite valency.

The construction of an isometric action of G on a connected *finite valency* graph Z quasi-isometrically quasi-conjugate to the original quasi-action is done in [40, §3.4] (this is the part where it is used that the tree is bushy). It is given by a homomorphism $G \to \text{Isom}(Z)$. By Theorem 19.31, $\text{Isom}(Z)$ has a continuous proper cocompact action on a tree T'. Note that the actions of $\text{Isom}(Z)$ on both Z and T' are quasi-isometrically conjugate to the left action of $\text{Isom}(Z)$ on itself, so there exists a quasi-isometry $Z \to T'$ which is quasi-equivariant with respect to the $\text{Isom}(Z)$-action, and therefore is quasi-equivariant with respect to the G-action. Finally, to get equivariance instead of quasi-equivariance, we use the same argument (based on the centre lemma, which holds in T') as in the proof of Theorem 19.28. □

19.4.8 Accessibility of compactly presented groups

Recall that a CGLC-group is *compactly presented* if it has a presentation with a compact generating set and relators of bounded length. If it has a compact open subgroup, it is equivalent to require that G has a continuous proper cocompact combinatorial action on a locally finite simply connected simplicial 2-complex (see Proposition 19.47).

Let us also use the following weaker variant: G is of type FP_2 mod 2 if G is a quotient of a compactly presented LC-group by a discrete normal subgroup N such that $\text{Hom}(N, \mathbf{Z}/2\mathbf{Z}) = 0$. If G has a compact open subgroup, it is equivalent (see Proposition 19.49) to require that G has a continuous proper co-compact combinatorial action on a locally finite connected 2-complex X such that $H^1(X, \mathbf{Z}/2\mathbf{Z}) = 0$.

By Lemma 19.34, if G_0 is not compact then G is (at most 2)-ended and thus is accessible by Corollary 19.39. Otherwise if G_0 is compact (i.e., G has a compact open subgroup), the restatement of the definition shows that as a consequence of Theorems 19.37 and 19.41, we have the following theorem, whose discrete version was the main theorem in [17].

Theorem 19.45   *Let G be a compactly generated locally compact group of type* FP_2 *mod 2 (e.g. G is compactly presented). Then G is accessible.*   □

Corollary 19.46   *Every hyperbolic locally compact group is accessible.*   □

Proposition 19.47   *Let G be a locally compact group with G_0 compact. Equivalences:*

(i) *G is compactly presented, that is, it has a presentation with a compact generating set and relators of bounded length;*

(ii) *G has a presentation with a compact generating set and relators of length ≤ 3;*

(iii) *G has a continuous proper cobounded combinatorial action on a simply connected simplicial 2-complex X.*

(iv) *G has a continuous proper cocompact combinatorial action on a locally finite simply connected simplicial 2-complex X.*

Proof   Trivially we have (iv)⇒(iii) and (ii)⇒(i).

(iii)⇒(ii). Let A be a connected bounded closed subset of X such that the G-translates of A cover X. Then if $S = \{g \in G \mid gA \cap A \neq \varnothing\}$, then S is compact by properness, and by [33, Lemma p. 26], G is generated by S and relators of length ≤ 3. So (ii) holds.

(i)⇒(iv). Let S be a compact symmetric generating subset with relators of bounded length; enlarging S if necessary, we can suppose that S contains a compact open subgroup K and that $S = KSK$. The Cayley–Abels graph (Proposition 19.7) is a graph with vertex set G/K and an unoriented edge from gK to $g'K$ if $g^{-1}g' \in S \smallsetminus K$. This is a G-invariant locally finite connected graph structure on G/K. Define a simplicial 2-complex structure by adding a triangle every time its 2-skeleton appears.

Define a simplicial 2-complex structure on the Cayley graph of G with respect to S in the same way. Then the projection $G \to G/K$ has a unique extension between the 2-skeleta, affine in each simplex (possibly collapsing a simplex onto a simplex of smaller dimension). The Cayley 2-complex of G is simply connected, because the relators of length ≤ 3 define the group.

It remains to check that the simplicial 2-complex on G/K is simply connected. Indeed, given a loop based at 1, we can homotope it to a combinatorial loop on the 1-skeleton. We can lift such a loop to a combinatorial path $(1 = g_0, g_1, \ldots, g_k)$ on G and ending at the inverse image of 1, namely K. If $k = 0$ there is nothing to do, otherwise observe that $g_{k-1} \in KS \subset S$. So we can replace g_k by 1, thus the new lifted path is a loop. This loop has a combinatorial homotopy to the trivial loop, which can be pushed forward to a combinatorial homotopy of the loop on G/K. So the simplicial 2-complex on G/K is simply connected. Thus (iv) holds. $\qquad\qquad\square$

Lemma 19.48 *Let G be a compactly presented locally compact group with G_0 compact. Then G has a continuous proper cocompact combinatorial action on a locally finite simply connected simplicial 2-complex X with no inversion (i.e. for each $g \in G$, the set of G-fixed points is a subcomplex) and with a main vertex orbit, in the sense that every $g \in G$ fixing a vertex, fixes a vertex in the main orbit, and if an element is the identity on the main orbit then it is the identity. Moreover, the main vertex orbit can be chosen as G/K for any choice K of compact open subgroup.*

Sketch of proof   The proof is essentially a refinement of the proof of (i)$\Rightarrow$(iv) of Proposition 19.47, so we just indicate how it can be adapted to yield the lemma. We start with the same construction, but using oriented edges (without self-loops), and then, each time we have three oriented edges (e_1, e_2, e_3) forming a triangle with three distinct vertices (with compatible orientations, i.e. the target of e_1 is the source of e_2, etc.), we add a triangle. Because of the double edges, this is not yet simply connected; so for any two adjacent vertices x, y we glue two bigons indexed by (x, y) and (y, x) (note that the union of these two bigons is homeomorphic to a 2-sphere). The resulting complex is simply connected. Then the action of G has the required property. The problem is that because of bigons, we do not have a simplicial complex. To solve this, we just add vertices at the middle of all edges and all bigons, split the bigons into four triangles by joining the centre to all four vertices (thus pairs of opposite bigons now form the 2-skeleton of a octahedron), and split the original triangles into four triangles by joining the middle of the edges. We obtain a simplicial 2-complex; the cost is that we have several vertex orbits, but then G/K is the "main orbit" in the sense of the proposition. $\qquad\qquad\square$

Proposition 19.49 *Let G be a locally compact group with G_0 compact. Let A be a discrete abelian group. Equivalences:*

(i) G *is isomorphic to a quotient* $\tilde{G}/N$ *with* $\tilde{G}$ *compactly presented, N a discrete normal subgroup of* $\tilde{G}$ *and* $\mathrm{Hom}(N,A) = 0$;

(ii) G *has a continuous proper cocompact combinatorial action on a locally finite simplicial 2-complex X with* $H^1(X,A) = 0$.

Proof Suppose (i) and let us prove (ii). We can suppose $G_0 = \{1\}$. Choose K to be a compact open subgroup with $K \cap N = \{1\}$. Consider an action α of $\tilde{G}$ on a 2-complex X as in Lemma 19.48 and main vertex orbit G/K. Then the action of N on X is free: indeed, if an element of N fixes a point, then it fixes a vertex in the main orbit and hence it has some conjugate in $N \cap K$, thus is trivial.

So $G = \tilde{G}/N$ acts continuously properly cocompactly on $N \backslash X$, which is a simplicial 2-complex as N acts freely on X. Now we have, fixing an implicit base-point in X

$$H^1(X,A) \simeq \mathrm{Hom}(\pi_1(N \backslash X), A). \tag{19.1}$$

Since $\pi_1(N \backslash X) \simeq N$, we deduce that $H^1(X,A) = 0$.

Conversely suppose (ii), denote by α the action of G on X and W its kernel. Fixing a base-vertex in X, let $\tilde{X}$ be the universal covering. Let H be the group of automorphisms of $\tilde{X}$ that induce an element of $\alpha(G)$. It is a closed subgroup and is cocompact on $\tilde{X}$; its projection ρ to $\alpha(G)$ is surjective. Let

$$\tilde{G} = \{(g,h) \in G \times H \mid \alpha(g) = \rho(h)\} \subset G \times H$$

be the fibre product of G and H over $\alpha(G)$. Then it contains $W \times \{1\}$ as a compact normal subgroup, and the quotient is canonically isomorphic to H. The kernel of the projection $\tilde{G} \to G$ is equal to $\{1\} \times \mathrm{Ker}(\rho)$, which is discrete and consists of deck transformations of the covering $\tilde{X} \to X$ and is in particular isomorphic to $\pi_1(X)$. Again using (19.1), we obtain that $\mathrm{Hom}(N,A) = \{0\}$. □

Remark 19.50 In case A is the trivial group, both (i) and (ii) of Proposition 19.49 hold. At the opposite, keeping in mind that any non-trivial abelian group B satisfies $\mathrm{Hom}(B, \mathbf{Q}/\mathbf{Z}) \neq 0$, the case $A = \mathbf{Q}/\mathbf{Z}$ of Proposition 19.49 characterizes locally compact groups with compact identity components that are of type FP_2; for $A = \mathbf{Z}/2\mathbf{Z}$ it characterizes CGLC-groups with compact identity components that are of type FP_2 mod 2.

19.4.9 Maximal 1-ended subgroups

In a locally compact group, let us call an M1E-subgroup a compactly generated, 1-ended open subgroup that is maximal among compactly generated, 1-ended open subgroups.

Lemma 19.51   *Let G be a 1-ended CGLC-group. Then every inversion-free continuous action of G on a non-empty tree with compact edge stabilizers has a unique fixed vertex.*

Proof   We can suppose that the action is minimal. If the tree is not reduced to a singleton, the action induces a non-trivial decomposition of G as a Bass–Serre fundamental group of a graph of groups with compact open edge stabilizers, contradicting that G is 1-ended. So G fixes a vertex v; if it fixes another vertex v', then it fixes every edge in between, contradicting that edge stabilizers are compact. $\square$

The following lemma is straightforward.

Lemma 19.52   *Let G be an LC-group and $p : G \to G/W$ the quotient by some compact normal subgroup. Then for every M1E-subgroup H of G, $p(H)$ is an M1E-subgroup of G/W, and for every M1E-subgroup L of G/W, $p^{-1}(H)$ is an M1E-subgroup of G.* $\square$

Lemma 19.53   *Let G be a compactly generated accessible LC-group and H a closed cocompact subgroup. Then*

(a) for every M1E-subgroup L of G, the intersection $L \cap H$ is an M1E-subgroup of H;

(b) every M1E-subgroup of H is contained in a unique M1E subgroup of G.

Proof   (a) Define $N = L \cap H$. Since L is open and H is cocompact, the intersection N is cocompact in L. So N is 1-ended. By Proposition 19.54, N is contained in an M1E-subgroup P of H. Let T be a tree on which G acts continuously with compact edge stabilizers. Then by Lemma 19.51, each of P, L and N fix a unique vertex in T, and since $P \supset N \subset L$, this unique vertex v is the same for all. Then $L \subset G_v \supset P$, where G_v is the stabilizer of v. Moreover, G_v is 1-ended and since L is M1E, we deduce that $G_v = L$. Thus $P \subset L$. Hence $P \subset L \cap H = N$, and therefore $P = N$.

(b). Let L be an M1E subgroup of H. By Proposition 19.54, L is contained in an M1E-subgroup M of G. Since H is cocompact in G and M is open, the intersection $H \cap M$ is cocompact in M; in particular, it is 1-ended and open in H. Since $L \subset H \cap M$ is an M1E-subgroup of G, we deduce that $L = H \cap M$. Hence L is cocompact in M, showing the existence. To prove the uniqueness, let M'

be another M1E-subgroup of G containing L. As in the previous paragraph, using the inclusions $M \supset L \subset M'$, using a tree with a continuous G-action with compact edge stabilizers and 1-ended vertex stabilizers and applying Lemma 19.51, we obtain that all fix a common vertex v, and since $M \subset G_v \supset M'$ and M, M' are M1E-subgroups, we obtain $M = G_v = M'$.   □

Proposition 19.54   *Let G be an accessible LC-group and H a closed, 1-ended compactly generated subgroup. Then H is contained in a M1E-subgroup. Moreover, H is an M1E-subgroup if and only if there is a continuous inversion-free action of G on a tree, with compact edge stabilizers, such that H is the stabilizer of some vertex.*

Proof   Assume that H is some vertex stabilizer in some inversion-free continuous action on a tree with compact edge stabilizers. Then H is open. Let L be a 1-ended subgroup containing H. By Lemma 19.51, L fixes some unique vertex v'. By uniqueness of the vertex fixed by H (again Lemma 19.51), we have $v = v'$. Since H is the stabilizer of v, we deduce that $H = L$. So H is an M1E-subgroup, proving one implication of the second statement.

Now start again with the assumptions of the proposition. Since G is accessible, it admits a continuous action on a tree T with no inversion, with compact edge stabilizers and (at most 1)-ended edge stabilizers. By Lemma 19.51, H is contained in a vertex stabilizer L. Then L is not compact, hence it is 1-ended. From the previous paragraph of this proof, we know that L is an M1E-subgroup, which proves the first statement. If moreover H is M1E, we deduce $H = L$, which establishes the converse implication of the second statement.   □

19.5   Commable groups, and commability classification of amenable hyperbolic groups

Let us recall some definitions and results from [13].

For a homomorphism between LC-groups, write *copci* as a (pronounceable) shorthand for *continuous proper with cocompact image*.

We say that a hyperbolic LC-group is *faithful* if it has no non-trivial compact normal subgroup. A simple observation [7, Lemma 3.6(a)] shows that any hyperbolic LC-group G has a maximal compact normal subgroup $W(G)$; in particular $G/W(G)$ is faithful. Note that if G is faithful then $G°$ is a connected Lie group.

19.5.1 Generalities

Let us say that two LC-groups G, H are *commable* if there exist an integer k and a sequence of copci homomorphisms

$$G = G_0 - G_1 - G_2 - \ldots - G_k = H, \qquad (19.1)$$

where each sign — denotes an arrow in either direction.

We sometimes use fancy arrows such as $\nearrow, \nwarrow$ to make it more readable. For instance, if there are three copci homomorphisms $G \leftarrow G_1 \to G_2 \leftarrow H$, we can write that G is commable to H through $\nwarrow \nearrow \nwarrow$, or that there is a commability $G \nwarrow \nearrow \nwarrow H$.

More generally, if $\mathscr{D}$ is a class of locally compact groups and $G, H \in \mathscr{D}$, we say that G, H are commable within the class $\mathscr{D}$ if the same condition holds with the additional requirement that the LC-groups G_i in (19.1) belong to $\mathscr{D}$.

We will especially consider for $\mathscr{D}$ the class of focal (resp. faithful, resp. focal and faithful) hyperbolic LC-groups.

Remark 19.55 To be compactly generated is invariant by commability among LC-groups. In particular, any two commable CGLC-groups are quasi-isometric. The converse is not true, even for finitely generated groups. Examples are $\Gamma_1 * \mathbf{Z}$ and $\Gamma_2 * \mathbf{Z}$, when Γ_1, Γ_2 are cocompact lattices in $\mathrm{SL}_3(\mathbf{R})$ that are not abstractly commensurable, according to an unpublished observation of Carette and Tessera (see §19.5.2).

Here is a possible other example, of independent interest. Let $\mathrm{BS}(m, n)$ be the Baumslag–Solitar group, defined by the presentation $\langle t, x \mid t x^m t^{-1} = x^n \rangle$. Then, by a result of Whyte [61], the groups $\mathrm{BS}(2, 3)$ and $\mathrm{BS}(3, 5)$ are quasi-isometric; he also makes the simple observation that these groups do not have isomorphic finite index subgroups (i.e. since they are torsion-free, are not commable within discrete groups). We do not know whether $\mathrm{BS}(2, 3)$ and $\mathrm{BS}(3, 5)$ are commable, but expect a negative answer. More generally, it would be of great interest to determine the commability and quasi-isometry classes (within locally compact groups) of any Baumslag–Solitar group $\mathrm{BS}(p, q)$. Their external quasi-isometry classification is addressed, within discrete groups, in [41].

Remark 19.56 If G and H are commable, we can define their commability distance as the least k such that there exists a commability such as in (19.1). We do not know if this can be greater than 4. Anyway, this notion forgets the direction of the arrows; for instance, to have a commensuration through $\nearrow \nwarrow$ or $\nwarrow \nearrow$ are very different relations.

Lemma 19.57 (Proposition 2.7 in [13]) *Any copci homomorphism between*

LC-groups $f : G_1 \to G_2$ satisfies $f^{-1}(W(G_2)) = W(G_1)$. In particular, if $W(G_1)$ is compact (e.g., if G_1 is hyperbolic), f thus factors through an injective copci homomorphism $f' : G_1/W(G_1) \to G_2/W(G_2)$, which is injective. ☐

Corollary 19.58 *Let $\mathscr{F}$ denote the class of focal hyperbolic LC-groups. Given faithful hyperbolic LC-groups G_1 and G_2,*

- *G_1 and G_2 are commable if and only if they are commable within faithful hyperbolic groups;*
- *assuming G_1 and G_2 focal, G_1 and G_2 are commable within focal hyperbolic groups if and only they are commable within faithful focal hyperbolic groups.* ☐

Remark 19.59 We can define *strict commability* in the same fashion as commability, but only allowing *injective* copci homomorphisms. Since a copci homomorphism between faithful focal hyperbolic LC-groups is necessarily injective, Corollary 19.58 shows that between faithful hyperbolic LC-groups, commability and strict commability are the same, and more generally that hyperbolic LC-groups G_1, G_2 are commable if and only if $G_1/W(G_1)$ and $G_2/W(G_2)$ are strictly commable. Similar consequences hold for commability within focal hyperbolic LC-groups.

19.5.2 Non-commable quasi-isometric discrete groups

Let us abbreviate totally disconnected to TD; thus compact-by-TD LC-groups are the same as Hausdorff topological groups with a compact open subgroup. Also, recall that maximal 1-ended (M1E) subgroups were introduced in §19.4.9.

Lemma 19.60 *Let G, H be compactly generated, accessible locally compact groups with infinitely many ends. Assume that G and H are commable. Then every M1E-subgroup of G is commable within compact-by-TD LC-groups to an M1E subgroup of H.*

Proof Since G and H have infinitely many ends, they are compact-by-TD as well as all LC-groups commable to them, and their closed subgroups. The result then follows, by an immediate induction, from Lemmas 19.52 and 19.53. ☐

Lemma 19.61 *Let Γ be a finitely generated group with no non-trivial finite normal subgroup. Assume that every compact-by-TD LC-group quasi-isometric to Γ is compact-by-discrete and has a maximal compact normal subgroup.*

Then a compact-by-TD LC-group H is commable to Γ within compact-by-TD LC-groups if and only if its compact normal subgroup W exists and H/W is commensurable (= strictly commable within discrete groups) to Γ.

Proof   Consider a sequence of copci homomorphisms

$$\Gamma = G_0 - G_1 - \ldots - G_k = H$$

between compact-by-TD LC-groups. When G is an LC-group, let $W(G)$ denote the union of all compact normal subgroups; when G admits a maximal compact normal subgroup, it is equal to $W(G)$. Note that $W(\Gamma) = \{1\}$ by the first assumption. Then any copci homomorphism $U \to V$ maps $W(U)$ into $W(V)$ (this follows from Proposition 19.57). It follows that the above sequence induces a sequence of homomorphisms $\Gamma - G_1/W(G_1) - \ldots - G_k/W(G_k) = H/W(H)$, which are copci homomorphisms since $W(G_i)$ is compact for all i, by the assumptions. All these groups being discrete, this means that Γ and $H/W(H)$ are commensurable. $\square$

Example 19.62 (Carette–Tessera)   Let S be a non-compact connected semisimple Lie group with trivial centre. Let Γ, Λ be non-commensurable cocompact lattices in S (this exists unless S is isomorphic to the product of a compact group with $SL_2(\mathbf{R})$). Then $\Gamma * \mathbf{Z}$ and $\Lambda * \mathbf{Z}$ are quasi-isometric but not commable, as we now show.

A consequence of Theorem 19.25 is that every LC-group quasi-isometric to S is compact-by-Lie. In particular, every compact-by-TD LC-group quasi-isometric to S is compact-by-discrete. Accordingly, the assumption of Lemma 19.61 is satisfied by Γ, and it follows this lemma that Γ and Λ are not commable within compact-by-TD LC-groups.

Since S is 1-ended, so are Γ and Λ, and therefore these are M1E-subgroups in $\Gamma * \mathbf{Z}$ and $\Lambda * \mathbf{Z}$. Thus, by Lemma 19.60, we deduce that $\Gamma * \mathbf{Z}$ and $\Lambda * \mathbf{Z}$ are not commable.

To show that $\Gamma * \mathbf{Z}$ and $\Lambda * \mathbf{Z}$ are quasi-isometric, we first need to know that Γ and Λ are bilipschitz (the result follows by an immediate argument). That they are bilipschitz follows from a result of Whyte [60] asserting that any two quasi-isometric non-amenable finitely generated groups are bilipschitz.

The next three subsections address commability within focal groups. See §19.5.6 for the link with commability.

19.5.3 Connected type

We say that a focal hyperbolic group is *focal-universal* if it satisfies the following: for every LC-group H, the group H is commable to G within focal groups if and only if there is a copci homomorphism $H \to G$.

Theorem 19.63 ([13]) *Let G be a focal hyperbolic LC-group of connected type. Then G is commable to a focal-universal LC-group $\hat{G}$, which thus satisfies: for every LC-group H, the group H is commable to G within focal groups if and only if there is a copci homomorphism $H \to \hat{G}$. Moreover, $\hat{G}$ is unique up to topological isomorphism.*

Corollary 19.64 *Any two commable focal hyperbolic groups of connected type are commable within focal groups through* $\nearrow^{\nwarrow}\diagdown$. $\quad\square$

Focal-universal groups of connected type give a canonical set of representatives of commability classes of focal hyperbolic LC-groups of connected type. However, another canonical set of representatives, sometimes more convenient (e.g. in view of the classification small dimension), is given by purely real Heintze groups.

Proposition 19.65 *Every focal-universal LC-group of connected type has a unique closed cocompact subgroup that is purely real Heintze. Every focal hyperbolic group of LC-type is commable to a purely real Heintze group, unique up to isomorphism.* $\quad\square$

Example 19.66 Here is (without proof) the classification of all purely real Heintze groups in dimension $2 \leq d \leq 4$. For each isomorphy class we give one representative.

- $d = 2$: the only group is the affine group $\mathrm{Carn}(\mathbf{R}) = \mathbf{R} \rtimes \mathbf{R}$, where the action is given by $t \cdot x = e^t x$. It is isomorphic to a closed cocompact subgroup in $\mathrm{Isom}(\mathbf{H}_{\mathbf{R}}^2)$.
- $d = 3$: the groups are the G_λ for $\lambda \geq 1$, as well as a certain G_1^u. All these groups are defined as a semidirect product $\mathbf{R}^2 \rtimes \mathbf{R}$. In G_λ, the action is given by $t \cdot (x,y) = (e^t x, e^{\lambda t} y)$, while in G_1^u it is given by $t \cdot (x,y) = e^t(x+ty,y)$. Note that $G_1 = \mathrm{Carn}(\mathbf{R}^2)$ is isomorphic to a closed cocompact subgroup of $\mathrm{Isom}(\mathbf{H}_{\mathbf{R}}^3)$.
- $d = 4$. It consists of semidirect products $\mathbf{R}^3 \rtimes \mathbf{R}$ and $\mathrm{Hei}_3 \rtimes \mathbf{R}$, where Hei_3 is the 3-dimensional Heisenberg group. More precisely, the semidirect products $\mathbf{R}^3 \rtimes \mathbf{R}$ consist of the $G_{\lambda,\mu}$ for $1 \leq \lambda \leq \mu$, with action given by $t \cdot (x,y,z) = (e^t x, e^{\lambda t} y, e^{\mu t} z)$, of the $G_\lambda^{(u)}$ ($\lambda > 0$) for which the action is given by $t \cdot (x,y,z) = (e^t(x+ty), e^t y, e^{\lambda t} z)$, and of G_1^u for which the action is given

by $t \cdot (x,y,z) = e^t(x+ty+t^2z/2, y+tz, z)$. If we use coordinates for the Heisenberg group so that the product is given by $(x,y,z)(x',y',z') = (x+x', y+y', z+z'+xy'-x'y)$ (these are not the standard matrix coordinates!), the corresponding Heintze groups are H_λ for $\lambda \geq 1$ with action given by $t \cdot (x,y,z) = (e^t x, e^{\lambda t} y, e^{(1+\lambda)t} z)$, and H_1^μ with action given by $t \cdot (x,y,z) = e^t(x+y, y, z)$. Note that $G_{1,1} = \mathrm{Carn}(\mathbf{R}^3)$ is isomorphic to a closed cocompact subgroup of $\mathrm{Isom}(\mathbf{H}_{\mathbf{R}}^4)$, and $H_1 = \mathrm{Carn}(\mathrm{Hei}_3)$ is isomorphic to a closed cocompact subgroup of $\mathrm{Isom}(\mathbf{H}_{\mathbf{C}}^2)$.

- For $d = 5$, we do not give the full classification, but just mention that the groups occurring are semidirect products $\mathbf{R}^4 \rtimes \mathbf{R}$, $(\mathrm{Hei}_3 \times \mathbf{R}) \rtimes R$, or $\mathrm{Fil}_4 \rtimes \mathbf{R}$, for various contracting actions we do not describe, where Fil_4 is the filiform 4-dimensional Lie group, whose Lie algebra has a basis (a, e_1, e_2, e_3) so that $[a, e_i] = e_{i+1}$ for $i = 1, 2$ and all other brackets between basis elements vanish. The Carnot type groups $\mathrm{Carn}(\mathrm{Hei}_3 \times \mathbf{R})$ and $\mathrm{Carn}(\mathrm{Fil}_4)$ are the smallest examples of Carnot type groups that are not isomorphic to closed cocompact groups of isometries of rank 1 symmetric spaces of non-compact type.

- For $d' \leq 6$, there are finitely many isomorphism classes of simply connected Carnot-gradable nilpotent Lie groups of dimension d' (namely $1, 1, 2, 3, 7, 21$ isomorphism classes for $d' = 1, \ldots, 6$), while for $d' \geq 7$ there are continuously many. Hence for $d \leq 7$ there are finitely many isomorphism classes of purely real Heintze groups of Carnot type, and continuously many for $d \geq 8$.

19.5.4 Totally disconnected type

Define, for every integer $m \geq 2$, FT_m as the stabilizer of a given boundary point in the automorphism group of an $(m+1)$-regular tree.

If G is any locally compact group having exactly two (opposite) continuous homomorphisms onto $\mathbf{Z}$ (e.g., any focal hyperbolic group not of connected type), for $n \geq 1$ we denote by $G^{[n]}$ the inverse image of $n\mathbf{Z}$ by any of these homomorphisms.

By *non-power* integer we mean an integer $q \geq 1$ that is not an integral proper power of any integer (thus excluding $4, 8, 9, 16, 25, 27, 32, 36, 49 \ldots$).

An easy observation (Proposition 19.74) is that any two focal hyperbolic LC-groups G_1, G_2 of totally disconnected type are always commable. However, they are not always commable within focal groups:

Proposition 19.67 ([13]) *Let G_1, G_2 be focal hyperbolic LC-groups of totally disconnected type. The following are equivalent:*

(i) G_1 and G_2 are commable within focal groups;

(ii) $\Delta(G_1)$ *and* $\Delta(G_2)$ *are commensurate subgroups of* $\mathbf{R}_+$, *i.e.* $\Delta(G_1) \cap \Delta(G_2)$
     has finite index in both.
(iii) *There exists a non-power integer* $q \geq 2$ *such that* $\Delta(G_1)$ *and* $\Delta(G_2)$ *are*
      both contained in the multiplicative group $\langle q \rangle = \{q^n : n \in \mathbf{Z}\}$;
(iv) *there is a commability within focal groups* $G_1 \nearrow^{\mathsf{K}}\!\!\diagdown \nearrow^{\mathsf{K}}\!\!\diagdown G_2$;
(v) *there exists a non-power integer* $q \geq 2$ *and an integer* $n \geq 1$, *such that*
     there is a commability within focal groups $G_i \nearrow^{\mathsf{K}}\!\!\diagdown \mathrm{FT}_q^{[n]}$ *for* $i = 1, 2$;
(vi) *there exists a non-power integer* $q \geq 2$ *such that there is a commability*
      within focal groups with $G_i \nearrow^{\mathsf{K}}\!\!\diagdown\nearrow \mathrm{FT}_q$ *for* $i = 1, 2$;
(vii) *there is a commability within focal groups* $G_1 \diagdown^{\mathsf{K}}\!\!\diagdown \nearrow^{\mathsf{K}}\!\!\nearrow G_2$;
(viii) *there exists an integer* $m \geq 2$, *such that for* $i = 1, 2$ *there is a commability*
       within focal groups $G_i \diagdown^{\mathsf{K}}\!\!\nearrow \mathrm{FT}_m$.

19.5.5 Mixed type

If G is a locally compact group, its locally elliptic radical $G^{\sharp}$ is its largest closed locally elliptic normal subgroup, where locally elliptic means that every compact subset is contained in a compact subgroup. If G is focal of mixed type, then $G/G^{\sharp}$ is focal of connected type.

Definition 19.68 ([13])   Let G be a focal hyperbolic LC-group. Consider the modular functions of G/G° and $G/G^{\sharp}$; by composition they define homomorphisms $\Delta_G^{\mathrm{td}}, \Delta_G^{\mathrm{con}} : G \to \mathbf{R}_+$, which we call restricted modular functions. Since $\mathrm{Hom}(G, \mathbf{R})$ is 1-dimensional, if G is not of totally disconnected type then Δ_G^{con} is non-trivial and hence there exists a unique $\varpi = \varpi(G) \in \mathbf{R}$ such that $\log \circ \Delta_G^{\mathrm{td}} = \varpi (\log \circ \Delta_G^{\mathrm{con}})$. Because of the compacting element in G, necessarily $\varpi(G) \geq 0$, with equality if and only if G is of connected type. If G is of totally disconnected type we set $\varpi(G) = +\infty$.

Definition 19.69   Let H, A be focal hyperbolic LC-groups, H being of connected type with a surjective modular function and A being of totally disconnected type. For $\varpi > 0$, define

$$H \overset{\varpi}{\times} A = \{(x, y) \in H \times A \mid \Delta_H(x)^{\varpi} = \Delta_A(y)\}.$$

This is a focal hyperbolic LC-group of mixed type, satisfying $\varpi(H \overset{\varpi}{\times} A) = \varpi$. If $q \geq 2$ is an integer and ϖ is a positive real number, define in particular

$$H[\varpi, q] = H \overset{\varpi}{\times} \mathrm{FT}_q.$$

Proposition 19.70 ([13])   *Let* G_1, G_2 *be focal hyperbolic LC-groups of mixed type. Equivalences:*

(i) G_1 and G_2 are commable;

(ii) the following three properties hold:

- G_1/G_1° and G_2/G_2° are commable within focal groups;
- $G_1/G_1^\sharp$ and $G_2/G_2^\sharp$ are commable within focal groups
- $\varpi(G_1) = \varpi(G_2)$;

(iii) there is a commability within focal groups $G_1 \nearrow\!\!\nwarrow\searrow\nearrow\!\!\nwarrow\searrow G_2$;

(iv) there exists a non-power integer $q \geq 2$, an integer $n \geq 1$, a focal-universal group of connected type H and a positive real number $\varpi > 0$ such that there is a commability within focal groups $G_i \nearrow\!\!\nwarrow\searrow H[q,\varpi]^{[n]}$ for $i = 1,2$;

(v) there exists a non-power integer $q \geq 2$, a focal-universal group of connected type H and a positive real number $\varpi > 0$ such that there is a commability within focal groups with $G_i \nearrow\!\!\nwarrow\searrow\nearrow H[\varpi,q]$ for $i = 1,2$;

(vi) there is a commability within focal groups $G_1 \searrow\nearrow\!\!\nwarrow\searrow\nearrow G_2$;

(vii) there exists an integer $m \geq 2$, a focal-universal group of connected type H and a positive real number $\varpi > 0$ such that for $i = 1,2$ there is a commability within focal groups $G_i \searrow\nearrow H[\varpi,m]$.

In view of Proposition 19.65, we deduce:

Corollary 19.71 *Every focal hyperbolic LC-group of mixed type G is commable to an LC-group of the form $H[\varpi,q]$, for some purely real Heintze group H uniquely defined up to isomorphism, a unique $\varpi \in \,]0,\infty[$, and a unique non-power integer q.* □

19.5.6 Commability between focal and general type groups

The following lemma from [13] is essentially contained in [7].

Lemma 19.72 *Let G_1,G_2 be focal hyperbolic LC-groups. Equivalences:*

(i) *there exists a hyperbolic LC-group G with copci homomorphisms $G_1 \to G \leftarrow G_2$.*

(ii) *there exists a focal hyperbolic LC-group G with copci homomorphisms $G_1 \to G \leftarrow G_2$.* □

Proposition 19.73 *Let G be a focal hyperbolic LC-group not of totally disconnected type. Let H be a locally compact group commable to G. Then the following statements hold.*

(a) *If H is focal, then G and H are commable within focal groups;*

(b) if H is non-focal, then there exists a rank 1 symmetric space of non-compact type X and continuous proper compact isometric actions of G and H on X.

Proof If G is quasi-isometric to any rank 1 symmetric space X of non-compact type, then by Theorem 19.25 we have copci homomorphisms $G \to \text{Isom}(X) \leftarrow H$; if H is of general type this proves (b); if H is focal then conjugating by some isometry we can suppose it has the same fixed point in ∂X as G, proving (a).

Assume otherwise that G is not quasi-isometric to any rank 1 symmetric space of non-compact type. To show the result, it is enough to check that the commability class of G consists of focal groups. Otherwise, there is a copci homomorphism $G_1 \to G_2$ between groups in the commability class of G, such that G_1 is focal and G_2 is not focal. By Theorem 19.1(b), it follows that G is quasi-isometric to some rank 1 symmetric space of non-compact type (which is excluded) or to a tree (which is excluded since ∂G is positive-dimensional). $\qquad\qquad\square$

Proposition 19.74 ([13]) *Two focal hyperbolic LC-groups G_1, G_2 of totally disconnected type are always commable through $\nearrow^{\kappa}\!\!\diagdown\, \nearrow^{\kappa}\!\!\diagdown$, and are commable to a finitely generated free group of rank ≥ 2 through $\nearrow^{\kappa}\!\!\diagdown$.*

Proof There is a copci homomorphism $G_i \to \text{Aut}(T_i)$ for some regular tree T_i of finite valency at least 3. For d large enough, $\text{Aut}(T_i)$ $(i = 1, 2)$ contains a cocompact lattice isomorphic to a free group of rank $k = 1 + d!$. Thus we have copci homomorphisms $G_1 \to \text{Aut}(T_1) \leftarrow F_k \to \text{Aut}(T_2) \leftarrow G_2$. $\qquad\square$

Special groups

Definition 19.75 By *special* hyperbolic LC-group we mean any CGLC-group commable to both focal and general type hyperbolic LC-groups. A CGLC-group quasi-isometric to a special hyperbolic group is called *quasi-special*.

See Remark 19.79 for some motivation on the choice of terminology. Taking Theorems 19.25 and 19.31 for granted, we have the following characterizations of special and quasi-special hyperbolic groups.

Proposition 19.76 *Let G be a CGLC-group. Equivalences:*

  (i) G is special hyperbolic;
 (ii) G is commable to the isometry group of a metric space X which is either a rank 1 symmetric space of non-compact type or a 3-regular tree.

We also have the equivalences:

(iii) G is quasi-special hyperbolic;

(iv) *G is quasi-isometric to either a rank 1 symmetric space of non-compact type or to a 3-regular tree;*

 (v) *G admits a continuous proper cocompact isometric action on either a rank 1 symmetric space or non-degenerate tree (see §19.4.1) (which can be chosen to be regular if G is focal).*

Moreover, if G is quasi-special and not special, then it is of general type and quasi-isometric to a 3-regular tree.

Proof   Let us begin by proving the equivalence (i)⇔(ii), not relying on Theorems 19.25 and 19.31.

(ii)⇒(i) is clear since, denoting by X the space in (ii), it follows that G is commable to both Isom(X) and Isom(X)$_\omega$ for some boundary point.

(i)⇒(ii) We can suppose that G is focal. If G is of totally disconnected type, then by the easy [15, Proposition 4.6] (which makes G act on its Bass–Serre tree), G has a continuous proper cocompact isometric action on a regular tree and thus (ii) holds. Assume that G is not of totally disconnected type. Let $G = H_0 - H_1 - H_2 \cdots - H_k$ be a sequence of copci arrows (in either direction) with $H_k = \text{Isom}(X)$, which is of general type. Let i be minimal such that H_i is of general type. Then $i \geq 1$, H_{i-1} is focal and necessarily the copci arrow is in the direction $H_{i-1} \to H_i$. So H_i is hyperbolic of general type and its identity component is not compact. Thus by [7, Proposition 5.10], H_i is isomorphic to an open subgroup of finite index subgroup in the isometry group of a rank 1 symmetric space of non-compact type, proving (ii).

Let us now prove the second set of equivalences.

(v)⇒(iv) is immediate, in view of Lemma 19.80.

(iv)⇒(iii) let X be the space as in (iv); then Isom(X) is special since it is of general type and the stabilizer of a boundary point is focal and cocompact.

(iii)⇒(v): if G is quasi-special, then it is quasi-isometric to a space as in (ii), so (v) is provided by Theorems 19.25 and 19.31, except the regularity of the tree in the focal case, in which case we can then invoke the easy [15, Proposition 4.6] (which makes G act on its Bass–Serre tree).

For the last statement, observe that (v)⇒(ii) holds in case in (v) we have a symmetric space or a regular tree.   □

Remark 19.77   In the class of hyperbolic LC-groups quasi-isometric to a non-degenerate tree (which is a single quasi-isometry class, described in Corollary 19.33), there is a "large" commability class, including

 (i) all focal hyperbolic groups of totally disconnected type (by Proposition 19.74);

(ii) all discrete groups (i.e. non-elementary virtually free finitely generated groups);

(iii) more generally, all unimodular groups quasi-isometric to a non-degenerate tree (because they have a cocompact lattice by [4]), including all automorphism groups of regular and biregular trees of finite valency;

(iv) non-ascending HNN extensions of compact groups over open subgroups: indeed the Bass–Serre tree is then a regular tree. Such groups are non-focal and are often non-unimodular.

It is natural to ask whether this "large" class is the whole class:

Question 19.78 Are any two hyperbolic LC-groups quasi-isometric to the 3-regular tree commable? Equivalently, does quasi-special imply special?

It is easy to check that any such group is commable to the Bass–Serre fundamental group of a connected finite graph of groups in which all vertex and edge groups are isomorphic to $\widehat{\mathbf{Z}}$ (the profinite completion of $\mathbf{Z}$). We expect a positive answer to Question 19.78.

Update. This question was settled positively in full generality by M. Carette in [8] after being asked in a previous version of this survey and in [13].

A thorough study of cocompact isometry groups of bounded valency trees is carried out in [39].

Remark 19.79 The adjective "special" indicates that special hyperbolic LC-groups are quite exceptional among hyperbolic LC-groups, although they are the best known. For instance, there are countably many quasi-isometry classes of such groups, namely one for trees and one for each homothety class of rank 1 symmetric space of non-compact type (and finitely many classes for each fixed asymptotic dimension), while there are continuum many pairwise non-quasi-isometric 3-dimensional Heintze groups.

Let us mention the following lemma, which is well known but often referred to without proof.

Lemma 19.80 *Let T be a bounded valency non-empty tree with no vertex of degree 1 and in which the set of vertices of valency ≥ 3 is cobounded. Then T is quasi-isometric to the 3-regular tree. In particular, if T is a bounded valency tree with a cocompact isometry group and at least three boundary points then it is quasi-isometric to the 3-regular tree T'.*

Proof A first step is to get rid of valency 2 vertices. Indeed, since there are no valency 1 vertices and by the coboundedness assumption, every valency 2 vertex v lies in a unique (up to orientation) segment consisting of vertices

$v_0, \ldots, v_n$ with v_0, v_n of valency ≥ 3, and each $v_1, \ldots, v_{n-1}$ being of valency 2, and n being bounded independently of v. If we remove the vertices $v_1, \ldots, v_{n-1}$ and join v_0 and v_n with an edge, the resulting tree is clearly quasi-isometric to T. Hence in the sequel, we assume that T has no vertex of valency ≤ 2.

Let $s \geq 3$ be the maximal valency of T. Denote by T^0 and T^1 the 0-skeleton and 1-skeleton of T. Let T_3 be a 3-regular tree, and fix an edge, called root edge, in both T and T_3, so that the n-ball $T(n)$ means the n-ball around the root edge (for $n = 0$ this is reduced to the edge). Let us define a map $f : T \to T_3$, by defining it by induction on the n-ball $T(n)$.

We prescribe f to map the root edge to the root edge, and, for $n \geq 0$, we assume by induction that f is defined on the n-ball of T^0. The convex hull of $f(T(n))$ is a finite subtree $A(n)$ of T_3. We assume the following property $\mathscr{P}(n)$: the subtree $A(n) \subset T_3^0$ has no vertex of valency 2, and the n-sphere of T^0 is mapped to the boundary of this subtree $A(n)$. Let v belong to the n-sphere of T. Then v has a single neighbour not in the $(n+1)$-sphere, and has a number $m \in [2, s-1]$ of other neighbours, in the $(n+1)$-sphere of T. The vertex $f(v)$ has a single neighbour w_0 in $A(n)$. (Choosing a root edge instead of a root vertex is only a trick to avoid a special step when $n = 0$.)

We use the following claim: in the binary rooted tree of height $k \geq 1$ and root o, for every integer in $m \in [2, 2^k]$ there exists a finite subset F of vertices of cardinal m with $o \notin F$ such that the convex hull of $F \cup \{o\}$ has exactly F as set of vertices of valency 1 and admits only o as set of vertices of valency 2. The proof is immediate: reducing the value of k if necessary, we can suppose $m \geq 2^{k-1}$, and then, writing $m = 2^{k-1} + t = 2t + (2^{k-1} - t)$, choose $2^{k-1} - t$ vertices of height $k - 1$, and choose the $2t$ descendants (of height k) of the remaining t vertices of height $k - 1$, to form the subset F, and it fulfills the claim.

Now choose $k = \lceil \log_2(s-1) \rceil$, consider the set

$$M_v = \{w \in T_3^0 : d(w, w_0) - 1 = d(w, f(v)) \leq k\}$$

(in other words, those vertices at distance $\leq k$ from $f(v)$ not in the direction of w_0); this is a binary tree, rooted at $f(v)$, of height k. Since $m \leq s - 1$, and $s - 1 \leq 2^k$, we have $m \leq 2^k$ and by the claim there is a finite subset $F_v \subset M \smallsetminus \{f(v)\}$ of cardinal m with the required properties. So the convex hull of $f(T(n)) \cup F_v$ admits, in M, only elements of F_v as vertices of valency 1, and no vertices of valency 1.

Noting that the M_v are pairwise disjoint (v ranging over the n-sphere of T), we deduce that the convex hull of $f(T(n)) \cup \bigcup_v F_v$ admits no vertex of valency 2 and admits $\bigcup F_v$ as set of vertices of valency 1. Now extend f to the $(n+1)$-ball by choosing, for every v in the n-sphere of T, a bijection between its set

322   Cornulier

of neighbouring vertices in the $(n+1)$-sphere and F_v (recall that they have the same cardinal by construction). Then $\mathscr{P}(n+1)$ holds by construction.

By induction, we obtain a map $f: T^0 \to T_3^0$; it is injective by construction, and more precisely $d(f(v), f(v')) \geq d(v, v')$ for all v, v'; moreover f is k-Lipschitz. In addition, if $A(n)$ is the convex hull of the image of $T(n)$, then an immediate induction shows that $A(n)$ contains the n-ball $T_3(n)$, and every point is at distance $\leq k$ to the image of f. Thus f is a quasi-isometry $T^0 \to T_3^0$. $\square$

19.5.7 A few counterexamples

Groups of connected type with no common cocompact subgroup
We have seen that any two commable focal groups of connected type are commable through $\nearrow\!\!\nwarrow$. This is not true with $\nwarrow\!\!\nearrow$. The simplest example is obtained as follows: start from the group $G = \mathbf{R} \rtimes \mathbf{R}$ (the affine group), $u = \log \circ \Delta_G$, $G_1 = u^{-1}(\mathbf{Z})$ and $G_2 = u^{-1}(\lambda\mathbf{Z})$ where λ is irrational. Since $\Delta(G_1) \cap \Delta(G_2) = \{1\}$, it is clear that G_1 and G_1 are commable through $\nwarrow\!\!\nearrow$.

Focal groups not acting on the same space
We saw that commable focal hyperbolic LC-groups of connected type are commable through $\nearrow\!\!\nwarrow$. This is not true in other types. We give here some examples of totally disconnected type; examples of mixed type can be derived mechanically by "adding" a connected part.

Recall from §19.5.4 that all FT_n, for $n \geq 2$, are commable, and that FT_m and FT_n are commable within focal groups if and only if m and n are integral powers of the same integer. In contrast, we have:

Proposition 19.81 ([13]) *If $2 \leq m < n$, there exists no hyperbolic LC-group G with copci homomorphisms $\mathrm{FT}_m \to G \leftarrow \mathrm{FT}_n$.* $\square$

Corollary 19.82 *If $m, n \geq 2$ are distinct, there is no proper metric space with continuous proper cocompact isometric actions of both FT_m and FT_n.* $\square$

Discrete groups not acting on the same space
This subsection is a little tour beyond the focal case, dealing with discrete analogues of the examples in Proposition 19.81.

Let us mention a consequence of Theorem 19.31, already observed in [40, Corollary 10] with a slightly different point of view. Let C_n be a cyclic group of order n. Recall that all discrete groups $C_n * C_m$, for $n \geq 2$ and $m \geq 2$ are quasi-isometric to the trivalent regular tree.

Proposition 19.83 *Let $\{p_1, q_1\} \neq \{p_2, q_2\}$ be distinct pairs of primes ≥ 3.*

Then the groups $C_{p_1} * C_{q_1}$ and $C_{p_1} * C_{q_2}$ are not isomorphic to cocompact lattices in the same locally compact group, and thus do not act properly cocompactly on the same non-empty proper metric space.

Lemma 19.84 *Let T be a tree with a cobounded action of its isometry group. Then it admits a unique minimal cobounded subtree T' (we agree that $\varnothing \subset T$ is cobounded if T is bounded). Moreover, $T = T'$ if and only if T has no vertex of degree 1.*

Proof Let T' be the union of all (bi-infinite) geodesics in T; a straightforward argument shows that T' is a subtree. Observe that any geodesic is contained in every cobounded subtree: indeed, any point of a geodesic cuts the tree into two unbounded components. It follows that T' is contained in every cobounded subtree; by definition, T' is Isom(T)-invariant.

Let us show that T' is cobounded. If T is bounded then $T' = \varnothing$ and is cobounded by convention. So let us assume that T is unbounded; then it is enough to show that $T' \neq \varnothing$, because then the distance to T' is invariant by the isometry group, so takes a finite number of values by coboundedness. To show that $T' \neq \varnothing$, it is enough to show that Isom(T) has a hyperbolic element: otherwise the action of Isom(T) would be horocyclic and thus would preserve the horocycles with respect to some point at infinity, which would prevent coboundedness of the action of Isom(T).

The last statement is clear from the definition of T'. ☐

If $p, q \geq 2$ are numbers and $m \geq 1$, define a tree $T_{p,q,m}$ as follows: start from the (p, q)-biregular tree and replace each edge by a segment made out of m consecutive edges. Note that if $p, q \geq 3$, the unordered pair $\{p, q\}$ is uniquely determined by the isomorphy type of $T_{p,q,m}$.

Lemma 19.85 *Let p, q be primes and let $C_p * C_q$ act minimally properly on a non-empty tree T with no inversion. Then T is isomorphic to $T_{p,q,m}$ for some $m \geq 1$.*

Proof This gives $C_p * C_q$ as Bass–Serre fundamental group of a finite connected graph of groups $((\Gamma_v)_v, (\Gamma_e)_e)$ with finite vertex groups. This finite graph X is a finite tree, because $\mathrm{Hom}(C_p * C_q, \mathbf{Z}) = 0$. Let v be a degree 1 vertex of this finite tree, and let e be the oriented edge towards v. Then the embedding of Γ_v into Γ_e is not an isomorphism, because otherwise the action on the tree would not be minimal (v corresponding to a degree 1 vertex in the Bass–Serre universal covering).

Since vertex stabilizers are at most finite of prime order, this shows that such edges are labeled by the trivial group. This gives a free decomposition of the group in as many factors as degree 1 vertices in X. So X has at most 2 degree 1 vertices and thus is a segment, and then shows that other vertices are labeled by the trivial group. If the number of vertices is $m + 1$, this shows that T is isomorphic to $T_{p,q,m}$.   □

Proof of Proposition 19.83   Note these groups have no non-trivial compact normal subgroup, so the second statement is a consequence of the first by considering the isometry group of the space.

Suppose they are cocompact lattices in a single locally compact group G. Then G is compactly generated and quasi-isometric to a tree, so by Theorem 19.31 acts properly cocompactly, with no inversion and minimally on a finite valency tree T. By Lemma 19.84, the action of $C_{p_i} * C_{q_i}$ is minimal for $i = 1, 2$. By Lemma 19.85, the tree is isomorphic to T_{p_i, q_i, m_i} for $i = 1, 2$, which implies $\{p_1, q_1\} = \{p_2, q_2\}$, a contradiction.   □

19.6 Towards a quasi-isometric classification of amenable hyperbolic groups

19.6.1 The main conjecture

We use the notion of commability studied in Section 19.5. Here is the main conjecture about quasi-isometric rigidity of focal hyperbolic LC-groups.

Conjecture 19.86   Let G be a focal hyperbolic LC-group. Then any compactly generated locally compact group H is quasi-isometric to G if and only if it is commable to G.

An LC-group H as in the conjecture is necessarily non-elementary hyperbolic. Thus the conjecture splits into two distinct issues:

• (internal case) when H is focal, in which case the conjecture can be restated as: two focal hyperbolic LC-groups are quasi-isometric if and only if they are commable; this is discussed in §19.6.2;

• (external case) when H is of general type; this is discussed in §19.6.3.

The commability classes of focal hyperbolic LC-groups having been fully described in §19.5 (except in the totally disconnected type), Conjecture 19.87 provides a comprehensive description. Note that unlike in the other two types, in the totally disconnected type the quasi-isometric classification is known (and trivial, since there is a single class) but the commability classification is still much more delicate (Question 19.78).

19.6.2 The internal classification

The main internal QI-classification conjecture

Let us repeat the internal part of Conjecture 19.86:

Conjecture 19.87 Let G be a focal hyperbolic LC-group. Then any focal hyperbolic LC-group H is quasi-isometric to G if and only if it is commable to G.

In other words, any two focal hyperbolic LC-groups are quasi-isometric if and only if they are commable. The conjecture is stated in a less symmetric formulation so that it makes sense to state that the conjecture holds for a given G. Note that any two focal hyperbolic LC-groups of totally disconnected type are commable by Proposition 19.74, so there is no need to discard them as in Conjecture 19.86.

Note that the "if" part is trivial. Thus the conjecture is a putative description of the internal quasi-isometry classification of focal hyperbolic groups (and thus of the spaces associated to these groups), using the description of commability classes, which is described in a somewhat satisfactory way in Section 19.5.

Conjecture 19.87 can in turn be split into the connected and mixed cases.

Internal QI-classification in connected type

The following was originally stated as a theorem by Hamenstädt in her PhD thesis [26], who told me to rather consider it as a conjecture:

Conjecture 19.88 Let H be a purely real Heintze group. If a purely real Heintze group L is quasi-isometric to H, then it is isomorphic to H.

In other words, the conjecture states that two purely real Heintze groups are quasi-isometric if and only if they are isomorphic. The non-symmetric formulation of the conjecture is convenient because it makes sense to assert that it holds for a given H. Note that the same statement holds true for commability, by Corollary 19.71.

Note that by Corollary 19.64 and Proposition 19.65, an equivalent formulation of the conjecture consists in replacing both times "purely real Heintze group" by "faithful focal-universal hyperbolic LC-group of connected type". We also have

Proposition 19.89 *Conjecture 19.87 specified to groups of connected type is equivalent to Conjecture 19.88.*

Proof Assume that Conjecture 19.87 holds for groups of connected type. If

H_1 is purely real Heintze and is quasi-isometric to H, by the validity of Conjecture 19.87, H and H_1 are commable; hence by Proposition 19.73 are commable within focal groups. By Theorem 19.63, there is a faithful focal-universal LC-group G and copci homomorphisms $H \to G \gets H_1$. Viewing these homomorphisms as inclusions, by Proposition 19.65, we obtain that $H = H_1$.

Conversely, suppose that Conjecture 19.88 holds. Let G_1, G_2 be quasi-isometric focal hyperbolic groups of connected type. By Proposition 19.65, they are commable to purely real Heintze groups H_1 and H_2, which are isomorphic by Conjecture 19.88. Hence G_1 and G_2 are commable. $\square$

The results of Section 19.5, or alternatively the more general Gordon–Wilson approach (see Remark 19.91), shows the following evidence in support of Conjecture 19.88.

Proposition 19.90 *Two purely real Heintze groups admit continuous simply transitive isometric actions on the same Riemannian manifold if and only if they are isomorphic.*

Proof If the groups are 1-dimensional, the only possibility is $\mathbf{R}$. Assume they have dimension ≥ 2; then they are focal hyperbolic of connected type and commable, hence isomorphic by the results of [13] (see Corollary 19.71). $\square$

Remark 19.91 The results of Gordon and Wilson [23] show that Proposition 19.90 holds in a much greater generality, namely for purely real simply connected solvable Lie groups (sometimes called real triangulable groups). Indeed, let H_1, H_2 be such groups and X the Riemannian manifold. Then H_1 and H_2 stand as closed subgroups in the isometry group of the homogeneous Riemannian manifold $\mathrm{Isom}(X)$. Gordon and Wilson define a notion of "subgroup in standard position" in $\mathrm{Isom}(X)$. In [23, Theorem 4.3], they show that any real triangulable subgroup is in standard position, and thus H_1 and H_2 are in standard position. In [23, Theorem 1.11], they show that in $\mathrm{Isom}(X)$, all subgroups in standard position are conjugate (and are actually equal, in case $\mathrm{Isom}(X)$ is amenable, see [23, Corollary 1.12]). Thus H_1 and H_2 are isomorphic.

Here are some partial results towards Conjecture 19.88.

- Conjecture 19.88 holds for H when $[H, H]$ is abelian, by results of Xie [63].
- Conjecture 19.88 holds for H when the purely real Heintze group H is co-compact in the group of isometries of a rank 1 symmetric space of non-compact type. This follows from Theorem 19.25 (combined, for instance, with Proposition 19.89).

- If purely real Heintze groups H_1, H_2 of Carnot type (see §19.2.7) are quasi-isometric, then Pansu's Theorem [46, Theorem 2] implies that H_1 and H_2 are isomorphic.
- Pansu's estimates of L^p-cohomology in degree 1 [47] provide useful quasi-isometry invariants.
- Carrasco [9, Cor. 1.9] proves the following: given a Heintze group $H = N \rtimes \mathbf{R}$, let $\mathfrak{n}^{\min}$ be the characteristic subspace relative to the smallest eigenvalue of some dilating element of $\mathbf{R}$, and $H^{\min} = \exp(\mathfrak{n}^{\min}) \rtimes \mathbf{R}$. Then he proves that if purely real Heintze groups H_1, H_2 are quasi-isometric, then $H_1^{\min}$ and $H_2^{\min}$ are isomorphic. He also proves that being of Carnot type is a quasi-isometry invariant among purely real Heintze groups.

Internal QI-classification in the mixed type

In mixed type, we can specify Conjecture 19.87 as follows; we refer to Definitions 19.68 and 19.69.

Conjecture 19.92 Let H be a nonabelian purely real Heintze group, $\varpi > 0$ a positive real number, and q a non-power integer, and define $G = H[\varpi, q]$. If (H', ϖ', q') is another such triple and G and $G' = H'[\varpi', q']$ are quasi-isometric then they are isomorphic.

Proposition 19.93 *Conjecture 19.87 specified to groups of mixed type is equivalent to Conjecture 19.92.*

Proof Suppose that Conjecture 19.87 specified to groups of mixed type holds. If G and G' are given as in Conjecture 19.92, then the validity of Conjecture 19.87 implies that G and G' are commable. By Corollary 19.71, we deduce that G and G' are isomorphic.

Conversely assume that Conjecture 19.92 holds. Let G be as in Conjecture 19.87, of mixed type, and let G' be a focal hyperbolic LC-group, quasi-isometric to G. Then G' is necessarily of mixed type (by Corollary 19.18). By Corollary 19.71, G and G' are respectively commable to groups of the form $H[\varpi, q]$ and $H'[\varpi', q']$, which by the validity of Conjecture 19.92 are isomorphic, so that G and G' are commable. $\square$

The following theorem indicates that a significant part of Conjecture 19.92 holds, and provides a full reduction to the connected type case Conjecture 19.88.

Theorem 19.94 *Let $G = H[\varpi, q]$ and $G' = H[\varpi, q']$, as in Conjecture 19.92, be quasi-isometric. Then the following statements hold:*

(a) [13] H and H' are quasi-isometric;
(b) [13] $\varpi = \varpi'$;

(c) (T. Dymarz [19]) $q = q'$.

In particular, if H satisfies Conjecture 19.88 then $H[\varpi,q]$ satisfies Conjecture 19.92. $\square$

Theorem 19.94 shows that Conjecture 19.92 can be reduced to Conjecture 19.88. However, the proofs of (a) and especially of (c) in [19] suggest that Conjecture 19.92 might be easier than Conjecture 19.88, because its boundary exhibits more rigidity.

Let us indicate an application from [13]. Let X be a homogeneous negatively curved manifold of dimension ≥ 2. For $t > 0$, let $X_{\{t\}}$ be obtained from X by multiplying the Riemannian metric by t^{-1} (thus multiplying the distance by $t^{-1/2}$ and the sectional curvature by t). For instance, $\mathbf{H}^2_{\{t\}}$ is the rescaled hyperbolic plane with constant curvature $-t$.

Theorem 19.95 ([13])   *If k_1,k_2 are integers ≥ 2 and t_1,t_2 are positive real numbers, then $X_{\{t_1\}}[k_1]$ and $X_{\{t_2\}}[k_2]$ are quasi-isometric if and only if k_1,k_2 have a common integral power and $\log(k_1)/t_1 = \log(k_2)/t_2$.* $\square$

In particular, when either $k \geq 2$ or $t > 0$ is fixed, the $X_{\{t\}}[k]$ are pairwise non-quasi-isometric.

The proof indeed consists in proving that if G is a focal hyperbolic LC-group with a continuous proper cocompact isometric action on $X_{\{t\}}[k]$, then $\varpi(G) = c\log(k)/t$, where the constant $c > 0$ only depends on X. In particular, the last statement of the theorem follows from the quasi-isometric invariance of ϖ, while the first statement follows from it as well as Dymarz' invariance of the invariant q, and for the (easier) converse, relies on Proposition 19.70.

19.6.3 Quasi-isometric amenable and non-amenable hyperbolic LC-groups

The external classification can be asked in two (essentially) equivalent but intuitively different ways:

- Which amenable hyperbolic LC-groups are QI to hyperbolic LC-groups of general type?
- Which hyperbolic LC-groups of general type are QI to amenable hyperbolic LC-groups?

These questions are equivalent but they can be specified in different ways. A potential full answer is given by the conjecture below.

Recall from Definition 19.75 that a hyperbolic LC-group is special if and only if it is commable to both amenable and non-amenable LC-groups and quasi-special if it is quasi-isometric to a special hyperbolic group. Such groups have a very peculiar form, see Proposition 19.76 for characterizations.

Conjecture 19.96   Let G be a hyperbolic LC-group. Then G is quasi-isometric to both an amenable and a non-amenable CGLC-group if and only if G is quasi-special hyperbolic.

Remark 19.97   The "if" part of the conjecture is clear. Conjecture 19.96 is a coarse counterpart to [7, Theorem D], which is transcribed here as Case (b) of Theorem 19.1 or as the equivalence (i)⇔(ii) of Proposition 19.76.

Proposition 19.98   *Conjecture 19.86 specified to H of general type is equivalent to Conjecture 19.96.*

Proof   Suppose that Conjecture 19.86 holds for H of general type. Let G be as in Conjecture 19.96; if G is quasi-isometric to the trivalent tree then it is quasi-special; otherwise G is then quasi-isometric, and hence commable by the (partial) validity of Conjecture 19.86, to both focal (not of totally disconnected type) and non-focal hyperbolic LC-groups. We conclude by Proposition 19.76 that G is special.

Conversely assume Conjecture 19.96 holds. Suppose that G, H are quasi-isometric hyperbolic LC-groups with G focal not of totally disconnected type and H of general type. By the validity of Conjecture 19.96, each of G and H is quasi-special, and hence special by Proposition 19.76. Thus again by Proposition 19.76(ii), G and H commable to the isometry group of a rank 1 symmetric space of non-compact type; since non-homothetic rank 1 symmetric spaces of non-compact type are not quasi-isometric, we get the same group for G and H and thus they are commable.   □

We now wish to specify Conjecture 19.96.

Conjecture 19.99   Let $H = N \rtimes \mathbf{R}$ be a purely real Heintze group of dimension ≥ 2. Then either H is special, or H is not quasi-isometric to any vertex-transitive finite valency graph.

Remark 19.100   The special role played by those purely real Heintze that are "accidentally" special makes the conjecture delicate.

Lemma 19.101   *Assume that H is a non-special purely real Heintze group of dimension ≥ 2. Then any CGLC group G quasi-isometric to H is either focal hyperbolic of connected type (as defined in §19.2.5), or is compact-by-(totally disconnected). If moreover H satisfies Conjecture 19.99, then G is focal hyperbolic of connected type.*

Proof If G is focal, its boundary is a sphere, it is of connected type. Otherwise G is of general type. By Theorem 19.25, H is not quasi-isometric to a rank 1 symmetric space of non-compact type and therefore G is compact-by-(totally disconnected). In particular, H is quasi-isometric to a vertex-transitive connected finite valency graph, and this is a contradiction in case Conjecture 19.99 holds.   □

Using a number of results reviewed above, we can relate the two conjectures.

Proposition 19.102 *Conjectures 19.96 and 19.99 are equivalent.*

Proof Assume Conjecture 19.96 holds. Consider H as in Conjecture 19.99, quasi-isometric to a vertex-transitive finite valency graph X. Then H is quasi-isometric to $\mathrm{Isom}(X)$; if the latter is focal, being totally disconnected, it is of totally disconnected type, hence its boundary is totally disconnected, contradicting Corollary 19.18. So $\mathrm{Isom}(X)$ is of general type; the validity of Conjecture 19.96 then implies that H is special.

Conversely assume Conjecture 19.99 holds. Let G be quasi-isometric hyperbolic LC-groups G_1, G_2, with G_1 of general type and G_2 focal; we have to show that G is special. By Corollary 19.22, G_2 is of either totally disconnected or connected type. In the first case, G_2 is special. Since by Proposition 19.76 being special hyperbolic is a quasi-isometry among CGLC-groups, we deduce that G is special. In the second case, by Proposition 19.65, G_2 is commable to a purely real Heintze group H. If by contradiction H is not special, since it satisfies Conjecture 19.99, by Lemma 19.101 G_1 is focal, a contradiction.   □

Remark 19.103 Another restatement of the conjectures is that the class of non-special focal hyperbolic LC-groups is closed under quasi-isometries among CGLC-groups.

Let us now give a more precise conjecture.

Conjecture 19.104 (Pointed sphere Conjecture) Let H be a purely real Heintze group of dimension ≥ 2. Let ω be the H-fixed point in ∂H. If H is not isomorphic to the minimal parabolic subgroup in any simple Lie group of rank one, then every quasi-symmetric self-homeomorphism of ∂H fixes ω.

The justification of the name is that the boundary ∂H naturally comes with a distinguished point; topologically this point is actually not detectable since the sphere is topologically homogeneous, but the quasi-symmetric structure ought to distinguish this point, at the notable exception of the cases for which it is known not to do so.

In turn, the pointed sphere Conjecture implies Conjecture 19.99. More precisely:

Figure 19.2 A pointed 2-sphere: the boundary of a generic hyperbolic semidirect product $\mathbf{R}^2 \rtimes \mathbf{R}$.

Proposition 19.105 *Let H satisfy the pointed sphere Conjecture. Then it also satisfies Conjecture 19.99.*

Proof Let H satisfy the pointed sphere Conjecture. If H is minimal parabolic, both conjectures are tautological, so assume the contrary. Assume by contradiction that H is QI to a vertex-transitive graph X of finite valency. The action of $\mathrm{Isom}(X)$ on its boundary is conjugate by a quasi-symmetric map to a quasi-symmetric action on ∂H. By assumption, this action fixes a point. So $\mathrm{Isom}(X)$ is focal of totally disconnected type, contradicting Corollary 19.18. $\square$

Theorem 19.106 (Pansu, Cor. 6.9 in [45]) *Consider a purely real Heintze group $N \rtimes \mathbf{R}$, not of Carnot type (see Definition 19.23). Suppose in addition that the action of $\mathbf{R}$ on the Lie algebra $\mathfrak{n}$ is diagonalizable. Then H satisfies the pointed sphere Conjecture.*

In case N is abelian, the assumption is that the contracting action of $\mathbf{R}$ is not scalar (in order to exclude minimal parabolic subgroups in $\mathrm{PO}(n,1) = \mathrm{Isom}(\mathbf{H}^n_{\mathbf{R}})$).

Theorem 19.107 (Carrasco [9]) *The pointed sphere Conjecture holds for all purely real Heintze groups that are not of Carnot type.*

In the particular case of Heintze groups of the form $H = N \rtimes \mathbf{R}$ holds with N abelian, this was previously proved by X. Xie [63] with different methods.

Theorem 19.106 covers all cases when H has dimension 3 (i.e. N has dimension 2), with the exception of the semidirect product $\mathbf{R}^2 \rtimes \mathbf{R}$ with action by the 1-parameter group $(U_t)_{t \in \mathbf{R}}$ where $U_t = \begin{pmatrix} e^t & te^t \\ 0 & e^t \end{pmatrix}$, which is dealt with specifically in [62].

In Xie's Theorem (the pointed sphere conjecture for $H = N \rtimes \mathbf{R}$ with N abelian), the most delicate case is that of an action with scalar diagonal part and non-trivial unipotent part; it is not covered by Pansu's Theorem 19.106. In turn, the first examples of purely real Heintze groups not covered by Xie's theorem are semidirect products $\mathrm{Hei}_3 \rtimes \mathbf{R}$, where Hei_3 is the Heisenberg group, with the exclusion of the minimal parabolic subgroup $\mathrm{Carn}(\mathrm{Hei}_3)$ (see Remark 19.24) in $\mathrm{PU}(2,1) = \mathrm{Isom}(\mathbf{H}_{\mathbf{C}}^2)$. If the action on $(\mathrm{Hei}_3)_{ab} = \mathrm{Hei}_3/[\mathrm{Hei}_3, \mathrm{Hei}_3]$ has two distinct eigenvalues, then Pansu's Theorem 19.106 applies. The remaining case of the pointed sphere Conjecture for $\dim(H) = 4$ is the one for which the action on $(\mathrm{Hei}_3)_{ab}$ is not diagonalizable but has scalar diagonal part, and is covered by Carrasco's theorem.

The remaining cases of the pointed sphere conjecture are now those Heintze groups of Carnot type $\mathrm{Carn}(N)$. Note that if N is abelian or is a generalized $(2n+1)$-dimensional Heisenberg group Hei_{2n+1} (characterized by the fact its 1-dimensional centre equals its derived subgroup), then $\mathrm{Carn}(N)$ is minimal parabolic. Therefore, the first test-cases would be when N is a non-abelian 4-dimensional simply connected nilpotent Lie group: there are two such Lie groups up to isomorphism: the direct product $\mathrm{Hei}_3 \times \mathbf{R}$ and the filiform Lie group Fil_4, which can be defined as $(\mathbf{R}[x]/x^3) \rtimes \mathbf{R}$ where $t \in \mathbf{R}$ acts by multiplication by $(1+x)^t = 1 + tx + \frac{t(t-1)}{2}x^2$.

19.6.4 Conjecture 19.88 and quasi-symmetric maps

An important tool, given a geodesic hyperbolic space and $\omega \in \partial X$, is the visual parabolic metric on the "parabolic boundary" $\partial X \smallsetminus \{\omega\}$. Note that unlike the visual metric, these are generally unbounded. It shares the property that any quasi-isometric embedding $f : X \to Y$ induces a quasi-symmetric embedding $\partial X \smallsetminus \{\omega\} \to \partial Y \smallsetminus \{\bar{f}(\omega)\}$. This follows from the corresponding fact for visual metrics and the fact that the embedding $\partial X \smallsetminus \{\omega\} \subset \partial X$ is quasi-symmetric [52, Section 5].

The parabolic boundaries and the quasi-symmetric homeomorphisms between those are therefore important tools in the study of quasi-isometry classification and notably Conjecture 19.88. Let us include the following simple lemma.

Lemma 19.108 *Let G, G' be focal hyperbolic LC-groups and ω, ω' the fixed points in their boundary. Suppose that there exists a quasi-isometry $f : G \to G'$. Then there exists a quasi-isometry $u : G \to G'$ such that $\bar{u}$ maps ω to ω'.*

Proof If $\bar{f}(\omega) = \omega'$ there is nothing to do. Otherwise, since G is transitive on $\partial G \smallsetminus \{\omega\}$, there is a left translation v on G such that $\bar{v}(\bar{f}^{-1}(\omega')) \neq \bar{f}^{-1}(\omega')$. Then $\bar{f}\bar{v}\bar{f}^{-1} = \overline{fvf^{-1}}$ is a quasi-symmetric self-homeomorphism of $\partial G'$ not fixing ω', so the group of quasi-isometries generated by fvf^{-1} and by left translations of G' is transitive on $\partial G'$. Thus after composition of f by a suitable quasi-isometry in this group, we obtain a quasi-isometry u such that $\bar{u}$ maps ω is mapped to ω'. □

If G is a focal hyperbolic LC-group of connected type, with no non-trivial compact normal subgroup, and N is its connected nilpotent radical, then the action of N on $\partial G \setminus \{\omega\}$ is simply transitive and the visual parabolic metric induces a left-invariant distance on N. When G is of Carnot type (see §19.2.7), this distance is equivalent (in the bilipschitz sense) to the Carnot–Caratheodory metric. In general the author deos not know how to describe it directly on N.

19.6.5 Further aspects of the QI classification of hyperbolic LC-groups

This final subsection is much smaller than it should be. It turns around the general question: how is the structure of a hyperbolic LC-group of general type related to the topological structure of its boundary?

The following theorem is closely related, in the methods, to the QI classification of groups quasi-isometric to the hyperbolic plane. It should be attributed to the same authors, namely Tukia, Gabai and Casson–Jungreis, to which we need to add Hinkkanen in the non-discrete case.

Theorem 19.109 *Let G be a hyperbolic LC-group. Then ∂G is homeomorphic to the circle if and only if G has a continuous proper isometric cocompact action on the hyperbolic plane.*

Proof The best-known (and hardest!) case of the theorem is when G is discrete. See for instance [30, Theorem 5.4]. It immediately extends to the case when G is compact-by-discrete (i.e. has an open normal compact subgroup).

Assume now that G is not compact-by-discrete. Consider the action $\alpha : G \to$ Homeo(∂G) on its boundary by quasi-symmetric self-homeomorphisms. Endow Homeo(∂G) with the compact-open topology, which (for some choice of metric on ∂G) is the topology of uniform convergence; the function α is continuous. Since G is non-elementary hyperbolic, the kernel of α is compact, and therefore by assumption the image $\alpha(G)$ is non-discrete. We can then apply Hinkkanen's theorem about non-discrete groups of quasi-symmetric self-homeomorphisms [28]: after fixing a homeomorphism identifying ∂G with the projective line, $\alpha(G)$ is contained in a conjugate of $\mathrm{PGL}_2(\mathbf{R})$. Thus after conjugating, we can view α as a continuous homomorphism with compact kernel from G to $\mathrm{PGL}_2(\mathbf{R})$. It follows that $H = G/\mathrm{Ker}(\alpha)$ is a Lie group. Since G is not compact-by-discrete, we deduce that the identity component H° is non-compact.

If H is focal, then it acts continuously properly isometrically cocompactly on a millefeuille space $X[k]$; the boundary condition implies that $k = 1$ (so $X[k] = X$) and X is 2-dimensional, hence is the hyperbolic plane. Otherwise H is of general type, and since H° is non-compact we deduce that H is a virtually connected Lie group; being of general type it is not amenable, hence not solvable and we deduce that the image of H in $\mathrm{PGL}_2(\mathbf{R})$ has index at most 2. Thus in all cases G admits a continuous proper cocompact isometric action on the hyperbolic plane. $\qquad\qquad\qquad\Box$

Conjecture 19.110 Let G be a compactly generated locally compact group. Suppose that G is quasi-isometric to a negatively curved homogeneous Riemannian manifold X. Then G is compact-by-Lie.

Proposition 19.111 *Conjecture 19.110 is implied by Conjecture 19.96.*

Proof Suppose Conjecture 19.96 holds. Let G be as in Conjecture 19.110; its boundary is therefore a sphere. If G is focal, then it is compact-by-Lie by Proposition 19.8. Otherwise it is of general type. Since $\mathrm{Isom}(X)$ contains a cocompact solvable group by [27, Proposition 1], we deduce from Conjecture 19.96 that G has a continuous proper isometric action on a symmetric space, which implies that it is compact-by-Lie. $\qquad\qquad\qquad\Box$

It is a general fact that if the boundary of a non-focal hyperbolic LC-group contains an open subset homeomorphic to $\mathbf{R}^n$ for some $n \geq 0$, then the boundary is homeomorphic to the n-sphere: see [30, Theorem 4.4] (which deals with the case of finitely generated groups and $n \geq 2$ but the proof works without change in this more general setting).

Question 19.112 Let G be a hyperbolic locally compact group whose boundary is homeomorphic to a d-sphere. Is G necessarily compact-by-Lie?

A positive answer to Question 19.112 would imply Conjecture 19.110, but, on the other hand would only be a very partial answer to determining which hyperbolic LC-groups admit a sphere as boundary, boiling down the question to the case of discrete groups. At this point, Question 19.112 is open for all $d \geq 4$. Still, it has a positive answer for $d \leq 3$ (and similarly Conjecture 19.110 has a positive answer for $\dim(X) \leq 4$), by the positive solution to the Hilbert–Smith conjecture in dimension $d \leq 3$, which is due to Montgomery–Zippin for $d = 1, 2$ and J. Pardon for $d = 3$.

Interestingly, in the Losert characterization of CGLC-groups with polynomial growth [35], the most original part of the proof precisely consists in showing that such a group is compact-by-Lie.

19.7 Beyond the hyperbolic case

Here is a far reaching generalization of Conjecture 19.88. We call *real triangulable Lie group* a Lie group isomorphic to a closed connected subgroup of the group of upper triangular real matrices in some dimension. Equivalently, this is a simply connected Lie group whose Lie algebra is solvable, and for which for each x in the Lie algebra, the adjoint operator $\mathfrak{ad}(x)$ has only real eigenvalues. The following conjecture appears in research statements the author wrote around 2010 as well as earlier talks.

Conjecture 19.113   Two real triangulable groups G_1, G_2 are quasi-isometric if and only if they are isomorphic.

Remark 19.91 provides some evidence: if G_1, G_2 admit isometric left-invariant Riemannian structures (or, equivalently, admit simply transitive continuous isometric actions on the same Riemannian manifold), then they are isomorphic. (This fails for more general simply connected solvable Lie groups: for instance both $\mathbf{R}^3$ and the universal covering of $\mathrm{Isom}(\mathbf{R}^2)$ admit such actions on the Euclidean 3-space.)

Let us mention the important result that for a real triangulable group, the dimension $\dim(G)$ is a quasi-isometry invariant, by a result of J. Roe [48].

A positive answer to Conjecture 19.113 would entail many consequences other than the hyperbolic case, including the (internal) quasi-isometry classification of polycyclic groups. A particular case is the case of polynomial growth.

Conjecture 19.114   Two simply connected nilpotent Lie groups G_1, G_2 are quasi-isometric if and only if they are isomorphic.

The latter conjecture, specialized to those simply connected nilpotent Lie groups admitting cocompact lattices, is equivalent to a more familiar (and complicated) conjecture concerning finitely generated nilpotent groups, namely: two finitely generated nilpotent groups are quasi-isometric if and only if they have isomorphic real Malcev closure (when a nilpotent Γ is not torsion-free and T is its torsion subgroup, its real Malcev closure is by definition the real Malcev closure of Γ/T). This is first asked by Farb and Mosher in [20] (before their Corollary 10).

Let us mention that two finitely generated nilpotent groups are commensurable if and only they have isomorphic *rational* Malcev closures. Thus any pair of non-isomorphic finite-dimensional nilpotent Lie algebras $\mathfrak{g}_1, \mathfrak{g}_2$ such that $\mathfrak{g}_1 \otimes \mathbf{R}$ and $\mathfrak{g}_2 \otimes \mathbf{R}$ are isomorphic provides non-commensurable quasi-isometric finitely generated nilpotent groups, namely $G_1(\mathbf{Z})$ and $G_2(\mathbf{Z})$, where G_i is the unipotent $\mathbf{Q}$-group associated to $\mathfrak{g}_i$, and some $\mathbf{Q}$-embedding $G_i \subset \mathrm{GL}_n$ is fixed. The smallest examples, according to the classification, are 6-dimensional. An elegant classical example is, denoting by Hei_{2n+1} the $(2n+1)$-dimensional Heisenberg group $(2n+1 \geq 3)$ and k is a positive non-square integer, $\mathrm{Hei}_{2n+1}(\mathbf{Z}) \times \mathrm{Hei}_{2n+1}(\mathbf{Z})$ and $\mathrm{Hei}_{2n+1}(\mathbf{Z}[\sqrt{k}])$ are not commensurable, although they are both isomorphic to cocompact lattices in $\mathrm{Hei}_{2n+1}(\mathbf{R})^2$.

The main two results known in this direction are, in the setting of Conjecture 19.114:

- the real Carnot Lie algebras (see Remark 19.24) $\mathrm{Carn}(G_1)$ and $\mathrm{Carn}(G_2)$ are isomorphic (Pansu [46]);
- (assuming that G_1 and G_2 have lattices) the real cohomology algebras of G_1 and G_2 are isomorphic (Sauer [49], improving Shalom's result [51] that the Betti numbers are equal).

Pansu's result supersedes a yet simpler result: for a simply connected nilpotent Lie group G, both the dimension d and the degree of polynomial growth rate δ are quasi-isometry invariant. This is obvious for the growth, while for the dimension this follows from Pansu's earlier result that the asymptotic cone is homeomorphic to $\mathbf{R}^{\dim G}$ [44]. This covers all cases up to dimension 4, since the only possible (d, δ) are (i, i) $(0 \leq i \leq 4)$, $(3, 4)$, $(4, 5)$, $(4, 7)$, are achieved by a single simply connected nilpotent Lie group $((i, i)$ are the abelian ones, $(3, 4)$ is the 3-dimensional Heinsenberg H_3 and $(4, 5)$ is its product with $\mathbf{R}$, and $(4, 7)$ is the filiform 4-dimensional Lie group, of nilpotency length 3).

In dimension 5, there are 9 simply connected nilpotent Lie groups up to isomorphism. They are denoted $L_{5,i}$ in [24] with $1 \leq i \leq 9$, and only two are not Carnot: $L_{5,5}$ and $L_{5,6}$. Their degrees of polynomial growth rate are 5, 6, 8, 6, 8, 11, 11, 7, 10 respectively. This only leaves three pairs not determined by the quasi-isometry invariance of (d, δ); the first is $L_{5,2}$ and $L_{5,4}$, where $L_{5,2} \simeq H_3 \times \mathbf{R}^2$ and $L_{5,4} \simeq H_5$ are distinguished by Pansu's theorem. The other two pairs we discuss below are $L_{5,3}$ and $L_{5,5}$ on the one hand, $L_{5,6}$ and $L_{5,7}$ on the other hand.

- For $L_{5,3}$ and $L_{5,5}$, the Betti numbers are $(1, 3, 4, 4, 3, 1)$, and $\mathrm{Carn}(L_{5,5}) \simeq L_{5,3}$ (they have growth exponent 8); thus they are distinguished neither by Pansu's theorem, nor by Shalom's theorem. A basis for the Lie algebra $\mathfrak{l}_{5,i}$ for $i = 3, 5$ is given as (a, b, c, d, e) with for both, non-zero brackets $[a, b] = c$, $[a, c] = e$, and for $\mathfrak{l}_{5,5}$ the additional non-zero bracket $[b, d] = e$.

Note that $\mathfrak{l}_{5,3}$ is the product of the 1-dimensional Lie algebra and the 4-dimensional filiform Lie algebra.

However, they are distinguished by Sauer's theorem: indeed, the cup product $S^2(H^2(\mathfrak{l}_{5,i})) \to H^4(\mathfrak{l}_{5,i})$ has rank 1 (in the sense of linear algebra) for $\mathfrak{l}_{5,3}$ and 2 for $\mathfrak{l}_{5,5}$. Thus the cohomology algebras are not isomorphic as graded algebras.

- For both $L_{5,7}$ and $L_{5,6}$, the Betti numbers are $(1,2,3,3,2,1)$. Moreover, we have $\mathrm{Carn}(L_{5,6}) \simeq L_{5,7}$ (they have growth exponent 11); thus they are distinguished neither by Pansu's theorem, nor by Shalom's theorem. A basis for the Lie algebra $\mathfrak{l}_{5,i}$ for $i = 7, 6$ is given as (a,b,c,d,e) with for both, non-zero brackets $[a,b] = c$, $[a,c] = d$, $[a,d] = e$, and for $\mathfrak{l}_{5,6}$ the additional non-zero bracket $[b,c] = e$. The Lie algebra $\mathfrak{l}_{5,7}$ is the standard filiform 5-dimensional Lie algebra.

Actually the graded cohomology algebras are isomorphic: both have a basis

$$(1, a_1, a_2, b_5, b_6, b_7, c_8, c_9, c_{10}, d_{13}, d_{14}, e_{15})$$

so that the product is commutative with unit 1, H^1 has basis (a_1, a_2), H_2 has basis (b_6, b_7, b_8), etc. and the non-zero products of basis elements except the unit are

- $a_i d_{15-i} = b_j c_{15-j} = e_{15}$;
- $a_1 b_7 = -2c_8$, $a_2 b_6 = c_8$, $a_2 b_7 = c_9$;
- $b_6 b_7 = d_{13}$, $b_7 b_7 = -2d_{14}$

(it is commutative because the product of any two elements of odd degree is zero). Thus $L_{5,7}$ and $L_{5,6}$, although non-isomorphic, are not distinguishable by either Pansu or Sauer's theorem. Thus they seem to be the smallest open case of Conjecture 19.114 (note that being rational, they have lattices; the rational structure being unique up to isomorphism, these lattices are uniquely defined up to abstract commensurability). Still, these can be distinguished by the adjoint cohomology: $H^1(\mathfrak{g}, \mathfrak{g})$ (the space of derivations modulo inner derivations) has dimension 5 for $\mathfrak{l}_{5,7}$ and 4 for $\mathfrak{l}_{5,6}$; however it is not known if this dimension is a quasi-isometry invariant for simply connected nilpotent Lie groups.

It would be interesting to show that the adjoint cohomology of the Lie algebra is a quasi-isometry invariant of simply connected nilpotent Lie groups (at least as a real graded vector space): indeed, as mentioned by Magnin who did comprehensive dimension computations in dimension ≤ 7 [36], it is much more effective than the trivial cohomology to distinguish non-isomorphic nilpotent Lie algebras.

Indeed, the classification of 6-dimensional nilpotent Lie algebras provides 34 isomorphism classes over the field of real numbers, 26 of which are not 2-nilpotent. Among those, 13 are Carnot (over the reals); among those Lie algebras with the same Carnot Lie algebra, all have the same Betti numbers (i.e. the same trivial cohomology as a graded vector space) with two exceptions, which provide, by the way, the smallest examples for which Shalom's result improves Pansu's (Shalom [51, §4.1] provides a 7-dimensional example, attributed to Y. Benoist).

For the interested reader, we list the 6-dimensional real nilpotent Lie algebras, referring to [24] for definitions, according to their Carnot Lie algebra. Since the Betti numbers and adjoint cohomology computations are done in [36] with a different nomenclature, we give in each case both notations.

- Nilpotency length 5:

    - Carnot: $\mathfrak{l}_{6,18}$ (standard filiform). It is the Carnot Lie algebra of three Lie algebras: $\mathfrak{l}_{6,i}$ for $i = 18, 17, 15$. In [36], they are denoted $\mathfrak{g}_{6,j}$ with $j = 16, 17, 19$ (in the same order). All have the Betti numbers $(1, 2, 3, 4, 3, 2, 1)$. However they can be distinguished by adjoint cohomology in degree 1 (of dimension 6, 5, 4 respectively).

    - Carnot: $\mathfrak{l}_{6,16}$. It is the Carnot Lie algebra of two Lie algebras: $\mathfrak{l}_{6,i}$ for $i = 16, 14$. Both have the Betti numbers $(1, 2, 2, 2, 2, 2, 1)$.

- Nilpotency length 4:

    - Carnot: $\mathfrak{l}_{6,7}$ (product of a 5-dimensional standard filiform with a 1-dimensional abelian one). It is the Carnot Lie algebra of five Lie algebras: $\mathfrak{l}_{6,i}$ for $i = 7, 6, 11, 12, 13$. In [36], they are denoted $\mathfrak{g}_{5,5} \times \mathbf{R}$, $\mathfrak{g}_{5,6} \times \mathbf{R}$, and $\mathfrak{g}_{6,j}$ for $j = 12, 11, 13$. The first four have the Betti numbers $(1, 3, 5, 6, 5, 3, 1)$, while the last one has the Betti numbers $(1, 3, 4, 4, 4, 3, 1)$ and thus the corresponding group can be distinguished by Shalom's theorem (incidentally, the first four are metabelian while the last one is not).

    - Each of the last three Carnot Lie algebras of nilpotency length 4 are only Carnot Lie algebras of itself. They are denoted $\mathfrak{l}_{6,21}(\varepsilon)$ for $\varepsilon = 0, 1, -1$; in [36] they are denoted $\mathfrak{g}_{6,14}$ and $\mathfrak{g}_{6,15}$ (twice, the last two having isomorphic complexifications).

- Nilpotency length 3:

    - Carnot: $\mathfrak{l}_{6,9}$. It is the Carnot Lie algebra of four Lie algebras (three over the complex numbers): $\mathfrak{l}_{6,9}$, $\mathfrak{l}_{6,24}(1)$, $\mathfrak{l}_{6,24}(-1)$ and $\mathfrak{l}_{6,24}(0)$. (In [36], these are $\mathfrak{g}_{5,4} \times \mathbf{R}$, $\mathfrak{g}_{6,5}$ (twice) and $\mathfrak{g}_{6,8}$.) All have the same Betti numbers

$(1,3,5,6,5,3,1)$. The dimension of the zeroth and first adjoint cohomology distinguishes them, except the two middle ones having isomorphic complexification.

- Carnot: $\mathfrak{l}_{6,25}$. It is the Carnot Lie algebra of two Lie algebras: $\mathfrak{l}_{6,i}$ for $i = 25,23$. (In [36], these are $\mathfrak{g}_{6,j}$ for $j = 6,7$.) They both have the Betti numbers $(1,3,6,8,6,3,1)$. The dimension of the first adjoint cohomology distinguishes them.

- Carnot: $\mathfrak{l}_{6,3}$. It is the Carnot Lie algebra of three Lie algebras: $\mathfrak{l}_{6,i}$ for $i = 3,5,10$. (In [36], these are $\mathfrak{g}_4 \times \mathbf{R}^2$, $\mathfrak{g}_{5,3} \times \mathbf{R}$ and $\mathfrak{g}_{6,2}$.) The first two have the Betti numbers $(1,4,7,8,7,4,1)$, while the last one has the Betti numbers $(1,4,6,6,6,4,1)$. The first two, still, can be distinguished by adjoint cohomology in degree 0 (i.e. they have centres of distinct dimension), and also in degree one.

- Each of the last four Carnot Lie algebras of nilpotency length three are only Carnot Lie algebras of itself. They are denoted $\mathfrak{l}_{6,19}(0)$, $\mathfrak{l}_{6,19}(1)$, $\mathfrak{l}_{6,19}(-1)$, and $\mathfrak{l}_{6,20}$ and are called in [36] $\mathfrak{g}_{6,4}$, $\mathfrak{g}_{6,9}$ for the two middle ones which have isomorphic complexification, and $\mathfrak{g}_{6,10}$. They have Betti numbers $(1,3,6,8,6,3,1)$ for the first one and $(1,3,5,6,5,3,1)$ for the last three ones.

• Nilpotency length ≤ 2. They are all Carnot and thus distinguished by Pansu's theorem. They are denoted $\mathfrak{l}_{6,26}$, $\mathfrak{l}_{6,22}(\varepsilon)$ for $\varepsilon = 0,1,-1$, and $\mathfrak{l}_{6,i}$ for $i = 8,4,2,1$, and in [36] they are denoted $\mathfrak{g}_{6,3}$, $\mathfrak{g}_{6,1}$, $\mathfrak{g}_3 \times \mathfrak{g}_3$ (twice), $\mathfrak{g}_{5,2} \times \mathbf{R}$, $\mathfrak{g}_{5,1} \times \mathbf{R}$, $\mathfrak{g}_3 \times \mathbf{R}^3$ and $\mathbf{R}^6$.

Note that in all the cases above distinguished neither by Pansu nor by Shalom's theorem, we have not computed the cup product in cohomology and thus have not checked in which cases they are distinguished by Sauer's theorem.

Acknowledgements. The author is grateful to thank Tullia Dymarz, Pierre Pansu and Romain Tessera for useful discussions, comments and corrections.

References for this chapter

[1] H. Abels. Specker-Kompaktifizierungen von lokal kompakten topologischen Gruppen. (German) Math. Z. 135 (1973/74), 325–361.

[2] H. Abels. On a problem of Freudenthal's. Comp. Math. 35(1977), no. 1, 39–47.

[3] S. Antonyan. Characterizing maximal compact subgroups. Arch. Math. 98 (2012), no. 6, 555–560.

[4] H. Bass and R. Kulkarni. Uniform tree lattices. J. Amer. Math. Soc. 3 (1990), 843–902.

[5] B. Bekka, P. de la Harpe and A. Valette. Kazhdan's Property (T). New Math. Monographs 11, Cambridge Univ. Press, Cambridge, 2008.

[6] S. Buyalo and V. Schroeder. Elements of asymptotic geometry. European Math. Soc. 2007.

[7] P-E. Caprace, Y. Cornulier, N. Monod and R. Tessera. Amenable hyperbolic groups. J. Eur. Math. Soc. (JEMS) 17 (2015), no. 11, 2903–2947.

[8] M. Carette. Commability of groups quasi-isometric to trees. ArXiv math/1312.0278 (2013).

[9] M. Carrasco Piaggio. Orlicz spaces and the large scale geometry of Heintze groups. Math. Ann. 368 (2017), no. 1-2, 433–481.

[10] A. Casson and D. Jungreis. Convergence groups and Seifert fibered 3-manifolds. Inv. Math. 118 (1994), no. 1, 441–456.

[11] R. Chow. Groups quasi-isometric to complex hyperbolic space. Trans. Amer. Math. Soc. 348 (1996), 1757–1770.

[12] Y. Cornulier. Dimension of asymptotic cones of Lie groups. Journal of Topology 1 (2008), 342–361.

[13] Y. Cornulier. Commability and focal locally compact groups. Indiana Univ. Math. J. 64(1) (2015), 115–150.

[14] Y. Cornulier and P. de la Harpe. Metric geometry of locally compact groups. EMS Tracts in Mathematics, 25. European Mathematical Society (EMS), Zürich, 2016.

[15] Y. Cornulier and R. Tessera. Contracting automorphisms and L^p-cohomology in degree one. Ark. Mat. 49 (2011), no. 2, 295–324.

[16] W. Dicks and M. Dunwoody. Groups acting on graphs. Cambridge University Press, Cambridge, 1989.

[17] M.J. Dunwoody. The accessibility of finitely presented groups. Inventiones Mathematicae 81 (1985), no. 3, 449–457.

[18] M.J. Dunwoody. An inaccessible group. Geometric group theory, Vol. 1 (Sussex, 1991), pp. 75–78, London Math. Soc. Lecture Note Ser., 181, Cambridge University Press, Cambridge, 1993.

[19] T. Dymarz. Quasisymmetric maps of boundaries of amenable hyperbolic groups. Indiana Univ. Math. J. 63(2), (2014), 329–343.

[20] B. Farb and L. Mosher. Problems on the geometry of finitely generated solvable groups. In "Crystallographic Groups and their Generalizations (Kortrijk, 1999)", Cont. Math. 262, Amer. Math. Soc. (2000).

[21] K. Fujiwara and K. Whyte. A note on spaces of asymptotic dimension one. Algebr. Geom. Topol. 7 (2007), 1063–1070.

[22] D. Gabai. Convergence groups are Fuchsian groups, Ann. of Math. 136 (1992), no. 3, 447–510.

[23] C. S. Gordon and E. N. Wilson. Isometry groups of Riemannian solvmanifolds. Trans. Amer. Math. Soc. 307 (1988), no. 1, 245–269.

[24] W. de Graaf. Classification of 6-dimensional nilpotent Lie algebras over fields of characteristic not 2. Journal of Algebra 309 (2007), 640–653.

[25] M. Gromov. Hyperbolic groups, Essays in group theory, Math. Sci. Res. Inst. Publ., vol. 8, Springer, New York, (1987), 75–263.

[26] U. Hamenstädt. Theorie von Carnot-Carathéodory Metriken und ihren Anwendungen, Dissertation, Rheinische Friedrich-Wilhelms-Univ. Bonn, 1986. Bonner Mathematische Schriften 180, 1987.

[27] E. Heintze. On homogeneous manifolds of negative curvature, Math. Ann. 211 (1974), 23–34.

[28] A. Hinkkanen. The structure of certain quasisymmetric groups. Mem. Amer. Math. Soc. 83 (1990), no. 422.

[29] C.H. Houghton. Ends of locally compact groups and their coset spaces. J. Australian Math. Soc. 17, 274–284.

[30] I. Kapovich and N. Benakli. Boundaries of hyperbolic groups. In: Combinatorial and geometric group theory (New York, 2000/Hoboken, NJ, 2001), 39–93, Contemp. Math. 296, Amer. Math. Soc., Providence, RI, 2002.

[31] B. Kleiner and B. Leeb. Rigidity of quasi-isometries for symmetric spaces and Euclidean buildings. Publ. Math. IHES 86 (1997), no. 1, 115–197.

[32] B. Kleiner and B. Leeb. Induced quasi-actions: A remark. Proc. Amer. Math. Soc. 137 (2009), 1561–1567.

[33] J-L. Koszul. Lectures on group of transformations. Tata Institute of Fundamental Research, 1965.

[34] B. Krön and R. Möller. Analogues of Cayley graphs for topological groups. Math. Z. 258 (2008), no. 3, 637–675.

[35] V. Losert. On the structure of groups with polynomial growth. Math. Z. 195 (1987), 109–117.

[36] L. Magnin. Adjoint and trivial cohomologies of nilpotent complex Lie algebras of dimension ≤ 7. Int. J. Math. Math. Sci. 2008, Art. ID 805305, 12 pp.

[37] N. Monod. Continuous bounded cohomology of locally compact groups, Lecture Notes in Mathematics, vol. 1758, Springer-Verlag, Berlin, 2001.

[38] D. Montgomery and L. Zippin, Topological transformation groups, New York, Interscience Publishers, Inc., 1955.

[39] L. Mosher, M. Sageev and K. Whyte. Maximally symmetric trees. Geom. Dedicata 92(1) (2002), 195–233.

[40] L. Mosher, M. Sageev and K. Whyte. Quasi-actions on trees I. Bounded valence. Ann. of Math. 158 (2003), 115–164.

[41] L. Mosher, M. Sageev and K. Whyte. Quasi-actions on trees II: Finite depth Bass–Serre trees. Mem. Amer. Math. Soc. 214 (2011), no. 1008.

[42] G. Mostow. Self-adjoint groups. Ann. of Math. 62(1) (1955), 44–55.

[43] G. Mostow. Quasi-conformal mappings in n-space and the rigidity of the hyperbolic space forms. Publ. Math. IHES 34 (1968), 53–104.

[44] P. Pansu. Croissance des boules et des géodésiques fermées dans les nilvariétés. Ergodic Theory Dynam. Systems 3 (1983), no. 3, 415–445.

[45] P. Pansu. Dimension conforme et sphère à l'infini des variétés à courbure négative, Ann. Acad. Sci. Fenn. Ser. A I Math. 14 (1989), no. 2, 177–212.

[46] P. Pansu. Métriques de Carnot-Carathéodory et quasiisométries des espaces symétriques de rang un. Ann. of Math. (1989), 1–60.

[47] P. Pansu. Cohomologie L^p en degré 1 des espaces homogènes. Potential Anal. 27 (2007), 151–165.

[48] J. Roe. Coarse cohomology and index theory on complete manifolds. Mem. Amer. Math. Soc. 104, 1993.

[49] R. Sauer. Homological invariants and quasi-isometry. Geom. Funct. Anal. 16(2) (2006), 476–515.

[50] J-P. Serre, Arbres, amalgames, SL_2. Astérisque 46, Soc. Math. France 1977.

[51] Y. Shalom. Harmonic analysis, cohomology, and the large-scale geometry of amenable groups. Acta Math. 192 (2004), 119–185.

[52] N. Shanmugalingam and Xiangdong Xie. A rigidity property of some negatively curved solvable Lie groups. Comment. Math. Helv. 87(4) (2012), 805–823.

[53] E. Specker. Endenverbande von Raumen und Gruppen. Math. Ann. 122 (1950), 167–174.

[54] J. Stallings. On torsion free groups with infinitely many ends. Ann. of Math. 88 (1968) 312–334.

[55] R. Tessera. Large scale Sobolev inequalities on metric measure spaces and applications. Rev. Mat. Iberoam. 24 (2008), no. 3, 825–864.

[56] R. Tessera. Vanishing of the first reduced cohomology with values in a L^p-representation. Ann. Inst. Fourier 59 no. 2 (2009), 851–876.

[57] C. Thomassen and W. Woess. Vertex-transitive graphs and accessibility. J. Combin. Theory, Ser. B 58 (1993), no. 2, 248–268.

[58] P. Tukia. On quasiconformal groups. Journal d'Analyse Mathématique 46 (1986), no. 1, 318–346.

[59] P. Tukia. Homeomorphic conjugates of Fuchsian groups. Journal für die reine und angewandte Mathematik (Crelles Journal) 391 (1988), 1–54.

[60] K. Whyte. Amenability, bi-Lipschitz equivalence, and the von Neumann conjecture. Duke Math. J. 99 (1999), no. 1, 93–112.

[61] K. Whyte. The large scale geometry of the higher Baumslag–Solitar groups. Geom. Funct. Anal. 11(6) (2001), 1327–1343.

[62] Xiangdong Xie. Quasisymmetric maps on the boundary of a negatively curved solvable Lie group. Math. Ann. 353(3) (2012), 727–746.

[63] Xiangdong Xie. Large scale geometry of negatively curved $\mathbf{R}^n \rtimes \mathbf{R}$. Geom. Topol. 18 (2014), 831–872.

20

Future directions in locally compact groups: a tentative problem list

Pierre-Emmanuel Caprace and Nicolas Monod

She went on saying to herself, in a dreamy sort of way, "Do cats eat bats? Do cats eat bats?" and sometimes, "Do bats eat cats?" for, you see, as she couldn't answer either question, it didn't much matter which way she put it.
(Lewis Carroll, *Alice's Adventures in Wonderland*, 1865.)

Abstract

The recent progress on locally compact groups surveyed in this book also reveals the considerable extent of the unexplored territories. We conclude this volume by mentioning a few open problems related to the material covered in the text and that we consider important at the time of this writing.

20.1	Chabauty limits	344
20.2	p-adic Lie groups	344
20.3	Profinite groups	345
20.4	Contraction groups	345
20.5	Compactly generated simple groups	346
20.6	Lattices	348
20.7	Commensurated subgroups and commensurators	349
20.8	Unitary representations and C*-simplicity	350
20.9	Elementary groups	351
20.10	Galois groups	353
	References for this chapter	353

We shall group problems along the themes indicated in the above table of contents; each problem is briefly discussed and accompanied by a list of relevant references for further reading.

20.1 Chabauty limits

Recall that the collection **Sub**(G) of all closed subgroups of a locally compact group G carries a natural topology, the **Chabauty topology**, for which it is a compact space.

Problem 20.1.1 *Let G be a locally compact group. Is the collection of closed amenable subgroups of G a closed subset of the Chabauty space* **Sub**(G)?

The answer is positive for a large number of natural examples, see [11] (which contains a detailed discussion of the problem) and [41]. A theorem of P. Wesolek ensures moreover that if G is second countable, then the set of closed amenable subgroups of G is Borel, see [1, Theorem A.1]. We don't know the answer to Problem 20.1.1 in the case of the Neretin group (see Chapter 8).

A closely related problem is the following.

Problem 20.1.2 *Let G be a locally compact group. Is the collection of closed locally elliptic subgroups of G a closed subset of the Chabauty space* **Sub**(G)?

The answer is positive for the Neretin group (as a consequence of [28, Corollary 3.6]), but open in general. For totally disconnected groups, it is equivalent to the following.

Problem 20.1.3 (C. Rosendal) *Let G be a tdlc group and $n > 0$ be an integer. Is the set $P_n(G)$ consisting of the n-tuples of elements of G contained in a common compact subgroup, closed in the Cartesian product G^n?*

For $n = 1$, the answer is positive by a theorem of G. Willis recalled in Chapter 9 above.

20.2 *p*-adic Lie groups

Every p-adic Lie group has a continuous finite-dimensional linear representation over $\mathbf{Q}_p$ given by its adjoint action on its Lie algebra. Hence a topologically simple p-adic Lie group is either linear or has a trivial adjoint representation. The linear topologically simple p-adic Lie groups are classified: they are all simple algebraic groups over $\mathbf{Q}_p$, and in particular compactly generated, see [15, Proposition 6.5]. We do not know whether other simple p-adic Lie groups exist:

Problem 20.2.1 *Is there a topologically simple p-adic Lie group whose adjoint representation is trivial? Can it be one-dimensional?*

A p-adic Lie group whose adjoint representation is trivial has an abelian Lie algebra, and is thus locally abelian. In particular it is elementary of rank 2 (in the sense of Wesolek), see Chapter 16 in this volume.

20.3 Profinite groups

Thanks to the major advances due to N. Nikolov and D. Segal reviewed in Chapter 5, the abstract algebraic structure of finitely generated profinite groups is now well understood. To what extent is the assumption of finite generation necessary in their theory? We mention some specific questions that could guide research in this direction.

Problem 20.3.1 (J. Wilson) *Can a pro-p group have a non-trivial abstract quotient that is perfect?*

Problem 20.3.2 *Can a hereditarily just-infinite profinite group have a proper dense normal subgroup?*

More problems on profinite groups are included and discussed in Chapter 3 of this volume.

20.4 Contraction groups

Contraction groups appear naturally in the structure theory of tdlc groups, in the presence of automorphisms whose scale is greater than one, see Chapter 10. Moreover, when the contraction group of an automorphism is closed, it is subjected to the far-reaching results from [23]. However, the following basic question remains open.

Problem 20.4.1 *Let G be a tdlc group and $\alpha \in \mathrm{Aut}(G)$ be a contracting automorphism, i.e. an automorphism such that $\lim_{n \to \infty} \alpha^n(g) = 1$ for all $g \in G$. Assume that the exponent of G is a prime power. Does it follow that G is nilpotent?*

The results from [23] ensure that G is solvable. Moreover, the problem is known to have a positive answer if G is a Lie group over a local field of arbitrary characteristic, see [22, Application 9.2].

20.5 Compactly generated simple groups

Let $\mathscr{S}$ denote the class of non-discrete compactly generated tdlc groups that are topologically simple. One of the recent trends in the structure theory of tdlc groups is the approach that relates the global properties of compactly generated simple tdlc groups with the structural properties of its compact open subgroups. This was initiated in [42] and [3] and further elaborated in [13] (see Chapters 17 and 18). Generally speaking, a property verified by all sufficiently small compact open subgroups is called **local**. One is thus interested in relating the local and global structures of groups in $\mathscr{S}$.

A basic question is to evaluate the number of local isomorphism classes: two groups are called **locally isomorphic** if they contain isomorphic open subgroups.

Problem 20.5.1   *Is the number of local isomorphism classes of groups in $\mathscr{S}$ uncountable?*

An equivalent way to think of the local approach to the study of the class $\mathscr{S}$ is to ask which profinite groups embed as a compact open subgroup in a group in $\mathscr{S}$. Despite important recent progress, our understanding of that problem remains elusive, as illustrated by the following.

Problem 20.5.2 (Y. Barnea, M. Ershov)   *Can a group in $\mathscr{S}$ have a compact open subgroup isomorphic to a free pro-p group?*

A basic observation from [13] ensures that it cannot be a free profinite group, since every group in $\mathscr{S}$ is locally pro-π for a finite set of primes π.

Another natural problem occurring in the realm of simple tdlc groups is the difference between topological simplicity (every closed normal subgroup is trivial) and abstract simplicity (every normal subgroup of the underlying abstract group is trivial). Examples show that a topologically simple tdlc group can fail to be abstractly simple, see [42] or the introduction of [13]. However, no compactly generated such example is known.

Problem 20.5.3   *Can a group in $\mathscr{S}$ have a proper dense normal subgroup?*

The following problem is closely related, see [13] and [9, Appendix B].

Problem 20.5.4   *Can a group G in $\mathscr{S}$ be such that $\mathrm{Inn}(G)$ is not closed in $\mathrm{Aut}(G)$?*

It has been proved in [12] that a topologically simple tdlc group having an element with a non-trivial contraction group has a smallest abstract normal subgroup, which is moreover simple. It is thus desirable to know whether all groups in $\mathscr{S}$ admit such an element. That property could fail if a group in $\mathscr{S}$ satisfied either of the following two conditions.

Problem 20.5.5 *Can a group in $\mathscr{S}$ have a dense conjugacy class? Can all its closed subgroups be unimodular?*

We believe that a better understanding of the class $\mathscr{S}$ as a whole necessitates to develop an intuition based on a larger pool of examples. The results from [13] show that all groups in $\mathscr{S}$ with a non-trivial centraliser lattice share some fundamental features with the full automorphism group of a tree. On the other hand, as long as the centraliser lattice is trivial, many of the tools developed in loc. cit. become inefficient or even useless.

Problem 20.5.6 *Find new examples of groups in $\mathscr{S}$ whose centraliser lattice is trivial.*

Another intriguing direction to explore is the relation between the algebraic structure of a group in $\mathscr{S}$ and its analytic properties, and in particular its unitary representations. In the classical case of simple Lie and algebraic groups, Kazhdan's property (T) is a landmark that is also a gateway to numerous fascinating rigidity phenomena. Some Kac–Moody groups in $\mathscr{S}$ also enjoy property (T). We do not know whether a group in $\mathscr{S}$ with a non-trivial centraliser lattice can have (T).

Problem 20.5.7 *Find new examples of groups in $\mathscr{S}$ satisfying Kazhdan's property (T).*

One of the main results from [13] is that a group in $\mathscr{S}$ with a non-trivial centraliser lattice is not amenable. The question of the existence of an infinite finitely generated simple amenable group was solved positively in [25]. Its non-discrete counterpart remains open.

Problem 20.5.8 *Can a group in $\mathscr{S}$ be amenable?*

If this question has a negative answer, then none of the topological full groups considered in [25] admits proper, infinite commensurated subgroups. Additionally, no non-virtually abelian finitely generated just infinite amenable group would appear as a lattice in a non-trivial way.

Amenable groups in $\mathscr{S}$ would be highly interesting since their behaviour would necessarily be very different from that of all currently known examples. On the other hand, a negative answer to the previous problem would have far-reaching consequences on discrete amenable groups.

For a more detailed discussion of the class $\mathscr{S}$ and open problems about it, we refer to [8].

20.6 Lattices

A fundamental impetus to the study of simple tdlc groups beyond the case of algebraic groups was the ground-breaking work of M. Burger and S. Mozes on lattices in products of trees, see [6] and Chapters 6 and 12. However, the mechanisms responsible for the existence or non-existence of lattices in general simple groups remain largely mysterious. The following vague problem consists in investigating that question.

Problem 20.6.1   *Which groups in $\mathscr{S}$ contain lattices? Which products of groups in $\mathscr{S}$ contain lattices with dense projections? For a group (or a product of groups) in $\mathscr{S}$, can one classify its lattices up to commensurability?*

Even in the basic case (considered by Burger–Mozes) of a product $G = G_1 \times G_2$ of two groups in $\mathscr{S}$, each acting properly and cocompactly on a regular tree, it is not clear how the existence of an irreducible cocompact lattice is reflected by the global structure of G (see the Basic Question on p. 5 in [7]).

Another fundamental problem, also naturally suggested by the work of Burger and Mozes, is the following.

Problem 20.6.2   *Let $G = G_1 \times \cdots \times G_n$ be the product of n non-discrete compactly generated topologically simple locally compact groups and $\Gamma \leq G$ be a lattice. Assume that the projection of Γ to every proper subproduct of G has dense image. Can Γ be non-arithmetic and residually finite if $n \geq 2$? Can Γ be simple if $n \geq 3$?*

Partial results when the factors are certain Kac–Moody groups in $\mathscr{S}$ have been established in [10]. The special case of Problem 20.6.2 where each G_i is a closed subgroup of the automorphism group of a locally finite tree T_i acting cocompactly (and even 2-transitively at infinity) is already highly interesting.

A related problem consists in finding 'exotic' lattices in the full automorphism group of a simple algebraic group over a local field.

Problem 20.6.3 ([16, Annexe A, Problem 1])   *Let $\mathbf{G}$ be a simply connected absolutely simple algebraic group over a local field k, of k-rank ≥ 2. Has every lattice finite image in $\mathrm{Out}(\mathbf{G}(k))$?*

In additive combinatorics, there is a long history of considering sets that are just-not-quite groups; the modern notion of **approximate groups** was introduced by T. Tao in [38]. In a recent preprint [5], M. Björklund and T. Hartnick consider certain subsets of a locally compact group G which they call **uniform approximate lattices**. They further investigate three different tentative definitions of (non-uniform) "approximate lattices".

Problem 20.6.4 (Björklund–Hartnick)   *Which is the "right" definition of approximate lattices?*

More specifically, we could ask for a definition such that (i) uniform approximate lattices are approximate lattices; (ii) a subgroup of G is an approximate lattice if and only if it is a lattice.

20.7 Commensurated subgroups and commensurators

The class of tdlc groups is closely related to the class of pairs (Γ, Λ) consisting of a discrete group Γ and a commensurated subgroup $\Lambda \leq \Gamma$, see for example [36, Section 3].

Problem 20.7.1 (Margulis–Zimmer conjecture)   *Let G be a connected semisimple Lie group with finite centre and $\Gamma \leq G$ be an irreducible lattice. Assume that the rank of G is at least 2. Prove that every commensurated subgroup of Γ is either finite or of finite index.*

See [36] for an extended discussion and partial results. We emphasise that the only known cases where that problem has been solved concern non-uniform lattices: there is not a single example of a uniform lattice for which the Margulis–Zimmer conjecture has been proved.

Instead of looking for commensurated subgroups in a given group, one can dually consider the largest group in which a given group G embeds as a commensurated subgroup. That group is called the **group of abstract commensurators** of G. It is defined in Section 6, Appendix B of [2], where the idea of the concept is attributed to J.-P. Serre and W. Neumann.

Problem 20.7.2 (A. Lubotzky)   *Let F be a non-abelian free group of finite rank. Is the group of abstract commensurators of F a simple group?*

That group of abstract commensurators is countable, but not finitely generated, see [4]. Several variants of that problem can be envisioned. The following one is due to A. Lubotzky, S. Mozes and R. Zimmer (Remark 2.12(i) in [29]): is the relative commensurator of F in the full automorphism group of its Cayley tree virtually simple? We refer to the appendix of [32] for partial answers to the latter question as well as Problem 20.7.2. It is also natural to ask whether the group of abstract commensurators of the profinite (resp. pro-p) completion of F is simple, or whether it is topologically simple with respect to the natural tdlc group topology that it carries (see [3]). The latter question is thus related to Problem 20.5.2.

20.8 Unitary representations and C*-simplicity

The problems in this section pertain to a general research direction consisting in relating the intrinsic algebraic/geometric/dynamical structure of a locally compact group with the properties of its unitary representations. Given the difficulty and depth of the theory of unitary representations of semi-simple groups over local fields, it is of course not realistic at this stage to hope for a meaningful general theory. However, recent breakthroughs suggest that some specific questions could be solved. We mention a few of them.

Problem 20.8.1   *Let G be a tdlc group. Characterise the C*-simplicity of G in terms of its Furstenberg boundary.*

For discrete groups, such a characterisation has been obtained recently by M. Kalantar and M. Kennedy: they proved in [26] that a discrete group is C*-simple if and only if its action on its Furstenberg boundary is topologically free.

The following problem was suggested by T. Steger, who reported that C. Nebbia asked it in the 1990s. We recall that a representation of a tdlc group is called **admissible** if the subspace of fixed points of every compact open subgroup is finite-dimensional.

Problem 20.8.2 (C. Nebbia)   *Let T be a locally finite leafless tree and G ≤ Aut(T) be a closed subgroup acting 2-transitively on the set of ends ∂T. Is every continuous irreducible unitary representation of G admissible?*

A classical criterion (see [18] or Thm. 2.2 in [14]) implies that a tdlc group all of whose continuous irreducible unitary representations are admissible, is of type I, i.e. all of its continuous unitary representations generate a von Neumann algebra of type I. Problem 20.8.2 thus leads us naturally to the following.

Problem 20.8.3   *Let T be a locally finite leafless tree. Is it true that G ≤ Aut(T) is of type I if and only if G is 2-transitive on the set of ends?*

Thus a positive solution to Problem 20.8.2 implies that the 'if' part of Problem 20.8.3 holds. The converse implication in Problem 20.8.3 was recently proved by C. Houdayer and S. Raum (see [24], which also contains an extensive discussion of Problem 20.8.3).

A topological group is called **unitarisable** if all its uniformly bounded continuous representations on Hilbert spaces are conjugate to unitary representations. Following work of B. Sz.-Nagy, it was observed in 1950 by M. Day, J. Dixmier, M. Nakamura and Z. Takeda that this property holds for amenable groups (their argument for discrete groups holds unchanged for topological groups). These authors raised the question whether, conversely, only amenable groups are unitarisable. This question, surveyed in [31], has led to deep results by G. Pisier. Despite modest contributions by other authors in more recent years [19], [30], it remains completely open.

The unitarisability question makes sense more generally for locally compact groups, see [21] for a partial result (beyond the locally compact setting, there are amenable topological groups without any uniformly bounded continuous representation [20]). However, even the most basic tools to study it appear to fall short in the non-discrete setting. For instance:

Problem 20.8.4 *Let G be a unitarisable tdlc group and H < G a closed subgroup. Must H also be unitarisable?*

We do not even know the following particular cases:

Problem 20.8.5 *Can a unitarisable tdlc group contain a discrete non-abelian free subgroup? Can it contain a non-abelian free subgroup as a lattice?*

20.9 Elementary groups

When trying to decompose a general locally compact group into 'atomic building blocks' by means of subnormal series, several families of subquotients appear to be unavoidable: discrete groups, compact groups and topologically characteristically simple groups, see Chapter 15 and [9]. The class of elementary groups was introduced and studied by P. Wesolek (see Chapter 16 and [40]) as a tool to investigate general tdlc groups by understanding which of them are exclusively built out of discrete and compact pieces. In that sense, those tdlc groups are the most elementary, whence the choice of terminology.

The most general decomposition results on arbitrary tdlc groups have been obtained in the past two years by C. Reid and P. Wesolek in a deep and far-reaching theory which highlights the key role played by elementary groups, see [33], [34] and references therein. We thank both of them for their suggestions about the present subsection; most of the problems selected here are due to them.

By definition, an elementary group is constructed by means of an iterative procedure involving more and more building blocks. The complexity of the resulting group is measured by an ordinal-valued function called the **rank**, see Chapter 16 and [40].

Problem 20.9.1   *Do there exist second countable elementary groups of arbitrarily large rank below ω_1?*

A basic fact is that a group in $\mathscr{S}$ is not elementary. Moreover, since the class of elementary groups is stable under passing to closed subgroups and Hausdorff quotients, it follows that an elementary group cannot have a subquotient isomorphic to a group in $\mathscr{S}$. We do not know whether the converse holds.

Problem 20.9.2   *Let G be a second countable tdlc group that is not elementary. Must there exist closed subgroup H,K of G with K normal in H such that H/K belongs to $\mathscr{S}$?*

Another fundamental problem is to understand which elementary groups are topologically simple. The known simple elementary groups include all discrete simple groups, as well as various topologically simple locally elliptic groups. All of them are thus of rank ≤ 2.

Problem 20.9.3   *Characterise the elementary groups that are topologically simple. Are they all of rank 2?*

The results from [13] show that for many groups in $\mathscr{S}$, the conjugation action of the group on its closed subgroups has interesting dynamics. To what extent is that feature shared by non-elementary groups? The following problems are guided by this vague question.

Problem 20.9.4   *Characterise the tdlc groups all of whose closed subgroups are unimodular. Are they all elementary?*

Problem 20.9.5   *Characterise the tdlc group G such that the only minimal closed invariant subsets of the Chabauty space $\mathbf{Sub}(G)$ are $\{1\}$ and $\{G\}$. Are they all elementary?*

We refer to [27] for very recent exemples.

Finally, a question of P. Wesolek blending the notion of elementarity discussed here with the classical notion of elementary amenability (of discrete groups) is as follows.

Problem 20.9.6 (P. Wesolek)   *Let G be an amenable second countable tdlc group. Must G be elementary?*

Notice that a positive answer to that question would imply a negative answer to Problem 20.5.8. Moreover, in case the answer to Problem 20.9.2 is positive, then Problems 20.5.8 and 20.9.6 are then formally equivalent.

A more specific sub-question of Problem 20.9.6 is:

Problem 20.9.7 (P. Wesolek) *Let G be a compactly generated tdlc group of subexponential growth. Must G be elementary?*

We refer to [17] for recent examples of tdlc groups of subexponential growth that are elementary, but not compact-by-discrete.

20.10 Galois groups

Let K/k be a field extension and $G = \mathrm{Aut}(K/k)$ its automorphism group, i.e. the set of those automorphisms of K acting trivially on k. Endow G with the topology of pointwise convergence. It is well known that if K/k is algebraic, then G is a profinite group. More generally, if K/k is of finite transcendence degree, then G is a tdlc group, which is moreover discrete if and only if K is finitely generated over k (see Chapter 6, §6.3 in [37]). Thus infinitely generated transcendental field extensions of finite transcendence degree provide a natural source of examples of non-discrete tdlc groups. While it is known that every profinite group is the Galois group of some algebraic extension (see [39]), the corresponding problem does not seem to have been addressed for non-compact groups.

Problem 20.10.1 *Which tdlc groups are Galois groups?*

Some natural field extensions moreover yield simple groups. This is for example the case if the fields k and K are algebraically closed of characteristic 0 as soon as K/k is non-trivial (hence transcendental): Indeed, by [35, Theorem 2.9] the Galois group $\mathrm{Aut}(K/k)$ has a characteristic open subgroup which is topologically simple.

Problem 20.10.2 *Which topologically simple tdlc groups are Galois groups? Which topologically simple tdlc groups continuously embed in Galois groups?*

References for this chapter

[1] BADER, U., DUCHESNE, B., AND LÉCUREUX, J. Amenable invariant random subgroups. *Israel J. Math. 213*, 1 (2016), 399–422. With an appendix by Phillip Wesolek.

[2] BASS, H., AND KULKARNI, R. Uniform tree lattices. *J. Amer. Math. Soc. 3*, 4 (1990), 843–902.

[3] BARNEA, Y., ERSHOV, M., AND WEIGEL, T. Abstract commensurators of profinite groups. *Trans. Amer. Math. Soc. 363*, 10 (2011), 5381–5417.

[4] BARTHOLDI, L., AND BOGOPOLSKI, O. On abstract commensurators of groups. *J. Group Theory 13*, 6 (2010), 903–922.

[5] BJÖRKLUND, M., AND HARTNICK, T.   Approximate lattices.   Preprint, arXiv:1612.09246, 2016.

[6] BURGER, M., AND MOZES, S. Lattices in product of trees. *Inst. Hautes Études Sci. Publ. Math. 92* (2000), 151–194 (2001).

[7] BURGER, M., MOZES, S. AND ZIMMER, R. J. Linear representations and arithmeticity of lattices in products of trees. *Essays in geometric group theory*, Ramanujan Math. Soc. Lect. Notes Ser., 9 (2009), 1–25.

[8] CAPRACE, P.-E. Non-discrete simple locally compact groups. Preprint, to appear in the Proceedings of the 7th European Congress of Mathematics, 2016.

[9] CAPRACE, P.-E., AND MONOD, N. Decomposing locally compact groups into simple pieces. *Math. Proc. Cambridge Philos. Soc. 150*, 1 (2011), 97–128.

[10] CAPRACE, P.-E., AND MONOD, N. A lattice in more than two Kac-Moody groups is arithmetic. *Israel J. Math. 190* (2012), 413–444.

[11] CAPRACE, P.-E., AND MONOD, N. Relative amenability. *Groups Geom. Dyn. 8*, 3 (2014), 747–774.

[12] CAPRACE, P.-E., REID, C. D., AND WILLIS, G. A. Limits of contraction groups and the Tits core. *J. Lie Theory 24*, 4 (2014), 957–967.

[13] CAPRACE, P.-E., REID, C. D., AND WILLIS, G. A. Locally normal subgroups of totally disconnected groups. Part II: Compactly generated simple groups. *Forum Math. Sigma 5* (2017), e12, 89.

[14] CIOBOTARU, C. A note on type I groups acting on d-regular trees. Preprint, arXiv:1506.02950, 2015.

[15] CLUCKERS, R., CORNULIER, Y., LOUVET, N., TESSERA, R., AND VALETTE, A. The Howe-Moore property for real and *p*-adic groups. *Math. Scand. 109*, 2 (2011), 201–224.

[16] CORNULIER, Y. Aspects de la géométrie des groupes. Mémoire d'habilitation à diriger des recherches, Université Paris-Sud 11, 2014.

[17] CORNULIER, Y. Locally compact wreath products. Preprint , arXiv:1703.08880, 2017.

[18] DIXMIER, J. *C*-algebras*. North-Holland Publishing Co., Amsterdam-New York-Oxford, 1977. Translated from the French by Francis Jellett, North-Holland Mathematical Library, Vol. 15.

[19] EPSTEIN, I., AND MONOD, N. Nonunitarizable representations and random forests. *Int. Math. Res. Not. IMRN 22* (2009), 4336–4353.

[20] GHEYSENS, M. Inducing representations against all odds. Thesis (Ph.D.)–EPFL, 2017.

[21] GHEYSENS, M., AND MONOD, N. Fixed points for bounded orbits in Hilbert spaces. *Ann. Sci. Éc. Norm. Supér. (4) 50*, 1 (2017), 131–156.

[22] GLÖCKNER, H. Invariant manifolds for analytic dynamical systems over ultrametric fields. *Expo. Math. 31*, 2 (2013), 116–150.

[23] GLÖCKNER, H., AND WILLIS, G. A. Classification of the simple factors appearing in composition series of totally disconnected contraction groups. *J. Reine Angew. Math. 643* (2010), 141–169.

[24] HOUDAYER, C., AND RAUM, S. Locally compact groups acting on trees, the type I conjecture and non-amenable von Neumann algebras. Preprint, arXiv:1610.00884, 2016.

[25] JUSCHENKO, K., AND MONOD, N. Cantor systems, piecewise translations and simple amenable groups. *Ann. of Math. (2) 178*, 2 (2013), 775–787.

[26] KALANTAR, M., AND KENNEDY, M. Boundaries of reduced C^*-algebras of discrete groups. *J. Reine Angew. Math. 727* (2017), 247–267.

[27] LE BOUDEC, A., AND MATTE BON, N. Locally compact groups whose ergodic or minimal actions are all free Preprint, arXiv:1709.06733, 2017.

[28] LE BOUDEC, A., AND WESOLEK, P. Commensurated subgroups in tree almost automorphism groups. Preprint, arXiv:1604.04162, 2016.

[29] LUBOTZKY, A., MOZES, S., AND ZIMMER, R. Superrigidity for the commensurability group of tree lattices. *Comment. Math. Helv. 69*, 4 (1994), 523–548.

[30] MONOD, N., AND OZAWA, N. The Dixmier problem, lamplighters and Burnside groups. *J. Funct. Anal. 258*, 1 (2010), 255–259.

[31] PISIER, G. *Similarity problems and completely bounded maps*, expanded ed., vol. 1618 of *Lecture Notes in Mathematics*. Springer-Verlag, Berlin, 2001. Includes the solution to "The Halmos problem".

[32] RADU, N. New simple lattices in products of trees and their projections. ArXiv preprint 1712.01091, 2017.

[33] REID, C. D., AND WESOLEK, P. R. The essentially chief series of a compactly generated locally compact group. ArXiv preprint 1509.06593, 2015.

[34] REID, C. D., AND WESOLEK, P. R. Dense normal subgroups and chief factors in locally compact groups. ArXiv preprint 1601.07317, 2016.

[35] ROVINSKY, M. Motives and admissible representations of automorphism groups of fields. *Math. Z. 249*, 1 (2005), 163–221.

[36] SHALOM, Y., AND WILLIS, G. A. Commensurated subgroups of arithmetic groups, totally disconnected groups and adelic rigidity. *Geom. Funct. Anal. 23*, 5 (2013), 1631–1683.

[37] SHIMURA, G. *Introduction to the arithmetic theory of automorphic functions*. Publications of the Mathematical Society of Japan, No. 11. Iwanami Shoten, Publishers, Tokyo; Princeton University Press, Princeton, N.J., 1971. Kanô Memorial Lectures, No. 1.

[38] TAO, T. Product set estimates for non-commutative groups. *Combinatorica 28*, 5 (2008), 547–594.

[39] WATERHOUSE, W. C. Profinite groups are Galois groups. *Proc. Amer. Math. Soc. 42* (1973), 639–640.

[40] WESOLEK, P. Elementary totally disconnected locally compact groups. *Proc. Lond. Math. Soc. (3) 110*, 6 (2015), 1387–1434.

[41] WESOLEK, P. A note on relative amenability. *Groups Geom. Dyn. 11*, 1 (2017), 95–104.

[42] WILLIS, G. A. Compact open subgroups in simple totally disconnected groups. *J. Algebra 312*, 1 (2007), 405–417.

Index

She said that indexing was a thing that only the most amateurish author undertook to do for his own book.

Kurt Vonnegut, *Cat's Cradle*, 1963

accessible
  diameter- —, 300
  graph, 300
  group, 302
action
  cobounded, 10
  geometric, 10
  locally bounded, 10
  metrically proper, 10
  micro-supported, 273
amenable, 5, 14, 172, 274, 278, 282, 288, 344, 347
  action, 175, 176
  geometrically —, 15
  metrically —, 282
  radical, 5, 224

Benjamini–Schramm topology, 198
Borel density, 193
boundary, 15, 94, 132, 175, 256, 271, 284
  Furstenberg–Poisson —, 175
branch
  group, 27
Burger–Mozes
  lattices, 179
  universal groups, 102, 124

C-stable, 269
  locally —, 270
Cayley–Abels graph, 7, 11, 230, 239, 283
centraliser lattice, 270
Chabauty space, 169, 187, 344
characteristic
  equivariant Euler —, 220
  subgroup, 22, 228, 237, 257
characteristically simple, 228, 237, 257
coarse

connectedness, 11
  equivalence, 10, 221
  simple connectedness, 12
cohomology
  abstract —, 87
  continuous —, 30, 87, 208, 211, 212
  L^2- —, 211, 212
  reduced —, 208
commable, 311
compact presentation, 13, 305
contraction group, 39, 52, 162, 277, 345
  nub, 167
core, 78, 126
  Tits —, 169

dense
  Borel —, 193
  geometrically —, 98
elementary
  group, 236, 240, 351
  hyperbolic group, 284
  non- — hyperbolic group, 278
  radical, 237
elliptic
  automorphism, 95, 149, 271
  locally —, 229, 316, 344
  locally — radical, 229, 255, 316
  purely —, 97
focal hyperbolic group, 278
Galois group, 19, 74, 353
geometrically dense, 98
Gleason–Yamabe theorem, 5
graph
  accessible —, 300
  Cayley–Abels —, 7, 11, 230, 239, 283
Gromov–Hausdorff topology, 197

Higman–Thompson groups, 139
hyperbolic
  automorphism, 95, 149, 271
  group, 276
    focal — —, 278
    purely — —, 97
independence property, 100, 122
invariant random subgroup, 189
irreducible lattice, 171
IRS, 189

just infinite, 171

Kazhdan property, 174, 347
Kazhdan–Margulis theorem, 194

L^2-Betti numbers, 205, 220
lattice, 171
  Burger–Mozes —, 179
  centraliser —, 270
  cocompact —, 171
  irreducible —, 171
  local decomposition —, 268
  non-existence thereof, 191, 348
  structure —, 260, 268
  uniform —, 171
Lazard theorem, 26, 32, 53
Levi factor, 55, 163
Lie group, 5, 11, 37, 45, 188, 189, 277
  p-adic —, 37, 45, 155, 344
local action, 103, 120
local decomposition lattice, 268
locally
  C-stable, 270
  elliptic, 229, 316, 344
  elliptic radical, 229, 255, 316
  equivalent, 267
  normal, 258, 267

micro-supported, 273
millefeuille, 286
minimal normal subgroup, 227, 232, 255
minimising subgroup, 146

Neretin group, 131, 191, 344
non-elementary
  group, *see* elementary group
  hyperbolic group, 278
normal subgroup
  dense —, 345, 346
  locally —, 258, 267
  minimal —, 227, 232, 255
  theorem, 171, 195
normal subgroup theorem, 171
nub, 167

parabolic subgroup (of an automorphism), 162
proximal (strongly), 272

quasi-centraliser, 261, 269
quasi-discrete, 251
quasi-product, 255
radical
  amenable —, 5, 224
  elementary —, 237
  locally elliptic —, 229, 255, 316
rank
  construction —, 248
  Hattori–Stallings —, 217
  of a topological group, 88
residually discrete, 241

scale function, 146, 163
simple
  abstractly —, 93, 101, 106, 109, 124, 125,
    128, 143, 169, 346
  characteristically —, 228, 237, 257
  lattice, 179
spheromorphism, *see* Neretin group
strongly proximal, 272
structure lattice, 260, 268

theorem
  Gleason–Yamabe, 5
  Kazhdan–Margulis, 194
  Lazard, 26, 32, 53
  normal subgroup, 171
  Tits simplicity, 101, 125, 141
  van Dantzig, 2, 230, 241, 258
  Zassenhaus, 188
tidy subgroup, 51, 147, 162
Tits core, 169
Tits independence property (P), 100, 122
Tits simplicity theorem, 101, 125, 141
topology
  Benjamini–Schramm —, 198
  Gromov–Hausdorff —, 197
ultrametric
  absolute value, 40
  field, 42
  inequality, 40
uniform lattice, 171
universal groups
  Burger–Mozes —, 102, 124
  Smith —, 124

van Dantzig theorem, 2, 230, 241, 258

Zassenhaus theorem, 188

Printed in the United States
By Bookmasters